Literature, Science, Psychoanalysis, 1830–1970

Literature, Science, Psychoanalysis, 1830–1970

Essays in Honour of Gillian Beer

EDITED BY
HELEN SMALL
AND
TRUDI TATE

UNIVERSITY PRESS

OXFORD
UNIVERSITY PRESS

Great Clarendon Street, Oxford OX2 6DP

Oxford University Press is a department of the University of Oxford.
It furthers the University's objective of excellence in research, scholarship,
and education by publishing worldwide in

Oxford New York

Auckland Bangkok Buenos Aires Cape Town Chennai
Dar es Salaam Delhi Hong Kong Istanbul Karachi Kolkata
Kuala Lumpur Madrid Melbourne Mexico City Mumbai Nairobi
São Paulo Shanghai Taipei Tokyo Toronto

Oxford is a registered trade mark of Oxford University Press
in the UK and in certain other countries

Published in the United States
by Oxford University Press Inc., New York

First published 2003

British Library Cataloguing in Publication Data
Data available

Library of Congress Cataloging in Publication Data
Data available

ISBN 0–19–926667–0

1 3 5 7 9 10 8 6 4 2

Typeset by Cambrian Typesetters,
Frimley, Surrey
Printed in Great Britain
on acid-free paper by
T.J. International Ltd,
Padstow, Cornwall

Contents

List of Illustrations

Introduction

HELEN SMALL

> In the case of Darwin the restless curiosity of the child to know the "what for?" the "why?" and the "how?" of everything seems never to have abated its force.
>
> (A. R. Wallace, 1873)[1]

> [T]here is delight for Freud in identification with the robust invention and intelligence of [Little] Hans himself . . . trammelled by the half-lies of his society but persistently breaking through them to the insight born in fantasy. Freud hears with the ears of a child . . .
>
> (Gillian Beer, 2002)[2]

Where curiosity is at work the desire for knowledge is never entirely distinct from the desire to be transformed by knowledge. To be curious is to be like a child, as Alfred Russel Wallace recognized when he contemplated the persistent restlessness, after sixty-two years, of Darwin's wish to understand the workings of the natural world. More specifically, to be curious is to be like a child who wants to grow up—who wants to know what adults know (or what they are assumed to know), and who claims, as by right, the power to contest their versions of the world. Whatever curiosity's ends—and the content and quality of knowledge are to be distinguished here from the disposition toward knowledge—it requires putting oneself out in the world. As Gillian Beer's new introduction to Freud's *Case Histories* (1909) reminds us, curiosity is the most pleasurable form of resistance to orthodoxy. It relishes the idea, and the risk, that one's perceptions, and therefore oneself, may be altered by what one discovers.

From this perspective, curiosity is the vital counter-drive to that self-annihilating strain George Levine has recently diagnosed as a condition of nineteenth-century scientific epistemology: 'dying to know'.[3] To be curious is to extend the

[1] *Quarterly Journal of Science* (January 1873); quoted in Francis Darwin (ed.), *The Life and Letters of Charles Darwin Including an Autobiographical Chapter*, foreword by George Gaylord Simpson, 2 vols. (New York: Basic Books, 1959), ii. 350.

[2] Introduction to Sigmund Freud, *'The Wolf Man' and Other Cases*, trans. Louise Adey Huish, The New Penguin Freud (London: Penguin, 2002).

[3] See George Levine, *Dying to Know: Scientific Epistemology and Narrative in Victorian England* (Chicago: University of Chicago Press, 2001). There have been numerous recent studies of curiosity and wonder in the history of science and literature. See especially Neil Kenny, *Curiosity in Early Modern Europe: Word Histories* (Wiesbaden: Harrassowitz, 1998), Barbara M. Benedict, *Curiosity: A Cultural History of Early Modern Europe* (Chicago: University of Chicago Press, 2001); Lorraine Daston

self. The most immediate risk it involves is not death so much as the possibility of extending oneself spectacularly beyond the point where nature or custom or the capacity for self-recognition would have one stop. Alice in Wonderland's flouting of the laws of grammar, 'curiouser and curiouser', catches its elasticating effects with perfect succinctness. The desire to know pushes beyond itself, exclaiming at its own improbable expansions: 'Now I'm opening out like the largest telescope that ever was!'[4]

To liken the curious adult to a child can, of course, be a pre-emptive move—as it was for Wallace, seeking to mollify the critics of evolutionary theory. If the impulse behind the writing of *The Expression of the Emotions in Man and Animals* (the work Wallace was reviewing) is essentially that of a child, the drive to know can be assumed to be innocent. Its 'unabated force' is an imperialism not of the will but of youthful instinct—or something very close to instinct. But to claim the child's dispensation on behalf of an adult may also be to make allowance for something other than mere innocence or absorption in the subject at hand. When Beer draws our attention to the childlikeness of Freud she is suggesting that there are imperatives at work within the grown adult that may run counter to our expectations of what it will mean to 'grow' or 'mature' or 'come of age' through the acquisition of knowledge (though they preserve the hope and trust that we do indeed grow). It suggests, that is, that growing may involve—not regression, but a conscious or unconscious refusal to let go of the curious needs and impulses we started out with.

There is a moment in *Darwin's Plots* (1983) when a very similar observation is offered as something like the germ of that book's path-breaking analysis of Darwinian evolutionary narrative.

> So basic is the concept of development to our culture and to our adult experience that we forget that it is a *learnt* concept. Evolutionary assumptions of irreversible change have reinforced our observation of individual growth. It is salutary to remind ourselves that the two concepts are not inevitably connected. When our youngest child was a little less than three years old, he would often say: 'When I'm a baby again' or 'Granny a little girl soon' or 'Daddy carry me till Mummy gets bigger' or 'When Mummy's a baby I'll do so-and-so'. The idea of irreversible onward growth and cessation of growth was not yet fixed—perhaps he was trying it out and testing it, but against what seemed to him a natural order in which growth was variable, reversible, spasmodic.[5]

and K. Park, *Wonder and the Order of Nature, 1150–1750* (Oxford: Clarendon Press, 1991); Barbara Maria Stafford, *Artful Science: Enlightenment, Entertainment, and the Eclipse of Visual Education* (Cambridge, Mass.: MIT Press, 1994); Nigel Leask, *Curiosity and the Aesthetics of Travel Writing, 1770–1840: 'From an Antique Land'* (Oxford: Oxford University Press, 2002).

[4] Lewis Carroll, *Alice's Adventures in Wonderland* (first published as *Alice's Adventures Underground*, 1866), in *Alice in Wonderland*, ed. Donald J. Gray, 2nd edn. (New York: W. W. Norton, 1992), 13. 'Curiouser', of course, puns on the puzzling nature of what Alice sees and her own desire to understand.

[5] Gillian Beer, *Darwin's Plots: Evolutionary Narrative in Darwin, George Eliot and Nineteenth-Century Fiction* (London: Routledge & Kegan Paul, 1983), 111.

Books have multiple origins, of course, but this is a particularly intimate testimony to one of the strands of experience that played a part in shaping Beer's curiosity about Darwinian resistance to linear developmental plotting. Like the quotation from Wallace—however different the assumptions about audience—it testifies to a pleasurable awareness that knowledge is often advanced most profitably by means of our childlike resistances to established wisdom.

The use of the child's perspective as a way of displacing, and thereby questioning, our inherited assumptions about the world has been one of the quiet hallmarks of Beer's work. It is there in her first published book, her 1970 study of *The Romance*, when she remarks on the childlike involvement in narrative required of us by that mode: an involvement none the less predicated on the adult's 'sense of relief and sophistication' at acquiescing in romance's 'peculiar vagrancy of imagination'.[6] It is there, too, when she homes in on George Meredith's use of the child's perspective, in *Harry Richmond* (1871), to dislodge the more conventional responses of adult minds.[7] It is central to her readings of Darwin, with their alertness to his valuation of 'the exception, the anomalous, even in minutest instances'.[8] Most of all it explains the constant appeal for her of Lewis Carroll's Alice, with her 'untrammelled' persistence in the face of the nonsense Carroll parades before her, her determination to go on asking questions of the world, and letting its logic be tested through her, no matter how obstructive the responses she receives.

In valorizing such childlike willingness to entertain new and perhaps heretical plots, Gillian Beer has always remained keenly aware of the forces working, both outside us and within us, to constrain the free play of imagination. 'Death was the significant absence in his economy' is her concluding comment on her young son's proto-Darwinian take on the world. Her writing is characteristically attuned to the tussle between our 'hunger for permanence', in Arthur Eddington's memorable phrase (quoted in her 1987 essay 'Problems of Description in the Language of Discovery'),[9] and the desire to leave our statements about the world open to challenge and correction and change. That keynote can be heard in the literary choices she has been attracted to across the course of her writing career to date (Virginia Woolf, George Meredith, George Eliot, Thomas Hardy, John Donne, Sir Philip Sidney, Samuel Richardson, Vernon Lee, Edith Simcox, Sylvia Townsend Warner, Angela Carter, Lewis Carroll), and in her special fondness for those scientific writers who have been most open to the unsettling powers of language and literature (Darwin, Tyndall, James Clerk Maxwell, W. K. Clifford, James Jeans). As she put it in the conclusion to 'Problems of Description', 'The utmost resourcefulness and probity of language are needed, both by scientists and poets,

[6] (London: Methuen, 1970), 8–9, 4.

[7] *Meredith: A Change of Masks. A Study of the Novels* (London: Athlone Press, 1970), ch. 2.

[8] *Darwin's Plots*, 42.

[9] *Nature of the Physical World* (1927), quoted in Gillian Beer, *Open Fields: Science in Cultural Encounter* (Oxford: Clarendon Press, 1996), 172.

to outwit the tendency of description to stabilize a foreknown world and to curtail discovery.'[10]

Beer's willingness to recognize such resourcefulness in past writers, and her ability to find it always in potential in the language, have made her a pioneering contributor to that branch of interdisciplinary studies which is concerned with the relations between literature, science, and psychoanalysis. She has given us some of the most compelling accounts we have of what productive interdisciplinary reading should look like. Refusing to see her task as one of presiding over other spheres of knowledge, whether from the historical vantage point of the present or from the analytic vantage point of a literary critical training, she has always insisted on the openness of all fields of knowledge, even where their institutional settings have served to 'disguise' it, and insisted too on the necessarily equivalent openness of interdisciplinary studies themselves:

Interdisciplinary studies do not produce closure. Their stories emphasise not simply the circulation of intact ideas across a larger community, but transformation: the transformations undergone when ideas enter other genres or different reading groups, the destabilizing of knowledge once it escapes from the initial group of co-workers, its tendency to mean more and other than could have been foreseen.

Problems remain. How thoroughly interdisciplinary is it possible to be? Are we lightly transferring a set of terms from one practice to another, as metaphor, *façon de parler*? Are we appropriating *materials* hitherto neglected for analysis of the kind we have always used? Or are we trying to learn new *methods* and skills fast, which others have spent years acquiring? . . . That superior glossing of other people's controversies is one hazard in interdisciplinary work; the opposite temptation is to succumb to the glamour of the horizon.[11]

There is something here of the 'asperity' of self-scrutiny she recognized in George Meredith, or that determination to make us turn the lens back on our own motives and powers that she has analysed so adeptly in George Eliot's writing.

This said, the monitory mode is not Gillian Beer's preferred style. Acutely alert as she is to the practical limitations and the risk of intellectual self-deception in interdisciplinary criticism, one of the consistent pleasures of reading her is the characteristic vibrancy of her prose—its exploratory reach, but also its lightness of touch and its alertness to the comic and wayward in its chosen subjects. Her recent work on Lewis Carroll is only the most immediate sign of that deep delight so much of her writing takes in the affirmative powers of comedy. (Her students and colleagues are all familiar with her ability to pronounce the word 'yes' so flexibly as to register everything from delighted accord to a gritty determination to rescue something profitable from the unlikely materials at hand.)

When Darwin wrestled with the problem of taxonomical metaphor in his 1837 *Notebook* (as so often in the course of his career) he played whimsically with the

[10] In George Levine (ed.), *One Culture: Essays in Science and Literature* (Madison, Wis.: Wisconsin University Press, 1987); rpt. in *Open Fields*, 149–72 (172).

[11] *Forging the Missing Link: Interdisciplinary Stories*, inaugural lecture as the King Edward VII Professor of English Literature, University of Cambridge (Cambridge: Cambridge University Press, 1992), rpt. in *Open Fields*, 115–45 (115).

idea of replacing the tree of life with a sample of coral.[12] Looking across the range of Gillian Beer's writing, it is tempting to reach for the same kind of improved metaphorical capacity. Far too many subjects have attracted her interest to be represented as branches from a single stem: women's writing, psychoanalysis, anthropology, rhyme, comic verse, wave theory, music, the Gothic (especially female Gothic), post-colonial literature, the idea of the university, the development of 'academies' and (relatedly) of periodical publishing. And the historical reach has been impressive, from the Ovidian and medieval starting points of *The Romance*, or her very recent writing on the Platonic dialogue form,[13] through to her extensive involvement with contemporary fiction as both critic and valued judge on the literary prize circuit. The essays collected here are concerned with three, closely related, fields of inquiry in which she has been particularly influential: literature, the sciences, and psychoanalysis, from the 1830s through to the late 1960s. Not all the essays cover all three fields, but separately and collectively they bear witness to the centrality of Gillian Beer's influence on this branch of interdisciplinary studies in Britain and America. To that end, *Darwin's Plots* and *Open Fields* are the most regular points of reference here, though others of her writings are also referred to.

Nigel Leask's essay 'Darwin's "Second Sun" ' takes a lead from Beer's curiosity about the literary genesis of Darwin's first major publishing success, the *Journal of Researches into the Geology and Natural History of the Various Countries Visited by H.M.S. Beagle* (1839) (republished, with substantial revisions, as *The Voyage of the Beagle* (1845)). Leask explores the crucial influence on the young Darwin of Alexander von Humboldt's writing about his voyage around the South American tropics some thirty years earlier. In the Prussian traveller and naturalist's *Personal Narrative* (1814–25), Darwin found an attractive Romantic model for combining travelogue, scientific observation, and philosophical speculation. But Leask also charts a gradual process by which Humboldt's powerful example gave the maturing Darwin a model of scientific writing *against* which to define his own voice: a voice that, as he reworked the *Journal of Researches* over the 1840s, became more sober, more wary of Humboldt's free combining of objectivist and subjectivist modes, less readily seduced by the 'eudaemonic bliss of the Humboldtian tropics', and slower to reach for the grand (unearned) generalization. In Darwin reading and rewriting Humboldt, Leask finds a powerful example of the process by which one of the most curious minds of the nineteenth century (to recall Wallace) found a means of expressing and channelling that curiosity into the forms of an authoritative scientific writing.

[12] 'The tree of life should perhaps be called the coral of life, base of branches dead; so that passages cannot be seen.—this again offers contradiction to constant succession of germs in progress.—«no only makes it excessively complicated.»' *Charles Darwin's Notebooks, 1836–1844: Geology, Transmutation of Species, Metaphysical Enquiries*, transcribed and ed. Paul H. Barrett *et al.* (Cambridge: Cambridge University Press, 1987), 177, B25–26. Cf. the frontispiece to *Darwin's Plots*, which reproduces one of Darwin's many attempts to sketch the multiple branchings of evolution.

[13] In her forthcoming book on *Alice in Wonderland*.

George Levine's 'And If It Be a Pretty Woman All the Better' takes another kind of prompt from Gillian Beer's writing about Darwin. Levine warms to what he describes as a characteristic willingness, on her part, to go beyond the evidence of 'cultural prejudice' as it makes itself heard in Victorian science, in order to credit the ways in which ideological bias has proved compatible with, at times instrumental in, the discovery of new insights into the natural world—insights that even now can command our agreement. Levine focuses on a work that has attracted recurrent criticism from scholars concerned (and too often content) to expose the 'interestedness' of Darwin's thinking: *The Descent of Man and Sexual Selection in Relation to Sex*. Taking issue with some of the most influential of those readings, Levine seeks to demonstrate the ways in which Darwin's thoroughly 'Victorian' attitudes to gender and sexuality helped to press him towards surprising and not at all standardly Victorian insights into the role of female sexual choice in the evolution of species. In the course of Levine's polemical rereading of the *Descent*, 'interest' emerges strongly revalued as a potential condition of, rather than a barrier to, advances in knowledge.

In 'Ordering Creation, or Maybe Not', Harriet Ritvo examines Darwin as one among many Victorian scientists wrestling with the central problem of taxonomy: how to find a language and a set of symbols for systematizing the natural world without doing violence to the intricacies of its variations. 'The Victorian search for synthesis', as Beer remarks in *Open Fields*, goes in tandem 'apparently contradictorily' with the pursuit of 'taxonomic refinement'.[14] Ritvo charts the many difficulties that faced Victorian systematizers, whose energy and directness in dealing with the intellectual and practical hazards of taxonomy belied the common impression (then and now) that their work was a 'crabbed' and 'arcane' legacy of the age of Linnaeus. In the vigorous debate about the virtues and vices of the quinary system, during the third quarter of the nineteenth century, 'Ordering Creation' finds a rich instance of the complexity and intensity with which scientific writers, both professional and amateur, highly specialist and generalist, sought to defend and refine the necessary labour of classification, on which so much other, better remembered scientific work depended.

My own essay draws on *Darwin's Plots* and on Beer's 1990 Bateson Lecture 'The Reader's Wager' to examine the relation between the individual and the group in mid-nineteenth-century history, science, and literature. 'Chances Are' begins with Henry Buckle's attempt in *The History of Civilisation in England* to devise a scientific approach to history, by which the apparent irregularity of its individual agents and events would be brought within the scope of science's 'taming of chance'. John Venn's critique of Buckle, in *The Logic of Chance* (1866), was among the most prominent and effective of the book's rebuttals, insisting that probability theory did not provide arguments against individual efforts in the domain of

[14] 'Parable, Professionalization, and Allusion in Victorian Scientific Writing', *AUMLA* (*The Journal of the Australasian Universities Language and Literature Association*), 74 (1990); rpt. in *Open Fields*, 196–215.

ethics. Thomas Hardy's *The Return of the Native* works a very similar vein as it explores the fraught value of individualism through the figure of Clym Yeobright. Pitting his intellectual resources against the ethos of individualism in the expectation of thereby improving the life of the social group into which he was born (but out of which he has been educated), Clym necessarily fails, but the exceptionality of his failure ironically enables his society to see itself more coherently, and establish more clearly a sense of its values. His own necessary ethical reorientation under the pressure of an experience of chance not adequately spoken to by the sciences of probability presents a telling example of what Bernard Williams has called 'moral luck': a concept whose valuation of the irregular individual case might, I suggest, be extended to the ways in which we think about intellectual history.

Sally Shuttleworth's essay is the first of three in the volume to focus specifically on the figure of the child. With an eye on Gillian Beer's alertness to non-congruences between literary and scientific knowledge, Shuttleworth draws our attention to the surprising lack of a sustained or coherent medical literature on the psychology of childhood until very late in the Victorian period. The decades that saw some of the most detailed and serious fictional explorations of the child's mind (she singles out for close attention Brontë's *Jane Eyre*, George Eliot's *Daniel Deronda*, Hardy's *Jude the Obscure*, James's *What Maisie Knew*) saw for the most part silence, uncertainty, and confusion from science. Shuttleworth is especially interested in the intractability of the question of children's mental responsibility in Victorian psychology, advice-manuals, and (increasingly, as the century progressed) evolutionary psychiatry. Could a child be insane? What was the importance, or otherwise, of a child's propensity for lying? What should one make of a child's anger? More abstractly: how far was childhood analogous to, or even a recapitulation of, the earlier developmental stages of the species? In such a context, Shuttleworth suggests, it is to literature that we must turn if we wish to hear the voices of children represented with psychological depth, complexity, and sympathy, rather than to a scientific and more broadly instructional writing as yet so ill-equipped and ill-disposed to listen to them.

As Shuttleworth notes, Gillian Beer's work has always preserved a special place for the literary. Her interdisciplinary readings are characterized by an awareness of literature's ability not just to reflect or parallel developments in the sciences, but on occasion to anticipate them by virtue of its willingness to let the imagination, and the language itself, be a guide. For that reason there is a special sympathy in Beer's writing with psychoanalysis, which never gave up its avowed reliance on metaphor, simile, and analogy, even as the physical and biological sciences became more resistant to such intuitive leaps. Rachel Bowlby's reading of Freud's 'Analysis of a Case of Phobia in a Five-year-old Boy', echoing Beer's fascination with the heuristic powers of children's curiosity, draws out the analogy ('more or less explicit' in Freud) between the sexual enlightenment sought by children and the enlightenment offered to the culture by Freudian psychoanalysis. Her title 'A Freudian Curiosity' plays on the proximity, indeed the identification, between the curiosity of children and the curiosity of the analyst, vividly expressed in Freud's response to 'Little Hans'. The 'enlightenment' of children in sexuality, Bowlby shows, is for Freud 'neither an innate knowledge nor a process of simply seeking or simply acquiring the facts'. It is fraught with evasions, obfuscations, and wilful dissimulation on the part of the adults, and on the part of the children themselves, who are bound to resist the knowledge that comes their way (though in Freudian theory that resistance plays itself out with significant differences for boys and girls). In the almost contemporaneous *Leonardo da Vinci* (1910), Bowlby locates a further recognition on Freud's part that the drive towards knowledge, for children, for adults, for analysts, is bound to end in 'failure or negative illumination'—an epistemological disappointment that psychoanalysis equips us to describe, and that it might also help us to begin to challenge.

Suzanne Raitt addresses the status of psychoanalysis as a science, or would-be science, at the start of her essay on *Beyond the Pleasure Principle* (1920). Taking note of Freud's insistence that psychology be included within the category of the natural sciences, she sees his inability to establish a physiological basis for psychoanalysis as a spur to the critical reflections in *Beyond the Pleasure Principle* on the nature of scientific language—especially the status of figuration and analogy in scientific writing. 'Was there such a thing as a transparently literal language, as positivist language would have us believe?' In the neglected opening pages to chapter 6 of that work Freud returns to scientific experiments conducted during the 1890s on prostista—unicellular organisms which replicate by self-dividing—as a way to opening up the question of the conceptual distinction between life and death. The possibility emerges in Freud's account of the experimental literature (as also in Oscar Wilde's 'The Picture of Dorian Gray') that, under certain circumstances, the distinction may exist 'only in the realm of representation'. By arguing that a conceptual divide so essential to the life sciences might be no more here than an effect of language, Freud was able to free psychoanalysis from the burden of defining its own, equivalent 'foundation in metaphor'.

In 'On Not Being Able to Sleep', Jacqueline Rose examines the peculiarly challenging problem sleep presented for Freud's general theory of psychology. Her rereading of *The Interpretation of Dreams* (partly in the light of fresh insights opened up by Jocelyn Crick's 1999 translation) focuses especially on chapter 7: 'the most psychological, at first glance . . . most unpsychoanalytic' section of the book, and the least discussed by Freud's critics. For Freud, dreaming—and the sleep that makes dreaming possible—led not towards enlightenment but 'out into the dark'. In the 'Metapsychological Supplement to the Theory of Dreams', sleep presents itself not as a necessary or indispensable uncertainty of the kind psychoanalysis was so ready to accept elsewhere, but as an intractable 'obstacle' to the psychoanalytic investigator. Sleep, in other words, is to Freud's search for a general psychology what night terrors are to the child afraid of the dark—the point at which fear ceases to be a metaphor for the limits of knowledge and 'becomes real'. Reading the *Interpretation of Dreams* alongside Proust, Rose finds in Freud a deep, unwanted uncertainty about the nature of the distinction between sleep and dreaming: 'does the dreamer sleep?' More pointedly, 'does the child sleep?' When we dream we return to a 'prehistoric' condition of immersion in the sensory—a return that is both fearful regression (Proust, outstripping Freud, would imagine regressing so far that one left oneself behind altogether), but also, Rose suggests, a possible way forward into a more creative future.

When Ella Freeman Sharpe wrote her sequel to Freud's *Interpretation of Dreams*, *Dream-Analysis: A Practical Handbook for Psycho-Analysts* (1937), she departed from Freud by insisting upon the concrete, corporeal nature of the dream medium. 'Metaphor may be what we get in exchange for ceasing to express emotions through the body,' Mary Jacobus writes, 'but the origin of metaphor in forgotten bodily experience leaves its visceral imprint on dream-language'. ' "Brownie" Sharpe and the Stuff of Dreams' reveals the extent to which Sharpe's 'literalism' troubles Lacanian and other accounts of the relation of the unconscious to language, insisting on the 'textile', rather than 'textual', production of dream-thoughts and phantasies. For Sharpe '(however improbably)', language was a form of 'outer-ance', substituted for the child's literal expressions of faecal matter; but her recurrent reference to 'the technologies of material production and visual projection' also gives her work a much more 'distinctively modernist inflection' than that line of theorizing alone would suggest. Reading Sharpe's 1940 essay on 'Metaphor' alongside *Dream-Analysis*, Jacobus is particularly drawn to Sharpe's account of symbol-formation. Founding the symbol always to some degree in the material object, especially the maternal body, Sharpe came to see symbolization as a mourning and letting go of the object, but therefore also to see dreaming (in a move remarkably reminiscent of Freud worrying around the distinction between dreaming and sleeping) as a means by which we 'plant images between ourselves and the faceless, dreamless state of death'.

Virginia Woolf has been one of the touchstones of Gillian Beer's writing since the publication of her 1979 essay 'Beyond Determinism'—and indeed well before then. (Meredith's kinship with Woolf, and her correspondingly high valuation of him, is a repeated note in *George Meredith: A Change of Masks.*)[15] In her introduction to the 1992 Penguin edition of *Between the Acts*, she draws the reader's attention to a preoccupation with violence and horror, more marked in that novel than anywhere else in Woolf's fiction. The revelation on which Beer's reading turns is Woolf's use (lost to modern readers before this edition) of *The Times*' reports from June and July 1938 about the rape of a 14-year-old girl by a Whitehall guard. The episode intrudes into the domestic scene of Woolf's novel: unwanted but undeniable evidence of a 'real' and brutal world that will soon invade her

[15] In Mary Jacobus (ed.), *Women Writing and Writing about Women* (London: Croom Helm, with the Oxford University Women's Studies Committee, 1979), 80–99. And see *Arguing with the Past: Essays in Narrative from Woolf to Sidney* (London: Routledge, 1989), *Virginia Woolf: The Common Ground* (Edinburgh: Edinburgh University Press, 1996). *Wave, Atom, Dinosaur: Woolf's Science* (London: Virginia Woolf Society of Great Britain, 2000). Also her edition of *The Waves* (Oxford: World's Classics, 1992), and her introduction and notes to *Between the Acts*, text edited by Stella McNichol (Harmondsworth: Penguin, 1992).

characters' lives more definitively in the form of war.[16] Trudi Tate's essay 'On Not Knowing Why' begins with a more muted but structurally similar intrusion of violence and militarism into the serenity of English lives in a Woolf novel: the scene in *To the Lighthouse* where Mr Ramsay's blusteringly approximate rendition of Tennyson's 'Charge of the Light Brigade' alarms Lily Briscoe by its show of masculine brutalism and (she imagines) triumphalism. Following Beer's example, Tate presses at the question of how the reality of war gets registered by a culture, and what it does to the artifices that sustain that culture's sense of its values. Returning to the original *Times*' reports of the loss of the British Light Cavalry Brigade at Balaklava, she charts the process by which William Howard Russell's revelation of the gross errors that led to the loss of some 300 or more soldiers' lives in the Crimea found more lasting re-expression in Tennyson's poem. 'The Charge of the Light Brigade' emerges from this reading not as 'simple minded' patriotism but as a 'subtle and even anguished reflection' upon warfare. The 'disciplined purposelessness' of poetry, Tate suggests, was the Light Brigade's best memorial, granting expression to a deeply ambivalent valuation of war that Woolf would tap into so cannily some seventy years on.

Kate Flint's essay 'Sounds of the City' pays further tribute to the major role Beer has played in establishing Woolf at the forefront of contemporary thinking about the transition from the Victorian period to modernity, about the role of gender in that process, and about the continuities and discontinuities between science and literature that were part of the process. 'Sounds of the City' begins with a wide survey of nineteenth-century writers' increasing tendency to characterize noise as nuisance, and, especially in scientific texts, to see a heightened sensitivity to noise as one of the indices of evolutionary progress. By the 1930s noise had long been seen as 'synonymous with urban modernity'. Much modernist literature pits itself antagonistically against it; much prefers to close its ears altogether. But in Virginia Woolf Flint finds a striking openness to the auditory richness of city life. Through readings of *Night and Day*, *Mrs Dalloway*, *The Waves*, *Jacob's Room*, and *The Years*, she charts a growing willingness on Woolf's part to use auditory impressions as a means of exploring the connections—sometimes willed, often unwilled, very often intensely associative and affective—that pertain between individuals in the modern city, and that help to establish its inhabitants' sense of community with each other in that potentially alienating space.

Maroula Joannou's title ' "Chloe Liked Olivia" ' also takes a lead from Woolf, using the celebrated line from *A Room of One's Own* as a prompt to exploring the less utopian lives that fiction was elsewhere granting to the Chloes and Olivias at work in the modern research laboratories of London. Joannou begins by analysing the novel that probably did more than any other literary work in its period to set the terms for later fictional treatments of women in science: H. G. Wells's *Ann Veronica* (1909). Based on Wells's affairs with two young women

[16] Introduction to *Between the Acts*, rpt. in *Virginia Woolf: The Common Ground*, 137.

students of science (one of whom became his wife), *Ann Veronica* established a three-stranded narrative form for the 'New Woman Scientist novel' (as Joannou dubs it): 'the narrative of scientific interest and discovery' brought into tension with 'the narrative of heterosexual romance' and a narrative of suffragette activism. As Joannou shows, both science and women's suffrage lose out in Wells's formulation, overwhelmed by a eugenic imperative to match these women with biologically fit partners. Fifteen years later, Wells's inability or refusal to imagine a career in the laboratory as compatible with 'good' sexual and domestic choices for women was one of the provocations that helped to shape two novels by women with a stronger commitment than Wells's to the suffragette cause and to the success of women in science. Edith Ayrton Zangwill's *The Call* (1924) and Charlotte Haldane's *Man's World* (1924) both employ some of the key motifs popularized by Wells's work: most obviously the male scientist's infatuation with a younger female scientist, and her involvement in suffragette politics. With very different emphases, both challenge Wells's prejudicial representation of science and suffrage, yet also end by confirming his 'recognition of deep, perhaps irreconcilable conflicts of desire at work when imagining what it might mean to be a woman scientist in [the] early twentieth-century'.

Like much of the work collected in this volume, Alison Winter's essay follows Gillian Beer's example in attending to literature's supplementary, at times contradictory rather than simply confirmatory, relation to the sciences. 'The Chemistry of Truth' deals with one the most physiologically and psychologically 'literal-minded' expressions of science's pursuit of truth in the first half of the twentieth century, and with the challenges literature posed to that always contentious ambition. Winter describes the history of science's search for a 'truth serum': a drug that could provide controlled and dependable access to memories that its subjects would otherwise, voluntarily or involuntarily, suppress. Tracing the beginnings of this pharmaceutical quest in Victorian reflex physiology, she goes on to chart the emergence of the first recognized lie-detection technologies. The anaesthetic 'scopolamine' was synthesized in Germany in the 1890s. Its power to enable the elicitation of true statements from semi-conscious patients was discovered accidentally in 1916, and from the 1920s was given extensive forensic application in America and Europe. The new barbiturates sodium amytal and sodium pentothal followed at the end of the 1920s and 'rapidly eclipsed' scopolamine, but were more prominently used in psychiatry than in police work. Like scopolamine they attracted a great deal of interest and considerable trust within scientific circles. In the absence of a sustained critique from within science itself, it was left to literature to raise the necessary questions about the uses—especially the potential political uses—of these chemical technologies. Winter looks at two dystopian novels of the 1940s, George Orwell's classic protest against the state's attempt to control the minds of its citizens, *Nineteen Eighty-Four* (published in 1949) and the less well known novel *Kallocain* by the Swedish writer Karin Boye (first translated into English in 1949). In Boye's and Orwell's futuristic imaginings, truth serum emerges as 'an extremely powerful tool of repression', the means to abolishing dissent in favour of 'verified false consciousness'. Behind this hitherto untold history, Winter concludes, there lurks another, larger story waiting to be explored, about the cultural history of memory and the technologies that have attempted to answer our need to make it more accessible, more controllable, than we have always known it to be.

To read J. D. Watson's enormously popular account of his and Francis Crick's epochal discovery of the structure of DNA, *The Double Helix* (1968), is to find confirmation of how persistent and how deep-set the gender conflicts analysed in Joannou's essay (and underlying Winter's account of women as early experimental subjects for truth serum) would remain in the public representation of science. Approaching *The Double Helix* as a 'Coming of Age' narrative, Evelyn Fox Keller reveals a 'rather remarkable parallelism' at work on at least three levels in Watson's story: 'between models on the one hand and physical/biological reality on the other; between replication and reproduction; and between Cambridge "popsies" [desirable women, by Watson's strikingly limited criteria] and actual (especially mature) women'. Tracing the emergence of those key oppositions across the course of this highly 'literary' history of scientific discovery, Keller exposes Watson's aestheticized valuation of the model over the reality, the feminine ideal over the real woman, and (most simply) the male over the female. *The Double Helix* cannot and should not be taken as representative of twentieth-century science more broadly, Keller warns, but its heroic narrative has exerted, and continues to exert, a remarkably strong pull on readers: 'That fact alone . . . warrants the effort of identifying and exhibiting the very particular uni-dimensionality of its underlying intellectual and emotional aesthetic.'

All the contributors to this book have found Gillian Beer's writing both an influence and an inspiration. In her encouragement of her students and fellow scholars, she has always had a fine eye for the unexpected in the work of others. That our work has taken so many different and productive directions is a tribute to her generosity and good influence.

We present these essays to her with admiration, and gratitude, and deep affection.

CHAPTER 1

Darwin's 'Second Sun': Alexander von Humboldt and the Genesis of *The Voyage of the Beagle*

NIGEL LEASK

> [Humboldt] like another Sun illumines everything I behold.
>
> (Charles Darwin)[1]

For modern readers of Charles Darwin, the 1859 *Origin of Species* inevitably casts its long shadow back over his earlier travel book, commonly known as *The Voyage of the Beagle*, first published in 1839, and again, with substantial revisions, in 1845. Gillian Beer aptly describes Darwin's book (actually published under the title *Journal of Researches into the Geology and Natural History of the Various Countries Visited by H.M.S. Beagle*) as 'a narrative whose peripeteia is just beyond the scope of the text, has not yet been performed'.[2] Darwin himself paid retrospective tribute to the importance of the *Beagle* voyage in developing his ideas about the transmutation of species and natural selection in the opening pages of the *Origin*. Scientific voyages constituted a *rite de passage* for many distinguished nineteenth-century naturalists such as J. D. Hooker, A. R. Wallace, and T. H. Huxley, and Darwin's was no exception. He later described the voyage '[as] the first real training or education

[1] *Charles Darwin's Beagle Diary*, ed. Richard Darwin Keynes (Cambridge: Cambridge University Press, 1988), 42. Entry for 28 February 1832, Bahia, Brazil. Henceforth *Diary* in text.

[2] 'Writing Darwin's Islands: England and the Insular Condition', in Timothy Lenoir (ed.), *Inscribing Science: Scientific Texts and the Materiality of Communication* (Stanford, Calif.: Stanford University Press, 1998), 119–39 (121). Darwin's was the third volume of *Narrative of the Surveying Voyages of His Majesty's Ships Adventure and Beagle, between the years 1826 and 1836, describing their examination of the Southern Shores of South America and the Beagle's Circumnavigation of the Globe* (London: Henry Colburn, 1839). The first volume contained the narrative of Captain Parker King, captain of the first survey voyage of 1826–30, and the second Captain Robert Fitzroy's account of the *Beagle's* second voyage in 1831–6. Unless otherwise specified, I refer here to the useful abridgement of 1839, *Voyage of the Beagle: Charles Darwin's Journal of Researches*, ed. with an intro. by Janet Browne and Michael Neve (Harmondsworth: Penguin Books, 1989). Henceforth '1839' in text. The success of Darwin's *Journal* led his publisher Henry Colburn to unshackle it from the expensive original four-volume set (where it bore the title *Journal and Remarks 1832–6*) and republish it in the same year as a separate volume, now entitled *Journal of Researches*.

of my mind ... everything about which I thought or read was made to bear directly on what I had seen and was likely to see; and this habit of mind was continued during the five years of the voyage'.[3]

Some scholars have gone further in estimating the importance of the voyage for Darwin's scientific breakthrough. Darwin's granddaughter Nora Barlow long ago proposed that Darwin adopted transmutationist views in mid-1835 while he was on board the *Beagle*, a suggestion that cast the voyage as something of a 'eureka experience'.[4] Subsequent Darwin scholarship has refuted this suggestion, preferring a date of no earlier than January or February 1837 for Darwin's first grasp of transmutationist principles. That is to say, a date *after* his return to England, during the period when he was working on the *Beagle Journal* and the accompanying scientific treatises. Sandra Herbert sums up current thinking when she writes that, not the voyage itself, but rather 'the project of writing the naturalistic report on ... the voyage ... brought [the transmutation of species] to the forefront of Darwin's attention'. Darwin's personal experience of the voyage was broadened and generalized by the act of redacting his *Diary* and notebooks, and by the addition of references to at least thirty-five other travel narratives, as well as to contemporary works of science.[5]

Critics eager to discern hints and anticipations of Darwin's later theories have tended to overlook the conventions of nineteenth-century travel writing that influenced the shape of both Darwin's *Beagle Diary* (written during the voyage itself and sent back home in sections to be read by his family and friends) and the *Journal of Researches*. Like most travel accounts, the redacted *Journal* follows the narrative structure of the itinerary, although Darwin did also (conventionally) reorder it with a view to providing a modicum of geographical coherence to the *Beagle*'s 'to-ings and fro-ings' whilst surveying the coast of South America. 'I intend making it in a journal form,' he wrote to W. D. Fox in March 1837, 'but following the order of places rather than of time, giving results of my geology and the habits of animals where interesting.'[6] He accordingly confined to single chapters accounts of places such as Tierra del Fuego and the Falkland Islands which were visited twice by the *Beagle* at an interval of a year or more. Darwin's 'first sight in his native haunt of a real barbarian' (1839: 375) in Tierra del Fuego is given a climactic emphasis, further heightened in 1845 by the episode's replacement in the order of the narrative *after* the Falklands, and immediately *prior* to the passage of the Magellan straits and entry into the Pacific. In fact this event occurred early in the voyage, in December 1832, before his extensive land expeditions in Argentina and the Banda Oriental. As John Tallmadge has indicated,

[3] *Charles Darwin, T. H. Huxley, Autobiographies*, ed. with an intro. by Gavin de Beer (London: Oxford University Press, 1974), 45. Henceforth '*Auto.*' in text.

[4] Cited by Sandra Herbert, 'The Place of Man in the Development of Darwin's Theory of Transmutation, Part I, To July 1837', *Journal of the History of Biology* 7/2 (1974), 215–58 (234).

[5] Ibid. 234, 249, 245.

[6] *The Correspondence of Charles Darwin*, ed. Frederick Burkhardt *et al.*, in progress (Cambridge: Cambridge University Press, 1985–), ii. 14. Henceforth '*Corr.*' in text.

in the extensively revised edition of 1845, published by John Murray in his popular 'Home and Colonial Library' series, Darwin further sought to consolidate a sense of geographical and circumnavigatory 'progress' in the *Beagle*'s meandering survey voyage.[7]

In conformity with the conventions of nineteenth-century travel writing, in all these versions Darwin's observations are still presented in unsystematic form, rather than aspiring to the tightly developmental structure of a scientific monograph, exemplified in the powerful build-up of analogical arguments in the *Origin of Species*. This presentation is evident in the relationship between itinerary narrative and scientific disquisition; achieving a successful balance here was seen as a major problem for eighteenth- and nineteenth-century scientific travel writing. Although Darwin did add scientific material from his zoological and geological notebooks whilst transforming the *Diary* into the *Journal* (only about half of the 182,000 words in the *Diary* were incorporated into the *Journal*, the final length of which was 223,000 words[8]), the passages of scientific discussion still take their place within the overarching literary structure of the itinerary narrative. Notably, anthropological passages (which I discuss below) were transferred from the *Diary* to the published *Journal* with minimal revision. Both *Diary* and *Journal* place scientific observations within a personal, political, and geographical context that exerted a profound influence on the development of Darwin's thinking. Gillian Beer speculates upon the manner in which the genre of travel writing exerted its secret ministry on Darwin's theoretical breakthrough; 'the instability of register across which he worked . . . in itself helped to break up the securities from which he set out and brought into question the assumptions and values that would have curtailed his inquiry'.[9]

The *Diary* also—and most famously—bears witness to Darwin's failure to register the significance of the distribution of finches and turtles on the Galapagos Islands, whilst exploring the volcanic Pacific archipelago. Because Darwin's acceptance of transmutation seems to have occurred whilst he was redacting his *Diary* in the process of composing his travel account, the incremental commentary added to his original field notes in the 1839 and 1845 editions of the *Journal* provides a fascinating glimpse of Darwinian theory in the making, even though he was still not willing to publicize the fact or mechanism of the transmutation of species until 1859, goaded on by Alfred Wallace's parallel discovery.[10]

[7] It was this edition that would become famous, going through at least 160 editions in English, and being translated into twenty-two languages. See John Tallmadge, 'From Chronicle to Quest: The Shaping of Darwin's *Voyage of the Beagle*', *Victorian Studies* 23/3 (1980), 324–45 (325). Note Tallmadge's useful chart of the geographical reordering of the 1845 *Journal* in relation to the *Diary*'s itinerary (332). The chief weakness of Tallmadge's study is his failure to register the importance of the intermediate, 1839 text in his study of Darwin's redaction of the *Diary* into the 1845 text. Thanks to Jim Secord for this reference.

[8] *Diary*, pp. xx–xxi.

[9] 'Writing Darwin's Islands', 120–1.

[10] In chap. 17, Darwin writes, in a footnote on the diversity of species either side of the Andes, that his 'whole reasoning . . . is founded on the assumption of the immutability of species' [1839: 249]. This

DARWIN AND HUMBOLDT

Far from being merely an empirical transcript of the traveller's 'voyage into substance' (as travel writers often claimed, for rhetorical purposes), nineteenth-century scientific travel accounts demand to be interpreted according to specific generic and textual parameters. Despite the seminal influence of Lyell's *Principles of Geology* on the development of Darwin's thinking during the voyage, I want to argue here that Alexander von Humboldt's *Personal Narrative of a Journey to the Equinoctial Regions of the New Continent* (1814–25)[11] was the principal intertextual influence on Darwin's own travel account.[12] (Humboldt's book was brilliantly translated into English from the original French by Helen Maria Williams in 1814–28.) Humboldt was of course far more than just a travel writer for Darwin and his contemporaries. Building on Susan Cannon's designation of British science in the 1820–50 period as 'Humboldtian' rather than 'Baconian', and her identification of Darwin and his scientific mentors John Herschel and Charles Lyell as pre-eminent 'British Humboldtians',[13] one can appreciate the full justice of Darwin's claim in a letter to J. D. Hooker of February 1845 that 'my whole course of life is due to having read & reread as a Youth [Humboldt's] *Personal Narrative*' (*Corr.* iii. 140). Humboldt's influence extends across the whole burgeoning field of nineteenth-century bio-geography that would produce, in the work of J. D. Hooker and A. R. Wallace, and of Darwin himself, the crucial evidence for adaptive evolution.[14] Darwin frequently cited Humboldt in the *Transmutation Notebook*, the *Origin of Species*, and in later works.[15] As is clear from a throwaway remark in chapter 12 of the *Origin*, he was in the habit of ransacking 'the oldest voyages' as well as contemporary travel accounts (such as Humboldt's) in search of data concerning the geographical distribution of plant and animal species. Along with his evident interest in the geological sections of the *Personal Narrative*, it is this concern that inspires many of his marginal annotations in his copy held in the Cambridge University Library, particularly those

footnote was retained in the 1845 edition. As Martin J. S. Rudwick has indicated, Darwin's research in the years when he was writing up the results of his voyage represents 'an almost unparalleled example of a major scientific theory that can be followed in detail through all the phases of its development', thereby throwing considerable light both upon 'the dynamic processes of cognitive growth on the individual level' and the complex interactions between private speculation and public science. 'Charles Darwin in London: The Integration of Public and Private Science', *Isis* 73 (1982), 186–206 (186).

[11] Henceforth *PN* in text.

[12] Here I dissent from Jonathan Smith's argument in *Fact and Feeling: Baconian Science and the Nineteenth-Century Literary Imagination* (Madison, Wis.: Wisconsin University Press, 1994), especially ch. 3 'Seeing through Lyell's Eyes: The Uniformitarian Imagination and *The Voyage of the Beagle*'.

[13] *Science in Culture: The Early Victorian Period* (New York: Dawson and Science History Pubs., 1978), 82, 85, 86.

[14] Peter J. Bowler, *Charles Darwin: The Man and his Influence* (Cambridge: Cambridge University Press, 1990), 140.

[15] Paul Barrett and Alain F. Corcos, 'A Letter from Alexander Humboldt to Charles Darwin', *Journal of the History of Medicine* (April 1872), 159–72 (161).

dating from the climacteric 1837–9 period.[16] Pencil notes in this copy, some of which are dated 'Aug 1872' (III. 14) and 'July 6, 1881/April 3 1882' (V, inside back cover), indicate that Darwin's 'reading and rereading' of Humboldt continued to within a fortnight of his death.

Humboldt's narrative clearly had different meanings for Darwin at different stages of his long intellectual career. The most obvious link between Darwin's reading of Humboldt and his work on transmutation is the latter's pioneering work on the geographical distribution of species, although I will suggest other possible influences below.[17] Darwin as aspiring young gentleman-naturalist was particularly impressed by Humboldt's 'aristocratic' overview, his romantic drive to integrate all the sciences with a holistic and aesthetic outlook in an era of increasingly 'mechanical' specialization, without forfeiting scientific rigour. Humboldt's project of 'global physics' sought to 'investigate the confluence and interweaving of all physical forces, and the influence of dead nature on the animate animal and plant creation'.[18] Ironically, the increasing disciplinary specialization of the sciences by the 1830s and the limitations of Darwin's own training at the time of the *Beagle* voyage impeded him from matching the vertiginous range of Humboldt's scientific research. Humboldt remarked, on the eve of his departure for America in 1799, 'I have had to instruct myself in every empirical discipline . . . We have botanists, we have mineralogists, but no physicists [*Physiker*], as Bacon called for in the *Sylva Sylvanum*.'[19] As Darwin admitted in his *Autobiography*, he simply couldn't match Humboldt's enormous range of scientific expertise, and although his post-Lyellian geological observations superseded those Humboldt had made thirty years earlier, he later confessed to having 'wasted much of his time' in zoology as a result of ignorance of anatomy and bad drawing skills (*Auto.* 44). Humboldtian 'global physics' represented an intellectual challenge that Darwin would surmount in his mature scientific work on natural selection, although not as a callow 22-year-old ship's naturalist (without publications to his name) on the voyage itself.

The mature Darwin would come to disagree with Humboldt on many specific scientific issues, such as the multiplicity of creation, the migration of species, geographical catastrophism, and the dynamics of volcanic action. Above all Humboldt was more concerned with the empirical mapping of laws and regularities of nature than with the 'curious' and teasing anomalies which fascinated his younger disciple.[20] He warned, in the *Personal Narrative*, that 'the causes of the

[16] Charles Darwin, *The Origin of Species*, ed. J. W. Burrow (based on 1st edn.) (Harmondsworth: Penguin Books, 1968), 382; Barrett and Corcos, 'Letter from Alexander Humboldt', 160–1.

[17] Cannon, *Science in Culture*, 87–8.

[18] Michael S. Dettelbach, 'Global Physics and Aesthetic Empire: Humboldt's Physical Portrait of the Tropics', in David Philip Miller and Peter Hanns Reill (eds.), *Visions of Empire: Voyages, Botany and Representations of Nature* (Cambridge: Cambridge University Press, 1994), 266.

[19] Ibid.

[20] Dennis Porter writes of Darwin's 'peculiarly sharp eye for the odd or anomalous, for whatever breaks the pattern or does not fit any paradigm', by way of contrast with Humboldtian regularities. *Haunted Journeys: Desire and Transgression in European Travel Writing* (Princeton, NJ: Princeton University Press, 1991), 152.

distribution of species are among the number of mysteries, which natural philosophy cannot reach. This science is not occupied in the investigation of the *origin of beings*, but of the laws according to which they are distributed' (*PN* v. 180, heavily marked in Darwin's copy. Italics mine). Darwin *appears* to allude to this passage in Humboldt in the introduction to the *Origin of Species*, where the question of origins is described as 'that mystery of mysteries, as it has been called by one of our greatest philosophers' (65). The editors of Darwin's early notebooks have indicated, however, on the strength of Darwin's entry for 2 December 1838, that the immediate source of this remark was not Humboldt but rather the astronomer John Herschel, in a letter of 20 February 1836 to Charles Lyell: 'Of course I allude to the mystery of mysteries, the replacement of extinct species by others.' Darwin had met Herschel in Cape Town in June 1836 on the *Beagle* voyage, but he probably first encountered the phrase in Charles Babbage's *Ninth Bridgewater Treatise* in May 1837, which printed Herschel's letter.[21] Even if Herschel and not Humboldt is Darwin's immediate source, his turn of phrase is itself highly Humboldtian, and would doubtless have resonated in Darwin's mind with the *Personal Narrative* passage quoted above. Susan Cannon has suggested that Humboldt surely didn't intend his remark about the 'mystery' of origins to be taken as a prohibition, but rather as a challenge. Especially in the light of Herschel's romantic excitement about 'unveiling a dim glimpse of a region of speculation' (as expressed in the preceding sentence of his 1836 letter to Lyell), Darwin would surely have been keen to take up that challenge. For Humboldt, the laborious task of quantitative description—the isothermal, isodynamic, isochronic project of a 'physics of the earth'—must precede further forays into origins, that penetralium of mysteries. Theories floated in the past lacked the sheer empirical groundwork necessary to solve the weighty problem of origins. His English followers such as Charles Lyell, John Herschel, and Darwin himself, took a less cautious view.[22]

Darwin's *Journal* suggests a complex 'anxiety of influence' in relation to his great Prussian precursor. Had not the young English naturalist actually outdone his master by observing the continental extremes of South America, not just limiting himself, as Humboldt had done, to the 'equinoctial regions'? And hadn't Lyell's *Principles of Geology* permitted him to make connections in theorizing the formation of the earth's surface impossible to the Prussian traveller with his Wernerian training at Freiberg Mining Academy?[23] Moreover, Darwin had

[21] *Charles Darwin's Notebooks, 1836–44: Geology, Transmutation of Species, Metaphysical Enquiries*, transcribed and ed. Paul H. Barrett *et al.* (Cambridge: Cambridge University Press, 1987), 413, E59. ('Herschel calls the appearance of new species. the mystery of mysteries. & has grand passage upon problem.! Hurrah.'). Thanks to Simon Schaffer for explaining this connection to me. Cf. also Cannon, *Science in Culture*, 89.

[22] In fact Humboldt had earlier written 'in vain would reason forbid man to form hypotheses on the origin of things; he is not the less tormented with these insoluble problems of the distribution of beings' (iii. 491). Also marked in Darwin's copy.

[23] 'It is a rare piece of good fortune for me,' Darwin wrote to his sister Caroline on 29 April 1836, 'that of the many errant Naturalists, there have been few or rather no geologists. I shall enter the field

successfully performed the circumnavigation to which Humboldt and his French *companion de voyage* Aimé Bonpland had aspired, but failed to achieve, when they were forced to return to Europe in 1804. Although Darwin lacked Humboldt's informed interest in anthropology and comparative linguistics, he had encountered Native Americans in their 'lowest and savage state' (1839: 375), whereas Humboldt had really only observed them cooped up in the missions of the Orinoco, or as otherwise 'half-civilized' by the Catholic religion and European manners of their Spanish colonizers. Of course, the thirty-odd years that divided the two expeditions had seen momentous political changes in Latin America and in European attitudes to the 'New Continent'. Whilst Humboldt had been inspired by the republican spirit of the Spanish colonies he visited, on the eve of their independence struggle against colonial Spain, Darwin wrote at a time of disenchantment in Europe's attitudes to Latin America, after the collapse of the 1820s mining boom. National independence (the price of which was often massive economic dependence on Northern capital) appeared to have collapsed into civil war and debilitating political and economic instability. Another disturbing aspect of independence was the survival of colonial slavery in Brazil, and racial genocide against indigenous populations, witnessed by Darwin at first hand in Argentina, Uruguay, and Chile.

An obstacle to the full appreciation of Humboldt's influence upon the young Darwin has been an apparent difference between their respective practices of travel and therefore—inevitably—their styles of observation. In reading the introduction to Humboldt's *Personal Narrative*, Darwin would have been confronted with a strong caveat regarding the scientific usefulness of circumnavigatory and coastal surveying voyages such as that undertaken by the *Beagle*: 'it is not by sailing along the coast, that we can discover the direction of the chains of mountains, and their geological constitution, the climate of each zone, and its influence on the forms and the habits of organised beings' (*PN* i. p. vii). Only by penetrating the interior of continents, Humboldt argued, could the travelling naturalist grasp the complicated nexus of organic and inorganic forces that he termed the 'physique du monde'.[24] For Humboldt as exponent of the 'interiority' of romantic travel, maritime exploration (the exemplary form of enlightenment inquiry into nature), while casting lustre on the names of such explorers as Cook, Banks, and Bougainville, hardly advanced the real study of nature, holistically conceived.

In fact, far from being antipathetic to his aspirations as a scientific traveller, this was grist to Darwin's mill. Captain Fitzroy's official brief from the Admiralty was to complete the coastal survey of South America initially undertaken by Captain King in 1826, and to establish a chronometric chain around the world,

unopposed' (*Corr.* i. 496). Perhaps he preferred to overlook Humboldt's geological pretensions on the grounds that they had been superseded by Lyell, despite his evident fascination with the copious geological sections of the *Personal Narrative*.

[24] Dettelbach, 'Global Physics and Aesthetic Empire', 266.

but Darwin (Fitzroy's 'gentleman companion') had no role to play in this work. Despite the importance of islands and archipelagos in Darwin's scientific work (as Gillian Beer has reminded us), his voyage on the *Beagle* was, according to Michael Neve and Janet Browne, 'not so much a journey at sea but a voyage on land'.[25] Although the *Beagle* voyage lasted nearly five years, Darwin was actually on shore for a total of three years and three months, sometimes for lengthy periods, as in his three-month sojourn near Rio de Janeiro, his four-month travels around the Rio Negro in Patagonia, or his equally extensive Andean excursion from Richard Corfield's house in Valparaiso, Chile.[26] Most of the gripping episodic passages in Darwin's *Journal* describe his inland excursions (Rio Negro, Rio Santa Cruz, the Andean expedition, etc.), which are transferred from his *Diary* with only minor changes.[27] It wasn't just the seasickness that tormented him on board ship that encouraged Darwin to spend so much of his time on land, but rather a Humboldtian interest in the geology and natural history of continental interiors. Darwin's intellectual interests can even be to some extent quantified in terms of the work of transcription he did during the voyage: his geological notes comprise 1,383 pages, his zoological notes only 368, and his 'personal' diary 779 pages.[28] (But Humboldt's own particular favourite, plant geography, seems to have been a minor interest for the young Darwin.)

Last but not least, we need to consider Darwin's knowledge of Humboldt's *Personal Narrative* as part of the formative intellectual experience of the voyage itself, and not just 'another travel account' upon which he drew in redacting his *Diary* for publication. We know that he had read and re-read Humboldt at Cambridge, at home in Shropshire, and on a field trip with Sedgwick to Wales, and would certainly have had the complete *Personal Narrative* on his book-shelf whilst writing up his *Journal* (and developing the transmutationist hypothesis) in England in 1837–8. But what kind of access did he have to Humboldt's text on the voyage itself? As we have seen, later in life Darwin singled out the lesson of combining theory and reading with personal observation as the steepest learning curve of the *Beagle* voyage (*Auto.* 44). The editors of Darwin's *Correspondence*[29] propose that he possessed only the first two volumes (bound in one) of Humboldt's seven-volume *Personal Narrative* on the *Beagle*, a proposal that clearly minimizes Humboldt's probable influence on the genesis of his thought in these years. This volume was inscribed on the inside flyleaf with the words 'J. S. Henslow to his friend Charles Darwin on his departure from England upon a voyage around the World, 21st Sept. 1831'.[30]

[25] *Voyage of the Beagle*, 16.

[26] Ibid.

[27] Tallmadge is mistaken in arguing that 'the time taken up on these journeys was a relatively small fraction of the 5 years Darwin spent with the *Beagle* expedition'. 'From Chronicle to Quest', 334.

[28] Howard E. Gruber and Valai Gruber, 'The Eye of Reason: Darwin's Development during the *Beagle* Voyage', *Isis* 53 (1962), 186–200 (189).

[29] Appendix IV to *Corr.* i: 'The Books on Board the *Beagle*'.

[30] The fact that the individual volumes in Darwin's complete set of the *Personal Narrative* in Cambridge University Library are from heterogeneous editions (i and ii, 3rd edn., 1822; iii, 2nd edn., 1822; iv, 1st edn., 1819; v, 1st edn., 1819; vi, 1st edn., 1826; vii, 1st edn., 1829) is surely not an objection to

That this was the sole volume of the *Personal Narrative* that accompanied Darwin on the voyage seems highly unlikely, for several reasons. It is improbable that Professor Henslow was so parsimonious as to present his young Cambridge protégé with the first only in a seven-volume set of Humboldt. Moreover, there was no shortage of space for travel books in the *Beagle*'s poop-cabin library. Captain Fitzroy explicitly invited Darwin to 'take your Humboldt', urging that 'there will be plenty of room for books' (*Corr.* i. 167), a permission that he would hardly have felt necessary to grant if he had imagined Darwin's Humboldt to consist of a single volume. As is evident from citations in the *Diary* and *Correspondence,* Darwin spent his time on board reading the standard contemporary South American travel accounts as part of his 'homework', books such as Basil Hall's *Extracts from a Journal, Written on the Coasts of Chili, Peru, and Mexico* (1824), Francis Head's *Rough Notes taken during some rapid journeys across the Pampas* (1826), and Frederick Beechey's *Narrative of a Voyage to the Pacific* (1831). As the editors of the *Correspondence* themselves acknowledge, books of travel and natural history were as much a part of the *Beagle*'s equipment as her twenty-four precise chronometers and surveying instruments (*Corr.* i. 553–7). And Fitzroy's narrative, no less than Darwin's, owes a considerable debt to Humboldt.[31]

The really clinching evidence that the *Beagle* carried a complete set of Humboldt's *Personal Narrative,* however, is Darwin's request, in a letter to his sister Catherine from Maldonado in May/July 1833, for 'the 8th volume of Humboldt'.[32] If it were the case that Darwin possessed only the first volume inscribed by Henslow, why wouldn't he have requested the second and third, rather than the eighth volume? The editors of the *Correspondence* misleadingly annotate Darwin's statement as follows: 'The eighth and ninth volumes of [Humboldt's] *Voyage* were devoted to zoology and comparative anatomy; they were not translated' (*Corr.* i. 260¹]. In so doing they confuse the volume numbers of the English translation of Humboldt's *Personal Narrative* with the volume numbers of the complete thirty-one-volume French *Voyage aux régions équinoxiales du nouveau continent.*[33] In fact Humboldt had mysteriously suppressed

their having been purchased together in 1831 as part of Henslow's gift; maybe the Cambridge bookseller had sold out of any single edition and simply cobbled together a complete set from disparate editions.

[31] I have examined the annotations in Darwin's copy of the *Personal Narrative* in the Cambridge University Library in the hope of finding concrete evidence, but (as mentioned above) many of his annotations appear to have been made at a later date. Perhaps, for all his 'reading and rereading' of Humboldt's *Personal Narrative* during the five-year voyage, Darwin didn't make pencil notes in his copy precisely because, in accordance with the surviving regulations for the *Beagle*'s poop-cabin library, they were part of the ship's library. See *Corr.* i. 554.

[32] In a letter dated 18 August of the previous year, his brother Erasmus had informed him 'The 8th volume of the Personal Narrative was not published', so it was clearly not the first time that Darwin had made the request (*Corr.* i. 258).

[33] Vols. viii–xiv of the *Voyage* were in fact the Latin botanical account, *Nova genera et species plantarum, quas in peregrinatione ad plagam aequinoctialem orbis novi collegerunt, descripserunt, partim adumbraverunt,* 7 vols. (Paris: Schoell, 1815–25). The *Relation historique* (French original of the *Personal Narrative*) composed vols. xxviii–xxxi of Humboldt's complete *Voyage.*

almost one half of the personal narrative of his 1799–1804 Spanish American expedition, truncating it (at the end of the third French folio volume, seventh in the English translation) at the point when he and Aimé Bonpland set off down the Magdalena river in Columbia bound for Ecuador, Peru, and eventually Mexico. Headed as he was for Chile, Peru, and the Pacific coast on the second stage of the *Beagle*'s surveying voyage, it is hardly surprising that Darwin was eager to procure the continuation of Humboldt's narrative, oblivious of the fact that the seventh volume of the English translation (published in 1829) had been the last.

DARWIN AND HUMBOLDTIAN LANDSCAPE AESTHETICS

As the paramount traveller and travel writer of the romantic era and friend of Schiller and Goethe, Humboldt was anxious to 'enlist poetry under the banner of science' (in the memorable words of Darwin's grandfather Erasmus).[34] Humboldt's travel writings (notably the *Personal Narrative* and the *Tableaux de la nature* (1828), which Darwin requested on 5 July 1832 (*Corr.* i. 247)), had established a tropical aesthetic for nineteenth-century European readers, selecting privileged geographical sites in the 'New Continent' for aesthetic evocation: fecund and picturesque tropical forests, boundless *llanos* and pampas, sublime and lofty *cordilleras*. Mary Louise Pratt argues that Humboldt sought 'to reframe bourgeois subjectivity, heading off its sundering of objectivist and subjectivist strategies, science and sentiment, information and experience'.[35] The young Charles Darwin showed his appreciation of Humboldt's romantic holism in praising the German traveller's 'rare union of poetry with science . . . when writing on tropical scenery' (*Diary*, 42). Darwin was profoundly swayed by both the subjectivist aspirations of Humboldtian aesthetics (evident in the landscape descriptions of the *Beagle Diary* and the *Journal*),[36] and the objectivist concerns of his travel writing, his ability to reflect upon and generalize his particular experiences. Humboldt seemed to overcome the invidious distinction between the 'sedentary naturalist' working in the metropolitan 'centre of calculation' and the field collector who merely supplied the former with raw material. As Darwin wrote home in self-deprecating mood in May 1833, 'I am only a sort of Jackall, a lions provider; but I wish I was sure there were lions enough' (*Corr.* i. 316). Humboldt would teach him how a 'Jackall' could transmute itself into a lion.

Darwin tells us that his desire to travel was originally stimulated by his reading

[34] Advertisement to 'The Loves of the Plants', Part II of *The Botanic Garden* (Lichfield: J. Jackson, 1789).

[35] Mary Louise Pratt, *Imperial Eyes: Travel Writing and Transculturation* (London: Routledge, 1994), 119.

[36] It is noteworthy that the elderly Darwin mentioned the fact that his delight in landscape scenery 'has lasted longer than any other aesthetic pleasure', not excepting the poetry of Shakespeare or Milton (*Auto.* 23).

of Humboldt's account of Teneriffe and of 'Tropical scenery & vegetation' (*Corr.* i. 120). Another famous Humboldtian passage of landscape description that influenced Darwin was his evocation of the traveller's first impression of tropical forest in equatorial America, in the third volume of the *Personal Narrative.* In the heart of a continent 'where everything is gigantic', Humboldt wrote, 'if [the traveller] feels strongly the beauty of picturesque scenery, he can scarcely define the various emotions which crowd upon his mind; he can scarcely distinguish what most excites his admiration, the deep silence of those solitudes, the individual beauty and contrast of the forms, or the vigour and freshness of vegetable life, which characterise the climate of the tropics' (*PN* iii. 36). When Darwin first experienced Brazilian forest scenery at first hand in February 1832 he reached for this description of the tropical sublime as an 'embarrassment of riches'; 'the mind is a chaos of delight, out of which a world of future & more quiet pleasures will arise.—I am at present fit only to read Humboldt; he like another Sun illumines everything I behold' (*Diary*, 42). The intertextual sun of Humboldtian landscape aesthetics here outshines even the brilliant light of the tropics as experienced for the first time by the young and impressionable 23-year-old naturalist.

The *Beagle* of course spent more of her time surveying coastlines in temperate and sub-Antarctic latitudes than plying the lacquered waters of the tropics. Under the grey skies and icy winds of southern Patagonia and Tierra del Fuego, Darwin often yearned for Humboldt's equinoctial regions, and his account of Fuegian forest scenery reads almost like a nightmarish inversion of the more favoured Humboldtian tropics: 'In the deep ravines, the death-like scene of desolation exceeded all description . . . so gloomy, cold, and wet was every part, that not even the fungi, mosses, or ferns could flourish. In the valleys it was scarcely possible to crawl along, they were so completely barricaded by the great mouldering trunks' (1839: 196). Even in his later years, the beauty of the Humboldtian tropics is mediated by the Patagonian sublime of terror in Darwin's imagination: 'the glories of the vegetation of the tropics rise before my mind at the present time more vividly than anything else. Though the sense of sublimity, which the great deserts of Patagonia and the forest-clad mountains of Tierra del Fuego excited in me, has left an indelible impression on my mind' (*Auto.* 46). To the post-*Origin* reader, of course, it is the sublime of terror rather the Humboldtian aesthetic of tropical beauty that seems the more appropriately Darwinian landscape, an image of nature as an immense battlefield of extinction and death, rather than a fecund, self-reproducing ecosystem. On further reflection, however, Humboldt's tropical aesthetic is as true to the spirit of *The Origin of Species* as the wastes of Patagonia: *overproduction* takes on as much importance as *extinction* in his Malthusian theory of natural selection.

John Tallmadge notes that, although Humboldt's influence on Darwin's published travel writing remained all-pervasive, he took some pains to 'conceal or at least de-emphasize' the *Diary*'s youthful idolatry in the act of preparing to publish his travelogue: 'references to Humboldt's landscape descriptions disappear, and we find instead dry, technical citations addressed to specific geological, zoological, or meteorological issues'.[37] This is more true of the 1845 than the 1839 text, where he wrote that all his ideas of tropical scenery 'were taken from the vivid descriptions in the *Personal Narrative* of Humboldt, which far exceed in merit anything I have read on the subject' (1839: 374). Nevertheless, Darwin's (undeniable) demotion of Humboldt in later years may initially have been prompted by Caroline Darwin's strictures on the stylistic lushness of her brother's Brazilian journal in a letter dated 28 October 1833; 'you had, probably from reading so much of Humboldt, got his phraseology & occasionally made use of the kind of flowery french expressions which he uses, instead of your own simple straight forward & far more agreeable style' (*Corr.* i., 345). To Caroline there evidently seemed something un-English about Darwin's Humboldtian enthusiasm. In this light it is ironic that Helen Maria Williams's English translation of Humboldt—the text with which Darwin was familiar—is actually more 'florid' than the French original, partly on account of Williams's background as a poet of sensibility. Nevertheless her fine translation was fully endorsed by Humboldt himself, who assisted and checked her work.[38]

Darwin's youthful fascination with Humboldt's tropical stylistics seems to have been motivated by an almost libidinous attraction to the exotic and 'unhomely', which in the *Diary* (written partly for the benefit of his family) he analysed as a form of psychological disturbance:

> The entire newness, & therefore absence of all associations . . . of tropical scenery . . . requires the mind to be wrought to a high pitch . . . I often ask myself why I cannot calmly enjoy this; I might answer myself by also asking, what is there that can bring back the delightful ideas of rural quiet & retirement, what that can call back the recollection of childhood & times past, where all that was unpleasant is forgotten. (*Diary*, 64)

Darwin returns to this kind of aesthetic self-analysis in later correspondence, suggesting in July 1834 that the sensuous fascination of tropical forest exceeds even the richness of Humboldt's descriptions, so that a 'Persian writer could alone do justice to it' (*Corr.* i. 397). The Humboldtian tropical aesthetic is here *orientalized*, represented as the antithesis of English domesticity and the prospect of a respectable ecclesiastical career, embodied by the vision of Darwin's famous 'quiet parsonage . . . [seen] even through a grove of Palms' (ibid. 227). But in a letter to Susan Darwin from Brazil in August 1836, on the last lap of the voyage back to England, Darwin's 'immature' Humboldtian sense of tropical wonder has given place to a more austere spirit of comparison, leading him to a centripetal nostalgia for home in the place of his earlier desire for the exotic; 'I can now walk soberly through a Brazilian forest; not but what it is exquisitely beautiful, but

[37] Tallmadge, 'From Chronicle to Quest', 337–8. But note that Darwin's acknowledgement remains unchanged in the 1845 text (1845:, 482).

[38] When Thomasina Ross retranslated and abridged Humboldt's *Personal Narrative* for Bohn's Library in 1852, she criticized her precursor H. M. Williams's work on the grounds that 'it abounds in foreign turns of expression', as well as lacking 'fluency of style'. *Personal Narrative*, trans. Thomasina Ross, 3 vols. (London: H. Bohn, 1852), i. intro.

now, instead of seeking for splendid contrasts; I compare the stately Mango trees with the Horse Chesnuts of England' (ibid. 503). In a letter to Caroline Darwin written about the same time, he ridiculed his tropical infatuation in an explicitly gendered patriotic metaphor: 'People are pleased to talk of the ever smiling sky of the Tropics: must not this be precious nonsense? Who admires a lady's face who is always smiling? England is not one of your insipid beauties; she can cry, & frown, & smile, all by turns' (ibid. 501).

DARWIN AND HUMBOLDTIAN ANTHROPOLOGY

Sandra Herbert insists, in an influential article, that 'the subject of man was not one of those lines of inquiry which drew Darwin to transmutationist conclusions'.[39] According to Herbert's argument, Darwin's anthropological remarks in the *Journal* should be treated as non-scientific, on the grounds that unlike geology or zoology 'the observations were not backed up by a collection, or by critical pressure brought to bear by the existence of an audience of professionals'.[40] The validity of this latter point is not in question; however, Darwin's own travelogue persistently highlights the importance of his anthropological experiences on the voyage for his later thinking about transmutation. Amidst all the scientific discussion, the *Journal* bears anecdotal testimony—often in the most vivid terms—to the cruelties of the Atlantic slave-trade as witnessed by Darwin in Brazil, the murderous battle for the Argentinian Pampas between Rosa's gaucho army and the Pampas Indians, the oppression of Indians in Chile and Peru, or the violent impact of Euro-expansion on native societies in Tierra del Fuego, Tahiti, New Zealand, and Australia.

Population pressure and the struggle for existence are of course crucial factors in Darwin's later theory of Natural Selection as presented in the *Origin*, and he often couches these scientific theories in the contemporary discourse of European colonialism. It's no accident that, as Janet Browne has indicated, 'the word most used by early nineteenth-century students of animal and plant distribution patterns [Humboldt and Darwin included] was "colonist". This was the muscular language of [Euro-] expansionist power.'[41] In the final chapter of *The Origin*, Darwin speculates 'we need feel no surprise at the inhabitants of any one country, although on the ordinary view supposed to have been specially created and adapted for that country, being beaten and supplanted by the naturalised productions from another land'.[42] Turning back twenty years to the 1839 *Journal*, at the conclusion of his famous chapter on the Galapagos archipelago, Darwin considers 'what havoc the introduction of any new beast of prey must cause in a country, before the instincts of the aborigines become adapted to the stranger's craft or

[39] Herbert, 'Place of Man', 217.
[40] Ibid. 226.
[41] 'Biogeography and Empire', in N. Jardine, J. A. Secord, and E. C. Spary (eds.), *Cultures of Natural History* (Cambridge: Cambridge University Press, 1997), 315.
[42] *Origin*, 445.

power' (1839: 290). Although the word 'aborigine' is here used in a metaphorical sense (he is thinking explicitly about turtles, mocking birds, and finches rather than native peoples), the idea is expressed in quite literal terms later in the book: 'The varieties of man seem to act on each other; in the same way as different species of animals—the stronger always extirpating the weaker' (ibid. 322).

In the absence of any 'anthropological notebook' (on a par with the geological or zoological notebooks), the discursive site for Darwin's account of the global dispossession of indigenous peoples was the travel narrative itself, the pre-disciplinary forebear of the modern ethnographic treatise. Once again, Darwin's reading of Humboldt's *Personal Narrative*, with its elaborate anthropological and statistical account of contemporary Native Americans, in part substituted for the absent institutional embodiment of anthropology in 1830s Britain noted by George Stocking.[43] (Although Sandra Herbert notes that Darwin's observations on the geographical distribution of human populations and the effects of colonization are considerably less systematic than Humboldt's.)[44] The year 1839 saw (in addition to the publication of Darwin's *Journal*) the anthropologist James Prichard appeal to the British Association for the Advancement of Science on behalf of the 'Aborigines Protection Society', thus (arguably) initiating the institutional study of anthropology in Britain.[45]

Darwin's most famous anthropological pronouncement was made in response to his first contact with coastal Yaghans 'in the savage state' during the *Beagle*'s first visit to Tierra del Fuego in December 1832. He had had the opportunity on the outward journey to cultivate the company of Jemmy Button, York Minster, and Fuegia Basket (to give their 'colonial' names), the three surviving Fuegians kidnapped by Fitzroy and brought back to England to be 'civilized' and Christianized in October 1830, now being returned to their homeland, although the *Diary* mentions them only in passing. Darwin could now compare them with 'uncivilized' Yaghans in their native context: 'It was without exception the most curious and interesting spectacle I had ever beheld. I could not have believed how wide was the difference, between savage and civilised man. It is greater than between a wild and domesticated animal, in as much as in man there is a greater power of improvement' (1839: 172).

Although Darwin's Fuegians—in contrast to Humboldt's Orinoco Mission Indians—were observed in their 'savage' state, one might detect here a critical response to Humboldt's remark on the Chayma Indians in the third volume of the *Personal Narrative*: 'The savage and the civilised man are like those animals of the same species, several of which rove in the forest while others, connected with

[43] George Stocking, 'What's in a Name?: The Origins of the Royal Anthropological Institute, 1837–71', *Man* 6 (1971), 369–90.

[44] Herbert, 'Place of Man', 229.

[45] Gillian Beer, *Open Fields: Science in Cultural Encounter* (Oxford: Clarendon Press, 1996), 55–70 (59–60) and Stocking, 'What's in a Name?' Stocking points out the late 'institutionalization' of anthropology in Britain, compared with France and Germany.

us, share in the benefits and evils that accompany civilization' (*PN* iii. 227). (The passage is marked by Darwin in his copy of Humboldt in the Cambridge University Library.) Although both naturalists employ an offensive zoological analogy in describing native peoples, Darwin adds to Humboldt's observation the idea that savage and civilized man are *more* different from each other than wild and tame animals, because men possess a 'greater power of improvement' than animals. (This remark may have been inspired by the Fuegians on board the *Beagle*, but in the light of Jemmy Button's subsequent recidivism after the Europeans' departure, should be taken to imply that the process was reversible.) More significantly, Darwin's straining of Humboldt's analogy leads him to contemplate the Fuegian savage as disrupting the taxonomic barrier between beast and man; 'their hideous faces bedaubed with white paint, their skins filthy and greasy, their hair entangled, their voices discordant, their gestures violent and without dignity . . . Viewing such men', he continued, 'one can hardly make oneself believe that they are fellow-creatures, and inhabitants of the same world' (1839: 178). Gillian Beer suggests that 'Darwin's encounters with the Fuegians in their native place gave him a way of closing the gap between the human and other primates, a move necessary to the theories he was in the process of reaching.'[46] Arguably Humboldt *couldn't* have made the sort of connection that would eventually give rise to the controversial argument of Darwin's *Descent of Man*, precisely because his thoughts about the relations between men and animals remained at the level of analogy rather than genealogy. That's to say, despite his distinction between progressive Europeans and 'fossilized' non-European cultures,[47] he wasn't, like Darwin, in the habit of *literally* applying zoological concepts to the analysis of indigenous peoples.

To be fair to Darwin, the scarcity of his remarks on the Fuegians in the 1839 *Journal* was partly impelled by the agreed division of labour whereby Captain Fitzroy took charge of anthropological description, whereas Darwin confined himself to natural history. After all, chapters 7–9 of Fitzroy's narrative are entirely dedicated to describing the anthropology of the 'horse Indians' of Patagonia and the 'canoe Indians' of Tierra del Fuego, dilating at great length upon their manners, politics, and religion, with a large debt to the Jesuit missionary Thomas Falconer. Fitzroy's appendix also contains a vocabulary of the Alikhoolip and Tekeenica languages, as well as the shockingly 'zoological' phrenological account of Fuegian Indians by Wilson, the *Beagle*'s surgeon. ('The Fuegian, like a Cetacious animal which circulates red blood in a cold medium, has in his covering an admirable non-conductor of heat.')[48]

[46] *Open Fields*, 67.

[47] See my *Curiosity and the Aesthetics of Travel Writing, 1770–1840* (Oxford: Oxford University Press, 2002), 271–5.

[48] *Narrative of the Surveying Voyages of His Majesty's Ships Adventure and Beagle, between the years 1826 and 1836, describing their examination of the Southern Shores of South America and the Beagle's Circumnavigation of the Globe*, 4 vols. (London: Henry Colburn, 1839), iii. 143. Henceforth *NSV* in text.

Fitzroy, it is true, compares the colour of the Fuegians to 'the Devonshire breed of cattle' (*NSV* ii. 134) and in one place describes them as 'satires upon mankind' (*NSV* ii. 138). Elsewhere, however, he speaks of 'ignorant, though rather intelligent barbarians' (*NSV* ii. 2–3) and reminds his readers, in a temporalizing trope, that 'Caesar found the Britons painted and clothed in skins, like these Fuegians' (*NSV* ii. 121). Fitzroy—who, like Darwin, was avidly reading Humboldt during the voyage—was fascinated by physiology as the visible 'index of the mental quality' (*NSV* ii. 640) wrought by the civilizing process on the savage individual, and his remarks on the power of mind over matter echo Humboldt's assumption that 'the features acquire the habit of mobility in proportion as the emotions of the mind are more frequent, more varied, more durable' (*PN* iii. 228). Ironically Fitzroy's Christian anthropology found a stronger support from Humboldt's Lamarckian argument than would Darwin's darker, more deterministic reading of bio-geographical pressure as unfolded in the *Origin of Species* and the *Descent of Man*.[49]

Despite Humboldt's animadversions concerning what he termed the 'moral inflexibility' of the Indian race, he insisted on studying them in relation to their history, culture, and above all, language: 'Almost everywhere the [Amerindian] idioms display greater richness, and more delicate gradations, than might be supposed from the uncultivated state of the people, by whom they are spoken' (*PN* ii. 213). Darwin's conventionally ethnocentric view of the Fuegians is unmitigated by any such interest in culture or language (did he not admit, in the *Autobiography*, that 'during my whole life I have been singularly incapable of mastering any language' (*Auto.* 12)?). In his view the linguistic abilities of these abject and allegedly cannibalistic Fuegians was limited to 'so many hoarse, guttural and clicking sounds' and the constantly reiterated plea 'yammerschooner', meaning 'give me' (1839: 173, 180). 'The language of these people' commented Darwin, 'according to our notions, scarcely deserves to be called articulate' (ibid. 173). Consistent with his zoological paradigm, the only praise Darwin has for the Fuegians is of their animal-like power of mimicry in repeating complex English phrases and sentences, attributing this *natural* (rather than cultural) trait to 'the more practiced habits of perception and keener senses, common to all men in a savage state' (ibid.). The 'perfect equality' of the Fuegians, their refusal to submit to the principles of social hierarchy or the accumulation of property, is taken to retard the progressive civilization that, Darwin believed, was premissed upon both. Busy (in the words of Michael Taussig) 'investing sense-data in the bank of the Self', Darwin misses the spirit of gift-

49 Martin Rudwick questions the historiographical tendency to see natural theology as having had 'a wholly baneful influence on scientific research . . . the interpretation of the temporal progress of life in terms of the temporal actualisation of a divine plan tended to discourage speculation about possible mechanisms for the emergence of "higher" forms of life; yet at the same time it provided an equally powerful incentive to search for further evidence of organic progress'. *The Meaning of Fossils: Episodes in the History of Palaeontology*, 2nd edn. (New York: Neale Watson Academic Publications, 1976), 155–6. The implications of this for anthropological research are suggestive, although clearly Fitzroy's main concern is the potential for converting the Fuegians to Christianity.

exchange underlying the Fuegian's highly-developed powers of mimicry, the fact that 'give me' contains the fully germinated social bond of reciprocity.[50]

Once again Darwin's Tierra del Fuego proffered itself as a dark anthropological sublime set against the eudaemonic bliss of the Humboldtian tropics. Where Darwin did concur with Humboldt was in his assertive attack on slavery, and his sympathy for the heroism of the beleaguered savage pitted against the juggernaut power of European colonial expansion. In chapter 5 of the *Journal* (ibid. 111, 113) he echoed Humboldt's story of the Piedra de la Madre (a heroic Indian mother who performed a feat of superhuman endurance to rescue her children from missionary slave-gatherers), and even more directly (although with reference to an African rather than a Native American) in describing the suicide of an escaped slave near Rio de Janeiro: 'In a Roman matron this would have been called the noble love of freedom: in a poor negress it is mere brutal obstinacy' (ibid. 59). After witnessing at first hand General Rosas' brutal campaign of genocide on the Pampas, Darwin believed that 'there will not, in another half-century, be a wild Indian northward of the Rio Negro' (ibid. 112). Humboldt's very different geographical and historical experience led him to argue for a gradual demographic increase of Native Americans, whose total population he estimated at over 6 million (*PN* ii. 213).

Although Darwin expressed horror at the unrelenting slaughter of the Pampas Indians, he did write in his notebook (in a passage marked for deletion in pencil, and dropped from the *Journal*), 'if this warfare is successful, that is if all the Indians are butchered, a grand extent of country will be gained for the production of cattle: and the valleys of the R. Negro, Colorado, Sauce will be most productive in corn' (*Diary*, 181). He had no illusions that 'white Gaucho savages' represented moral virtue, any more than Argentina's ruling class ('the absence of gentlemen by profession appears to an Englishman something strange' (1839: 145)). Nevertheless, both were the blind instruments of progressive and expanding civilization. He was more sanguine about the virtues of British liberal colonization in the missions of Tahiti and New Zealand, the rapidly rising metropolis of Sydney, and the tarmacked roads of New South Wales: 'Here, in a less promising country, scores of years have effected many times more, than the same number of centuries have done in South America. My first feeling was to congratulate myself that I was born an Englishman' (ibid. 318). In an appropriately naturalistic metaphor, Darwin celebrated the fact that 'little embryo Englands are springing into life in many quarters' (ibid. 358). This kind of jingoism is rarely found in Humboldt's text, on account of the Prussian traveller's cosmopolitan background and non-affiliation with any particular national programme of colonization. With his experience of *mestizaje* in the long-established colonies of Spanish America, Humboldt marvelled at the will to survive and cultural adaptability of Native Americans, marked by their demographic increase. Darwin on the other hand,

[50] Michael Taussig, *Mimesis and Alterity: A Particular History of the Senses* (London: Routledge, 1993), 97.

with his dark experience of Pampas genocide, Fuegian abjection, and the extirpation of Tasmanian aborigines, perhaps understandably wrote of a 'mysterious agency' at work in the colonized world: 'wherever the European has trod, death seems to pursue the aboriginal . . . The varieties of man seem to act on each other, in the same way as different species of animals—the stronger always extirpating the weaker' (ibid. 322). It is hard to accept, *pace* Sandra Herbert, that these words are unconnected with the notion of the 'struggle for survival' germinating in Darwin's mind as he edited his *Diary* after his return to England, at the same time as he reread Thomas Malthus's *Essay on Population.* Once again Humboldt's *Personal Narrative* seems to have acted as a kind of springboard for the development of Darwin's thinking, despite or perhaps because of the absence of any institutional pressure shaping its anthropological sections, along the lines of contemporary geology or zoology.

HUMBOLDTIAN METHODOLOGY AND THE COMPOSITION OF THE 1845 *JOURNAL OF RESEARCHES*

An enthusiastic, elderly Humboldt described Darwin's *Journal* as 'one of the most remarkable works that, in the course of a long life, I have had the pleasure to see published'.[51] In a long, fulsome letter to Darwin dated 18 September 1839, which delighted and flattered the young naturalist, Humboldt praised him for having completed the programme of his own research: 'You told me in your kind letter that when you were young the manner in which I studied and depicted nature in the torrid zones contributed towards exciting in you the ardour and desire to travel in distant lands. Considering the importance of your work, Sir, this might be the greatest success that my humble work would bring. Scientific contributions are of value only if they give rise to better ones . . .'[52] Humboldt magnanimously listed Darwin's scientific achievements in 'enlarging and correcting' his own work, while lavishing praise on his poetic skill in offering 'the charm of a happy inspiration' in evoking 'the wild fertile earth of the torrid zone'.[53] He even provided a list of page numbers indicating the passages that had moved him most. It perhaps comes as no surprise that these coincide with passages in the *Journal* that were themselves most influenced by Humboldt's 'tropical aesthetic'.[54]

What Humboldt doesn't mention here is the debt that Darwin's book owed to the discursive form of his own travel account, particularly his decision to include long theoretical dissertations that punctuate the narrative, and increasingly seem

[51] Quoted in Barrett and Corcos, 'Letter from Alexander Humboldt', 160.

[52] Ibid. 163–5.

[53] Ibid. 165.

[54] For example, Darwin's description of the view from the top of the Andes: 'I felt glad I was alone: it was like watching a thunderstorm, or hearing a chorus of the Messiah in full orchestra' (394); and of 'the free scope given to the imagination' by the 'boundless' wastes of the Pampas (605), which evokes Humboldt's poetic description of the Venezuelan *llanos* in both the *Personal Narrative* and *Aspects of Nature*.

to qualify a rival desire to include subjective and aesthetic evocations of tropical nature and culture. Humboldt was in fact somewhat chary about the book that had made his reputation as a travel writer, as well as one of the most widely celebrated naturalists of his generation. The preface to the *Personal Narrative* contained a remarkable series of methodological strictures about travel writing that appear to have influenced Darwin when he came to transform his *Diary* into the form of the 1839 *Journal*. Humboldt described how he had originally planned to avoid 'what is usually called the historical narrative of a journey . . . [rather to] publish the fruits of my enquiries in works merely descriptive . . . I had arranged the facts, not in the order in which they successively presented themselves, but according to the relation they bore to each other' (*PN* i. p. xxxviii). Although he believed that the personality of the traveller contributed an aesthetic and emotional interest to travel writing, Humboldt's objectivist orientation made him none the less suspicious of the whole project of 'personal narrative', as well as deeply distrustful of the 'egotism' that characterized much popular post-romantic travel writing.

Part of Darwin's priority in publishing his *Journal* was, in John Tallmadge's words, to solve the 'problem of narrative integration by . . . developing a recognisable story line and an appealing voice for himself as a story teller'.[55] However (a fact which Tallmadge overlooks), in 1839 the question of literary integrity was subordinate to Darwin's quest for scientific recognition, of particular importance given the 'supplementary' role of the first publication of Darwin's *Journal*.[56] Considering the unprofessional natural history of Captain King's narrative ('it abounds with Natural History of a very trashy nature', wrote Darwin dismissively (*Corr.* ii. 80)) and the often diffuse and anecdotal nature of Fitzroy's volume, it became incumbent upon Darwin to fashion his narrative persona as that of a highly professional naturalist. By so doing he could win scientific recognition as well as 'transforming his voyage from a self-indulgent personal excursion (a "wild scheme" and "useless undertaking", in his father's harsh words) to a responsible quest for truth'.[57] The twin goals of professional credibility and moral self-justification weren't of course always commensurate with general literary interest: his cousin Elizabeth Wedgwood feared that his captivating *Diary* (the instalments of which had delighted her family) would become dry as dust in the process of redaction into a naturalist's *Journal and Remarks*. She exhorted Darwin to risk a bit of overlap with Fitzroy's narrative so as not to exclude 'too much of [the] Journal, especially about the people, which is always a more interesting subject than the place' (*Corr.* ii. 5).

Humboldt assumed model status in Darwin's professional self-fashioning. Taking his cue from the Prussian traveller's preference for synchronic over diachronic description, Darwin admitted his preference for ordering his narrative

[55] Tallmadge, 'From Chronicle to Quest', 330.

[56] The best account of Darwin's strategic manœuvring within the field of professional science is Rudwick's 'Charles Darwin in London'.

[57] Tallmadge, 'From Chronicle to Quest', 340.

by 'place' rather than 'time' (ibid. 14). Against Elizabeth Wedgwood's advice, he also followed Humboldt's principle of suppressing many circumstantial details of a personal nature, as well as references to Fitzroy and his fellow travellers, in order to preserve 'its peculiar character, that of a work of science' (*PN* vii. 472).[58] Despite the appeal of Humboldtian landscape aesthetics, Darwin's choice of Humboldt as his literary model distinguishes his *Journal* from the more egotistical 'literary' style of travel writing that was becoming popular in the 1830s. This style is exemplified in the *Quarterly Review*'s praise for the 'manly, unaffected style, rough and racy' and 'considerable skill and effect' in 'describ[ing] manners and scenery' of Capt. Basil Hall's *Extracts from a Journal, Written on the Coasts of Chili, Peru, and Mexico*.[59] Darwin disagreed: carrying Hall's travelogue on his excursions in Chile and Peru, he complained in his *Diary* that 'it appears to me that all Capt Hall's beautiful descriptions require a little washing with a Neutral tint—it may partly destroy their charms, but I am afraid will add to their reality' (*Diary*, 336).

Humboldt's strictures on travel writing in the introduction to his *Personal Narrative* seemed to offer Darwin a formula for achieving the correct balance between scientific generalization and integrated narrative.[60] 'When we have followed the traveller step by step', Humboldt had written in the third volume, 'in a long series of observations modified by the localities of a place, we love to stop, and raise our views to general considerations' (*PN* iii. 138). Examples of this shift from narrative to generalization abound in Humboldt's text: the long dissertation on volcanoes that follows the first volume's account of Humboldt's and Bonpland's ascent of the Pico de Teide on Teneriffe, the bizarre essay on geophagy that follows on Humboldt's observations of earth-eating Otomac Indians in the fourth volume, or (on a massive scale) the geognostic essay on South America that occupies most of the second part of volume vi. A similarly abrupt transition from personal narrative to scientific generalization informs Darwin's 1839 *Journal*, as in the zoological account of the Agouti in chapter 4, the generalized dissertation on the feeding habits of large quadrupeds in chapter 5, the excerpts from Robert Owen's paper on the Toxodon in chapter 8, and most spectacular of all, the lengthy account of the formation of coral islands in chapter 22, which breaks into the journal narrative with the sentence, 'I will now give a sketch of the general results at which I have arrived' (1839: 342).

In the *Journal*'s penultimate paragraph, Darwin even ventured a Humboldtian critique of the limitations of mere 'itinerary narrative' and the difficulty of presenting scientific information in diachronic rather than synchronic form: 'As the traveller stays but a short time in each place, his descriptions must generally consist of mere sketches, instead of detailed observation. Hence arises, as I have found to my cost, a constant tendency to fill up the wide gaps of knowledge, by

[58] Ibid. 336. By contrast, the *Diary* 'presents [Darwin] in a different light . . . part of a group, constantly interacting with others and recording their responses with lively candour' (ibid. 335).

[59] April 1831, 145, quoted in Tallmadge, 'From Chronicle to Quest', 330.

[60] See Smith, *Fact and Feeling*, 117.

inaccurate and superficial hypothesis' (ibid. 377). Darwin surely had in mind the sort of 'unprofessional' field-naturalist, or the naval and military traveller, who padded out his journal with half-baked speculations, exemplified (although Darwin is too tactful to say so) by Captain King's volume. Better, then, to follow Humboldt in punctuating the itinerary and 'raising our views to general considerations' by means of interpolated scientific dissertations.

However much Darwin's emulation of Humboldtian travelogue might have succeeded in securing his scientific credit, it wasn't without its own discursive risks. I have argued elsewhere that the *Personal Narrative* is ultimately flawed in its attempt to combine subjective affect and objective observation, aesthetics and science. Reluctant to begin his personal narrative, Humboldt terminated it just over halfway through the description of his South American journey.[61] Darwin's 'reading and rereading' of Humboldt's seven volumes would have doubtless alerted him to the dangers that he faced in emulating Humboldt's narrative manner, as would the reviewers' initial responses to his own travelogue upon its publication in 1839. He doubtless would have noted Humboldt's failure to sustain the difficult balance between subjective and objectivist discourses. Whereas Humboldt's first, 1814 volume (I refer to the English translation in Darwin's hands) is largely a conventional personal narrative combining episodic and anecdotal material with prolix dissertations on volcanoes and ocean currents, the 1818 and 1819 volumes contain an increased amount of generalized description. The balance shifts dramatically in the final two volumes, with their long dissertation on geology, and statistical account of Cuba. Humboldt's *Personal Narrative* gradually refined itself, in the course of its protracted and spasmodic publication, into a 'geographical narrative', a digest of scientific, statistical, and anthropological generalizations, rather than a travel account properly speaking. Humboldt literally wrote himself (and his itinerary) out of the picture, a fact that may have determined his abandonment of the whole project at the end of the seventh volume. As I argued above, the fact that Darwin probably carried all seven volumes of the *Personal Narrative* with him on the *Beagle*, rather than just the first two volumes inscribed by Henslow, makes it all the more probable that he was fully aware of its awkward metamorphosis into 'geographical narrative'.

Although reviewers were generally enthusiastic about Fitzroy and Darwin's four-volume 1839 *Journal*, some were ambivalent regarding the 'supplementary' nature of Darwin's volume. The *Quarterly*, it is true, while applauding Fitzroy's 'entertaining and interesting' account, commended Darwin's Humboldtian gift of combining state-of the-art scientific inquiry with the powers of 'a first-rate landscape painter with the pen'.[62] Writing in *The Edinburgh Review* as the leading contemporary authority on Latin American travel narrative, however, Basil Hall expressed reservations about Darwin's vigorous 'faculty of generalisation' and his

[61] *Curiosity and the Aesthetics of Travel Writing*, 281–98.
[62] *Quarterly Review* lxv (December 1839/March 1840), 194–233 (212, 233).

theoretical boldness.[63] Even more revealing was the *Athenaeum's* critique of the discursive 'division of labour' between Fitzroy's and Darwin's narratives. Not only had Fitzroy and King encroached on Darwin's territory by venturing their own clumsy observations on natural history, the review argued, but the generic form of the volume had obliged Darwin to 'connect together the results of his investigations by a separate web of personal narrative', thereby obstructing the free development of his various scientific topics.[64] Collaborative authorship of the *Beagle* volumes had obliged Darwin to abridge, in the course of which he had omitted too many details: as a result, the *Athenaeum* opined (echoing Hall's criticism in the *Edinburgh*) that his *Journal*, 'exhibits a predominating spirit of bold generalisation of which the world, not without justice, is exceedingly mistrustful'. The reviewer had no particular quarrel with Darwin's theories per se; but feared that 'denuded of elemental facts, we doubt not that many of the theories therein contained will meet with a less general concurrence than they are really entitled to do'.[65]

Although (strangely enough) no reviewer singled out the question of Darwin's debt to Humboldt, the *Athenaeum*'s criticism of Darwin echoed strictures levelled at Humboldt two decades earlier. Whilst the *Edinburgh* had lavished praise on his intellectual comprehensiveness and lack of egotism, John Barrow, in his 1816 review of the *Personal Narrative* in the *Quarterly*, complained bitterly of Humboldt's 'constant attempts at generalisation . . . a word or name suggests a hundred different ideas, and transports him to as many different places . . . in the meantime, the subject under discussion is lost sight of'.[66] This sort of criticism was voiced again and again, at least until the increasingly technical and experientially 'denuded' form of the later volumes of the *Personal Narrative* induced 'reviewers' fatigue' by the mid-1820s.[67] Moreover, Humboldt's scientific speculations were to some extent buttressed by the fact that the *Personal Narrative* was published after most of the specialist volumes of his *Voyage*; Darwin, by contrast, was in 1839 publishing his 'bold generalisations' *without* the textual support of the scientific volumes of the *Beagle* voyage, which largely remained to be published. In a letter of May 1837 he had foreseen this danger, suspecting that he had 'begun at the wrong end, I ought to have published detailed Geology, & Zoology first; & then all general views might have come out in as perfect a form, as the subject permitted' (*Corr.* ii. 21).

Prompted by his own earlier reservations and the reviewer's remarks about unsubstantiated generalizations, as well as frustration with his publisher Henry Colburn, Darwin sold the rights of his *Journal* to John Murray for £150 and set about reworking—and to some extent de-Humboldtizing—the *Journal* for a

[63] *Edinburgh Review* cxl (April/July 1839), 467–93 (489).

[64] *Athenaeum* 607 (15 June 1839), 446.

[65] Ibid.

[66] *Quarterly Review* (January 1816), 401–2.

[67] See Nicholas Rupke, 'A Geography of Enlightenment: The Critical Reception of Humboldt's Work', in David N. Livingstone and Charles Withers (eds.), *Geography and Enlightenment* (Chicago: Chicago University Press, 1999), 319–39.

second, 1845 edition. This new edition does register many of the developments in Darwin's scientific thinking in the intervening years—most famously he drops the theory of the senescence of species from chapter 9 (chapter 8 in 1845), and adds significant new material, including a suggestion concerning the mutability of species, to chapter 19 on the natural history of the Galapagos islands (chapter 17 in 1845). Nevertheless, even a brief comparison between the two editions, or a glance at Darwin's commentary on his revisions in the *Correspondence*, suffices to put in question Tallmadge's statement that 'he began revising his text, chiefly with an eye to correcting and expanding the scientific discussions'.[68] In fact Darwin dropped several rather technical scientific passages (such as the dissertation on the Agouti in chapter 4, and the extracts from Owen's paper on the Toxodon in chapter 8), as if conscious that they embodied Humboldtian scientific over-generalization of the sort criticized by the *Athenaeum*. When scientific passages were extended, they were also popularized, as in the account of coral reefs in chapter 23 (20 in 1845). Here the original passage in 1839, based on a paper read to the Geographical Society in May 1837, was replaced by a popular treatise omitting several of the more technical passages, and now illustrated with woodcuts and diagrams, offering the general reader a précis of Darwin's 1842 book *On the Structure and Distribution of Coral Reefs*. A major difference between the 1839 and 1845 texts is, of course, the fact that the latter now followed, rather than anticipated, the several specialized scientific publications deriving from the voyage. In his new 1845 introduction, Darwin was therefore able to refer the 'naturalist' (as opposed to the 'popular reader') 'for details to the larger publications, which comprise the scientific results of the Expedition', distinguishing between the readerships for specialized science and popular travel writing.

Just as revealing in this light are the passages that Darwin added *de novo*. In a letter to John Murray of 17 March 1845, he proposed 'to shorten a little . . . the geology & natural History & add something to my notices on the Fuegian savages &c &c' (*Corr*. iii. 158).[69] Perhaps remembering his cousin Elizabeth Wedgwood's plea for 'people' over 'places', he added to chapter 10 a lengthy passage describing the 'civilized' Fuegians on board the *Beagle*, as well as an account of their religion and manners, and a narrative of the ill-fated missionary settlement at Woollya. Given the fact that, by his own admission,[70] Darwin lacked any anthropological notes, these episodes were largely borrowed from his own unpublished *Diary*, supplemented by Fitzroy's narrative, from which the *Journal* was now of course completely unshackled. It is ironic that it is this account, in Darwin's best-selling 1845 *Journal*, rather than Fitzroy's now unread original, upon which most subsequent accounts of the *Beagle*'s Fuegian adventures depend.[71] Darwin's bid to

[68] Tallmadge, 'From Chronicle to Quest', 328.

[69] Writing to Lyell in July of the same year, he wrote of his improvements to the new edition: 'I have added a good deal about the Fuegians & cut down into half that mercilessly long discussion on climate & glaciers &c—' (*Corr*. iii. 214).

[70] See *Corr*. iii. 158.

[71] Until, that is, Darwin's *Beagle Diary* was first published in 1933 in Nora Barlow's edition.

'improve and popularise' his *Journal* was a resounding success, obviating the strictures of the reviewers and overcoming the 'denuded' Humboldtian habit of generalization that the *Athenaeum* had regretted. As J. D. Hooker wrote on 14 September, 'How doubly interesting you have made it to the common reader! I was so taken with the account of the poor Fuegians that I sat up I cannot say how long' (*Corr.* iii. 254).

One senses Darwin's increasing distance from Humboldt by mid-century as the former's star rose and the Prussian naturalist's waned. Darwin was 'a little disappointed' when he met the great man himself at Sir Roderick Murchison's in 1843 (*Auto.* 63), finding him unduly garrulous, and in 1845 he was likewise 'rather disappointed' by his reading of the first part of *Cosmos*, which contained 'so much repetition of the Personal Narrative, & I think no new views' (*Corr.* iii. 261). Darwin's newly integrated narrative persona, emancipated from the composite authorship of the 1839 edition, now eschewed the supplementary and limiting role of 'ship's naturalist' in search of professional legitimization. Although his book still made much of Humboldt's tropical aesthetic in its landscape descriptions, the success of Darwin's 1845 *Journal of Researches* may ultimately have depended upon its willingness to suppress other aspects of Humboldt's influence as a scientific travel writer. The 1845 text is inflected with the confident, expansionist mood of British imperialism rather than Humboldtian 'planetary consciousness', which in the end gives it a very different feel. In his adaptation of the genre of Humboldtian 'personal narrative' to the very different conditions of Victorian popular science, Darwin skilfully translated the global romanticism of his model into the popularizing idiom of the British 'civilizing mission'. Nor is this all: we should remember that in the following decade, Darwin's *Origin of Species* would owe something of its epoch-making importance to the challenge it posed to the Humboldtian ban on inquiry into origins, and its speculative rejection of the German savant's preference for painstaking empirical investigation over bold hypotheses.

CHAPTER 2

'And If It Be a Pretty Woman All the Better'—Darwin and Sexual Selection

GEORGE LEVINE

As Gillian Beer has argued, Darwin's theory of sexual selection puts back into the theory of evolution the intention and cultural direction that seemed so alarmingly absent in Darwin's first formulation *On the Origin of Species by Means of Natural Selection*. While, in the *Origin of Species*, Darwin was careful not to discuss the human species or culturally difficult issues (except, inevitably, through metaphor), in *The Descent of Man*, he met humanity, morality, and culture head on, not only to attempt to explain them naturalistically but to introduce cultural determinants into biological developments. It was a striking and difficult move, one whose effect was yet more dubious and controversial than the theory of natural selection itself. Intention, that central motif of natural theology, went out with the *Origin*, only to return with *The Descent*. And, of all things, it returned primarily by way of female choice. There is no way to talk about sexual selection without bringing to the surface all those cultural issues that Darwin tried hard to keep back, at least to hold in abeyance, as he made the case for his primary argument about evolution and its mechanisms.

Sexual selection is a minefield, not only for scientists, but for cultural critics as well. Nothing is easier than to fall back on the critique that Darwin's theory is really only a naturalizing of cultural prejudices about women, as it has been claimed that natural selection is nothing but a naturalizing of capitalism. Yet where natural selection will often withstand such ideological critique, if only because the 'nothing but' is clearly inadequate to the vast accumulations of knowledge that support it, to expose the degree to which sexual selection drew on or reinforced Victorian prejudices about women is often taken as its virtual elimination as a serious biological explanation.

But I want to argue here that the ideological critique, in both cases, has no purchase on the theory itself. It might indeed help us to understand the history of the development of the theory, but it cannot do much to validate or invalidate it. However embedded *The Descent of Man* might be in uninterrogated cultural assumptions, the theory of sexual selection is an astonishingly brilliant idea, teased out of a mixture of cultural assumptions and intense observation and careful thinking. These cannot be disentangled, and their entanglement (a deeply Darwinian concept itself) is part of the reason that ideological critique cannot, by itself, dislodge a theory, though it may give good reasons to like or dislike it. Good cultural theory might best take sexual selection not as a simple reflex of cultural prejudice but as a fascinating commentary upon it.

This, I take it, is the kind of approach Beer took when she turned her attention to sexual selection in her deservedly influential study, *Darwin's Plots*. Unfortunately, while her approach to Darwin has happily transformed almost all later cultural study of Darwin and his relation to literature and language, treatment of the question of sexual selection has not, for the most part, profited from her approach.[1] It is one of the distinctive and most satisfying qualities of Beer's remarkable treatment of Darwin in relation to culture that she talks freshly and revealingly about the ways in which cultural (dare one say, 'non-scientific') forces operate on Darwin's thinking without feeling obliged to judge him or his theories negatively from the perspective of our current ideological positionings. Indeed, she takes such judgmental critique as a symptom of an intellectual arrogance that closes down the possibility of learning from the past. Darwin is troublesome and troubling, never more so than as Beer has read and used him critically. One learns from her about how Darwin's thought developed in intimate contact with literary and cultural forces that are normally ignored in strictly scientific discussion, and about how his thought percolated through the culture—but in every case with a sense of the instability and richness of his language, and never with a sense that as a wealthy male citizen of his own time, he can be satisfactorily understood as representatively retrograde ideologically. She simply will not accept the all too common assumption that Darwin's implication in the values and ideals of his own culture somehow closed off the possibility that his work extends beyond the limits of that culture to make genuine discoveries and to criticize it. As she put it, Darwin's 'writing intensified and unsettled long-used themes and turned them into new problems'.[2]

It is easy to stick Darwin in an ideological box, hard to do what Beer did in recognizing cultural implication and at the same time refusing to relax into negative ideological placing. I hope that in returning to the subject of Darwin's handling of the question of sexual selection, I can follow the lines of Beer's

[1] Angelique Richardson, in a series of essays, has done extremely valuable work on the question of sexual selection in relation to Hardy and several interesting late nineteenth-century novelists. See, for example, 'Some Science Underlies All Art: The Dramatization of Sexual Selection and Racial Biology in Thomas Hardy's *A Pair of Blue Eyes* and *The Well-Beloved*', *Journal of Victorian Culture* 3/2 (1998), 302–28. Richardson is highly sensitive to questions of gender and race throughout, but her readings of Darwin and his adaptation by the writers in question are sensitive to the complexities of the arguments.

[2] Gillian Beer, *Darwin's Plots: Evolutionary Narrative in Darwin, George Eliot and Nineteenth-Century Fiction* (London: Routledge & Kegan Paul, 1983), 213. Revised edn., Cambridge University Press, 2000.

groundbreaking analyses into the very heart of cultural complicity. As I try this, I feel the pressures of conflicting allegiances. While in my studies of Darwin I have been much impressed by and have learned much from the sort of criticism that has demonstrated the connections between Darwin's theories and his ideological assumptions, about politics and gender, in particular, I have also always found useful and important an approach to Darwin that reads him very closely, follows out the lines of the history of the ideas he uses and develops, and emphasizes both the ways in which he developed his theory and the nature and originality of his arguments. Such approaches ought not to be incompatible, and in Gillian Beer they are not. But in much current practice they tend not only to be opposed but often to be quite hostile to each other. The first approach leans towards primary emphasis on Darwin's ideological complicity so that his work can begin to look like little more than an elaborate apology for some very bad Victorian habits. The other leans toward hagiography, an abstracted celebration of original genius.

While good cultural criticism does not reduce science to ideology, good intellectual history does not leave ideas and genius hanging abstractly out there without context. The extraordinary historical work of Adrian Desmond and James Moore—which often does, indeed, combine very close examination of language with broad contextual and ideological placing—is perhaps the best example of historical and cultural criticism that is ideologically driven but that is conscientiously committed to getting Darwin right, understanding how he thought, and watching both the cultural processes that gave shape to some of his ideas and the unique intellectual and personal qualities that allowed him to produce his theory as he manœuvred through social conditions in which he felt extremely comfortable but whose very foundations his theory threatened to challenge.

Desmond and Moore argue that the theory locked into place as Darwin found a form for it compatible with the economics of laissez-faire, and strong historical study like theirs seems to me to have established once and for all that the theory of natural selection has close and documentable ties to laissez-faire economics.[3] Desmond, especially in an earlier book, had demonstrated that early nineteenth-century evolutionism had largely radical and even revolutionary roots.[4] And, no doubt, Darwin had to work hard to disentangle his version of the theory from what he would have regarded as politically tainted ones. But the biggest political bang of natural selection was in the fact that it disallowed intelligence and direction in the development of species—and this point Darwin argued with scientific rigour, requiring no overt intrusion of political assumptions. The politics of sexual selection, however, spinning on the notion of female choice, were out there

[3] Adrian Desmond and James Moore, *Charles Darwin* (London: Michael Joseph, 1991). For a careful early analysis of the relation of Darwin's thinking to contemporary political economy, see Sylvan Schweber, 'Darwin and the Political Economists: Divergence of Character', *Journal of the History of Biology* 13 (1980), 195–289.

[4] *The Politics of Evolution: Morphology, Medicine, and Reform in Radical London* (Chicago: University of Chicago Press, 1989).

overtly and immediately. Here, contemporary critics go straight for the 'culture', and Darwin does not emerge looking very good.

Rosemary Jann's strong and tightly argued essay on sexual selection remains a point of departure for any discussion of the subject, and in my attempt to complicate matters in what I hope will be a Beer-like way, I take it as my point of departure as well.[5] I would add to it an important essay of about twenty-five years ago by Eveleen Richards, who claimed in what was then a groundbreaking argument that 'Darwin's re/construction of human evolution was pervaded by Victorian sexist ideology.'[6]

Those essays have been important in demonstrating the role of Darwin's very unscientific assumptions about the place of women in the development of the theory of sexual selection in *The Descent of Man*. There is no need to follow the details of their strong arguments, but surely it is difficult now to imagine picking up *The Descent of Man* without being struck by its thorough saturation in cultural assumptions. Darwin's anthropomorphizing is a central feature of his work, and virtually every one of his animals and insects seems to behave in very Victorian ways. Much of *The Descent* is anecdotal. Right at the start, Darwin confesses that 'This work contains hardly any original facts in regard to man',[7] and the absence of originality in these factual matters is striking. For the most part, Darwin summarizes his earlier notes and the views of others, often without the extraordinary care and rigour that mark his arguments in the *Origin*. Yet care and rigour do mark the arguments of the largest parts of the book—the sections dealing with sexual selection, largely of plants and animals. It is when he gets to talking about people that Darwin turns, it would seem almost defensively, to the views of others (he notes on the first page that he didn't talk about humans in the *Origin* because he feared doing so would 'add to the prejudices against my views' (1)). *The Descent* is often disappointing in parts even to Darwin enthusiasts like me because so much of its discussion of human behaviour seems to depend on commentaries by second-rate minds on Darwin's earlier first-rate work. Much of it relies on the materials provided by contemporary ethnologists and anthropologists, who were themselves ideologically saturated at a moment when the new social sciences were being born. Even in the *Descent* Darwin's most interesting ideas about the human get there only by indirection. His discussion of plants and animals, meticulously derived from close observation and carefully accumulated work, is saturated with his sense of the human, and what emerges is an extraordinary mental *tour de force*.

In arguing this, I do not mean to suggest that the Darwin whose personality gets expressed in the language of this book is anything but the respectable middle-

[5] Rosemary Jann, 'Darwin and the Anthropologists: Sexual Selection and its Discontents', *Victorian Studies* 37 (1994), 286–306.

[6] Eveleen Richards, 'Darwin and the Descent of Woman', in David Oldroyd and Ian Langham (eds.), *The Wider Domain of Evolutionary Thought* (Dordrecht: D. Reidel, 1983), 61.

[7] Charles Darwin, *The Descent of Man, and Selection in Relation to Sex*, reprint of the 2-vol. 1871 edn. with an intro. by John Tyler Bonner and Robert M. May (Princeton: Princeton University Press, 1981), i. 3. The 2nd edn. (London: John Murray, 1882) is cited in the text as (1882).

class gentleman (who, by the way, never allowed himself to be pictured, as Janet Browne has shown, as a man marked by his work in any way[8]). The cultural critiques by Richards and Jann were necessary and largely right. The respectable middle-class gentleman emerges from just about everything Darwin wrote. There is the homeowner who puttered about in his garden, who modestly took into account all possible objections to his ideas, who steered clear of controversy even while creating one, and who treated respectfully all who helped and all who disagreed.

But what do the arguments of Richards and Jann mean for the validity of the theory of sexual selection? What might Jann have meant when, near the end of her essay, she wrote: 'Acknowledging the extent to which the imposition of order on events is necessarily dependent on the ideological position of the observer who interprets data and fashions stories from them need not rule out the possibility of satisfying scientific standards of proof and logic in our reconstructions of the past' (304). A remark in Richards's essay raises a similar question: 'Darwin's conclusions on the biological and social evolution of women', she says, 'were as much constrained by his commitment to a naturalistic or scientific explanation of human mental and moral characteristics as they were by his socially derived assumptions of the innate inferiority and domesticity of women' (61). That sentence seems rather oddly phrased. On the one hand, it suggests surprise that scientific explanation might account for anything in Darwin's scientific theory. On the other, it suggests that the focus of Richards's arguments will be where it is not, that is, on the demonstration of the 'scientific constraints'. Nevertheless, it is extremely interesting and important that in two excellent essays implying the primacy of cultural explanation, both writers insist on the necessity of scientific constraints. The question that arises from that insistence, however, is what can be meant by such constraints and how such constraints can even be imagined. Is the suggestion that there are some issues that are constrained by the intrinsic nature of their materials *as opposed to* constraint by cultural forces? This, of course, is the fundamental argument of those who have taken up the cudgels for science in the recent lamentable 'science wars'. But beyond what I take to be an oversimple division between scientific and cultural restraints, one might ask the question, what if, as some cultural historians insist, *everything* is culturally constrained in some way or other? What might be the consequences of this assumption in study of particular scientific arguments? What might this mean for the theory of sexual selection, in particular?

That is to say, after all the evidence is in about the cultural forces that shaped Darwin's arguments, there remains the question of whether Darwin was right about sexual selection itself. 'Sexual selection', writes Fiona Erskine, 'is intrinsically

[8] Janet Browne, 'I Could Have Retched All Night: Charles Darwin and His Body', in Christopher Lawrence and Steven Shapin (eds.), *Science Incarnate: Historical Embodiments of Natural Knowledge* (Chicago: University of Chicago Press, 1998), 240–87.

anti-feminist.'[9] There are two questions to ask about such a claim: does it also imply a claim about the *validity* of the theory? And can sexual selection in all of its complications be contained by such an ideological implication? No doubt, when Darwin came to write about humans he largely reinforced, as Erskine says, his deeply rooted and widely shared views about the inferiority of women—though he had long supported women's education. But subtle critical analysis such as Jann's forces an unpacking of the question, for her essay locates contradictions within Darwin's own arguments that seem to make it impossible—except through the wildest chance—that they were entirely right (and it is widely agreed now that they were *not* entirely right). But there remains to this day a likelihood that Darwin's fundamental ideas of sexual selection were correct, and it would be a very bad mistake, solely because Darwin's sexism emerges so clearly when he talks about humans, to dismiss the theory. In fact to do so would be in effect to leave oneself open to the possibility that it is not only Darwin but nature that is intrinsically sexist. We are past the stage, either in cultural studies or in science, of having to take Darwin whole cloth. We should also be past the stage of thinking that an argument ends once it is demonstrated that a position is ideologically constructed. If it is universally the case (and the need to invoke the word 'universal' in such a matter is an ironic commentary on the futility of the 'everything is political' argument) that everything is ideological, then the important work of analysis and understanding must begin *after* the ideological work is recognized. Nothing is proved by proving that a thing is 'ideological'.

This is not the place, nor is there world enough and time, to discuss what it might mean to hold to scientific standards of proof and logic when we also believe that every discourse, including our own, must be marked by local cultural perspective. While I believe that there are legitimate standards of proof—perhaps not so systematically ordered and recognized as is sometimes argued—that resist the pressures of cultural forces, my argument about Darwin's peculiar genius here depends on an at least provisional acceptance of the general view that his ideas about sexual selection (and even natural selection) were significantly informed by his cultural assumptions. Obviously, on the understanding that cultural perspectives are pervasive, an answer to the question of what constitutes scientific standards of proof and logic would inevitably be determined by those very perspectives.

However such views are interpreted, some recognition of constraint in interpretation is as necessary to commentators on Darwin as Darwin felt them to be on himself. Darwin was certainly constrained by other things besides his ideological assumptions. He was alert to the possibility of the kind of criticism that cultural critics have been levelling at him as when, near the end of *The Descent of Man*, he plainly asserts that 'The views here advanced, on the part which sexual

[9] Fiona Erskine, '*The Origin of Species* and the Science of Female Inferiority', in David Amigoni (ed.), *Charles Darwin's Origin of Species: New Interdisciplinary Essays* (New York: St Martin's Press, 1995), 95–121.

selection has played in the history of man, want scientific precision' (ii. 383). The question of canons of scientific validity was serious for Darwin. He knew the degree to which his various rich but not totally substantiated arguments about natural and sexual selection required much more, particularly with regard to their roles in the development of humans, before they could be firmly established. He worried, legitimately, and not simply in the throes of ideological blindness, about whether he could produce enough evidence, and his whole work is marked by rhetorical admissions that if such and such were to be the case it would be fatal to his theory. He certainly knew that his discussion of humans was yet more speculative than any other part of his work. But surely, having read Darwin's work and his letters and notebooks, the most hardened critic would have to concede that his highest priority was not to enforce his sexist assumptions or his preferred economic theories but to get it right. In his own speculativeness, of course he believed what he argued while at the same time recognizing the vulnerability of his position.

On the subject of sexual selection, Darwin knew in how small a minority he was. But then if he were being ideologically complicit in insisting on his theory, what about the complicities of those who disagreed with him? On the one hand, there was A. R. Wallace, who was more Darwinian than Darwin, though politically he roamed well to what we might now call Darwin's left. On the other, there was St. George Mivart, the strongly Roman Catholic scientist, whose religious and political positions were obviously to the right of Darwin's and who talked of 'vicious feminine caprice' that made the idea of female choice helping establish permanent evolutionary changes absurd to him.[10] We might of course rub out the ideological differences among those who rejected Darwin by claiming that on at least one issue they were all in it together, trapped in the culture's prejudices about gender, but such erasure makes important distinctions impossible and suggests that nobody at the time could have either opposed or supported Darwin's theory without being guilty of the same sin—hardly an intellectually profitable argument. One of the more interesting questions for cultural study of science, then, would be whether rejection of Darwin's theory meant rejection of the culturally pervasive ideological assumptions now attributed to it (which would be odd). But if it did not mean that, what would that conclusion suggest about the direct relation between Darwin's theory and any particular ideological assumptions? Particularities are what would be needed in discussion of the various positions adopted by those opposed to Darwin, and what would it signify that those various positions all implied the *same* cultural assumptions? I hope I am not being merely naïve in claiming that in the unlikely event that Darwin had been presented with evidence that led to the conclusion that women were not inferior to men, he would have accepted it as 'fatal to his theory'. Such a concession could not have been greater for him than the one his discoveries of sexual selection

[10] For an interesting discussion of these two writers' relation to sexual selection, see Helena Cronin, *The Ant and the Peacock* (Cambridge: Cambridge University Press, 1991), esp. 172.

forced upon him, that there was some intelligence in the motors of evolutionary development after all. Darwin's theory, as Gillian Beer has shown so well, is aimed in part at disrupting the natural theological understanding of organisms, that they were intentionally designed and directly adapted to their positions in the world. Recall that Darwin makes his claims about the work of female choice in evolutionary development in the same book in which he confesses to having mistakenly assumed in his earlier work that 'every detail of structure, excepting rudiments, was of some special, though unrecognised, service' (i. 153). *The Descent of Man* presents a theory of evolutionary change that is self-consciously antagonistic to the idea that everything has a purpose, that all 'details of structure' are of use. Sexual selection, however, is a theory that depends on the assumption that only by recognizing that dimorphic details are indeed of some 'service', after all, can one make sense of racial and sexual difference.

Whatever Darwin's own views about intention, any discussion of the ideological implications of his arguments ought to take account of perhaps the most striking fact about the theory: virtually nobody believed it. To be sure, there is the interesting and extravagant book by Grant Allen, called *Physiological Aesthetics*. There is also strong evidence that Hardy read and was influenced by *The Descent of Man*. But any consideration of the cultural influence of the idea of sexual selection must take into account the fact that within biological science sexual selection had no serious place for many years, until in fact quite recently. As Helena Cronin has shown in her extensive survey of Darwin's theories, 'There has been very little discussion of sexual selection through the years and it is only now being taken seriously historically and scientifically.'[11] Obviously, Darwin's importance insured that the theory would generate discussion of the possible role of sexual relations in evolutionary development. But Wallace, who was totally committed to the view that natural selection provided fully explanatory power, was outspoken in his rejection of the theory, explaining what Darwin read as sexual—the often astonishing dimorphism of male and female organisms—entirely through natural selection. Unlike natural selection, which was also not scientifically accepted until well into the twentieth century, sexual selection does not seem to have seeped into the culture very deeply, though sexism, scientific and otherwise, was pervasive. Until very recently it was possible to treat the theory as Gertrude Himmelfarb has done, as public evidence not only of Darwin's recognition that his theory of natural selection had failed, but also of his intellectual shallowness.[12]

That the ideological work Jann and Richards detect was actually going on, I do not doubt. But that by way of Darwin the theory of sexual selection had much influence on Victorian culture is unlikely. Cynthia Russett claims that although the view that of the two sexes the male was the more variable was indeed used to do some dirty ideological work, that idea of the greater variability of the male did not at all depend on the theory for its support. She argues that 'The elaborate

[11] Ibid. 115.

[12] Gertrude Himmelfarb, *Darwin and the Darwinian Revolution* (London: Chatto & Windus, 1959).

edifice of female conservatism and male progressivism, female mediocrity and male genius, that was presently erected on the foundations of variability, does not derive immediately from *The Descent of Man*.'[13]

I have already intimated an idea that I want to make the basis of what I now have to say about the theory of sexual selection, this having to do with the question of the relation of ideology to scientific ideas. Oscar Kenshur has labelled the view that particular scientific theories have particular ideological implications 'ideological essentialism'. He makes the point this way: 'the ideological essentialist wants to be able to find intrinsic political significance in abstract theories that at first glance seem lofty and disinterested'.[14] The alternative to this view, the one I have already urged, Kenshur calls 'ideological contextualism'. The latter seems to me the only sensible one in this regard. The quickest glance at the uses to which Darwin's theory has been put would make the case because although natural selection is often linked to certain forms of social Darwinism emphasizing brutal competition, it has served an extraordinary number of other purposes, even to the extent of being taken as a justification for socialism, nihilism, and religion.[15] What remains after the uses and the misuses is always the theory again, to be appropriated and misappropriated.

And the basic theory of sexual selection, although after ideological exposures and 150 years it may be difficult to recognize, remains fertile and disruptive. But what I want to emphasize here is that its brilliance and originality are intimately connected with its cultural sources. Having identified the gender prejudices of the culture that play into Darwin's imagination of 'sexual selection', one will find that the theory itself forces a break with just those prejudices that produced it, and Darwin's reversal of his argument from animals to humans is a particularly good sign that his thought outleaped the culture that helped form it. How could he have come up with the idea that the aesthetic sense derives from animal sexuality? The difficulty of the conception is multiplied when one understands that it required that he recognize that female choice was at the root of it all. The scientific culture of his own moment found this idea particularly difficult to swallow. Almost nobody, whether sympathetic to Darwin or not, wanted to believe it. Cronin describes a representative position, taken by Darwin's opponent, St. George Mivart, which claimed that female birds simply could not have the sensibility to respond to the aesthetic appeals of ornamented males. So, he claimed,

[13] Cynthia Russett, *Sexual Science: The Victorian Construction of Womanhood* (Cambridge, Mass.: Harvard University Press, 1989), 92.

[14] Oscar Kenshur, *Dilemmas of Enlightenment: Studies in the Rhetoric and Logic of Ideology* (Berkeley: University of California Press, 1993), ch. 1: 'Ideological Essentialism'.

[15] In *Philosophy and Social Hope* (London: Penguin, 1999), Richard Rorty similarly argues that large philosophical theories are not linked to particular political positions. 'Both the orthodox and the postmoderns still want a tight connection between people's politics and their view on large theoretic (theological, metaphysical, epistemological, metaphilosophical) matters' (18). 'People on the left keep hoping for a philosophical view which cannot be used by the political right, one which will lend itself only to good causes. But there never will be such a view; any philosophical view is a tool which can be used by many different hands' (23).

'the female does not select; yet the display of the male may be useful in supplying the necessary degree of stimulation to her nervous system'.[16] Females are, after all, very coy.

Mivart's cultural prejudices against Darwin's arguments are clear enough, but I want to emphasize that the idea of female choice derived for Darwin as directly from his own unarticulated cultural assumptions as did his transference of choice to the males in humans. Darwin's working through of his theory suggests that discovery of the presence of cultural assumptions at work in scientific arguments need not undermine those arguments; in Darwin's case at least the presence of cultural assumptions made possible some good and innovative science, having the potential for implications counter to the very assumptions on which he unselfconsciously drew.

Notoriously, Darwin did not work on the 'Baconian principles' he claimed to use. The 'hypothetico-deductive' method, as it has been called admiringly by interesting commentators such as Michael Ghiselin,[17] begins with something like a guess. The 'guess' in the *Origin* that family resemblances among species are literal, was saturated with cultural assumptions. Darwin's science *needed* the ideological assumptions that cultural critics and scientists alike, for different reasons and in different ways, regard as evidence of bad science. The hypothesis, the guess, that underlies sexual selection, not only indicates the way in which Darwin's ideas participate in the culture's ideologies; it suggests to me that cultural assumptions are inevitable for anyone not in the dreamed of 'nowhere' of absolute objectivity and universality.

Predisposition to the two ostensibly opposed approaches I mentioned at the start has led me to try to understand, in Darwin's case at least, not so much how cultural prejudices and assumptions expose sexism or imperialism or other modes of complicity, but how they managed also to be creative, to produce a theory such as sexual selection. Let me begin with a quick glance at an aspect of Darwin's sexism.

There is a well-known passage in Darwin's *Autobiography*, in which he discusses the failure of his aesthetic sense, when he says that he can no longer read poetry, that Shakespeare positively nauseates him, but that

> novels, which are works of the imagination, though not of a very high order, have been for years a wonderful relief and pleasure to me, and I often bless all novelists. A surprising number have been read aloud to me, and I like all if moderately good, and if they do not end unhappily—against which a law ought to be passed. A novel, according to my taste, does not come into the first class unless it contains some person whom one can thoroughly love, and if it be a pretty woman all the better.[18]

[16] Cronin, *The Ant and the Peacock*, 157.

[17] *The Triumph of Darwinian Method* (1969), with a new Preface (Chicago: Chicago University Press, 1984).

[18] Charles Darwin, *The Autobiography of Charles Darwin, 1809–1882; with original omissions restored*, ed. Nora Barlow (London: Collins, 1958), 139.

Lamenting his fall from the higher pleasures of aesthetic culture, Darwin rather unselfconsciously falls back on the source of the aesthetic, which he had himself identified in *The Descent of Man*. There Darwin had argued that the 'sense of beauty' was not 'peculiar to man' (i. 63). Careful to point out that the sense of beauty is not single, he had shown himself aware that 'high tastes', as he calls them, depend 'on culture and complex associations' (i. 64). But what he did not quite call lower tastes are fundamental to complex animals, and they grow from regard for the 'pretty'. The assumptions of Victorian culture are obviously at work here and Darwin includes in his argument in the *Descent* some rather unpleasant (retrospectively) material about the inferiority of 'savages', whose aesthetic tastes are often, from his point of view, inferior to those of birds. In the context of his autobiographical comments, however, when he self-mockingly describes his love of happy endings and pretty heroines, Darwin implies that he has fallen back onto a primitive sense of the beautiful—a sense, by the way, that in the *Descent* he had shown common in 'barbarous tribes' where female choice was still the practice. Pleasure in the novel, itself largely recognized as a feminine form, was surely an indication of a lapse away from high culture back towards primitive feeling, and the attraction of 'pretty' women marks an appeal to fundamentally primitive desires. Higher culture and a higher level of evolutionary development shifted the power of choice to men, but in the *Autobiography* Darwin sadly concedes his own fall from that higher culture.

The 'pretty' attracts him as it attracts the birds, and he does not try to explain it. In the *Descent* he confesses: 'Why certain bright colours and certain sounds should excite pleasure, when in harmony, cannot, I presume, be explained any more than why certain flavours and scents are agreeable; but assuredly the same colours and the same sounds are admired by us and by many of the lower animals' (i. 64). Beer points out how, in his discussion of the female human, Darwin steadily 'gives primacy to beauty' over intelligence. Beauty is the key concept, one that entices him throughout and one that is obviously conditioned by Victorian expectations.

Darwin's continued enthusiasm for pretty girls in novels can serve as a reminder of how he arrived at the theory of sexual selection. The pretty girls of Darwin's unreflective, lower pleasures, are a condition for his theory. As he puts it succinctly in the *Origin*, 'when the males and females of any animal have the same general habits of life, but differ in structure, colour, or ornament, such differences have been mainly caused by sexual selection'.[19] What makes this insight possible is Darwin's assumption that those differences in appearance are noticeable and attractive to the opposite sex; not only that, but once these things are noticed, the opposite sex can make a choice about them, just as Victorian gentlemen (and maybe an occasional country girl) do.

One possible inference from this obvious reliance on cultural prejudices is that there must be something wrong with the whole theory. Another, the one I want

[19] *On the Origin of Species*, ed. J. W. Burrow (London: Penguin, 1985), 137.

to make here, is that Darwin's experience and sharing of those Victorian tastes made clear to him a problem and suggested a resolution that is really entirely anthropomorphic and at the same time almost certainly right. There have often been serious critiques of Darwin's arguments about non-human organisms, that they are too anthropomorphic. Certainly they are, and the flood of metaphors that does such work in the development of his theory reveals the degree to which his assumptions about human culture helped shape his scientific arguments. The obviousness of Darwin's assumptions about Victorian prettiness threatens to obscure the fact that there is nothing inevitable about the theory of sexual selection that he derived from them. Even those most sympathetic to Darwin's theories to this day have trouble distinguishing between the effects of sexual and those of natural selection. Characteristics developed by sexual selection have no consequences in natural selection; they are not conditions for the survival of the organisms. Indeed, if the theory of sexual selection is correct, certain elements of organisms necessary for sexual selection are dangerous in the world of natural selection—the most obvious, of course, being the often striking colours and displays of breeding males.

The theory is at times yet more counter-intuitive than natural selection was. Victory in sexual selection depends, as Darwin says, 'not on the general vigour' of the male but on its 'having special weapons' (*Origin*, 136). Those weapons are not matters of life and death but are necessary in the struggle with other males of the species to win the female and produce the most progeny. Since most females are not as ornate as most males, and since they survive in nature as well as the males, it follows that the ornamentation has another purpose, and Darwin inferred that the purpose was to win the female. What can all that finery be for?

Darwin unembarrassedly tries a thought experiment, a device he used brilliantly in the *Origin*, and, it would seem inevitably, it spins around a pretty girl:

> With respect to female birds feeling a preference for particular males, we must bear in mind that we can judge of choice being exerted, only by placing ourselves in imagination in the same position. If an inhabitant of another planet were to behold a number of young rustics at a fair, courting and quarrelling over a pretty girl, like birds at one of their places of assemblage, he would be able to infer that she had the power of choice only by observing the eagerness of the wooers to please her, and to display their finery. (ii. 122)

Much of the *Descent* depends on placing *ourselves* in imagination in the condition of some other being. When we do, even as we try to imagine otherness, the other gets to be rather like us, rather Victorian. The visitor from another world notices the 'pretty girl' rather as a Victorian bird watcher might watch courting birds. The only way to explain this behaviour is to see it as competition for a female, and one who, given the excess of suitors, has the power to exercise choice. Without choice there would be no dimorphism in birds. Victorian as it distinctly is, this guess is a remarkably good one—rather, *because* of its self-evident Victorianism. Like a good novelist, Darwin here makes something rich and creative of his ideological luggage by way of an imaginative leap.

It is worth pausing to notice, too, that in slipping into his thought experiment and creating a hypothetical encounter between a pretty country girl and competing male suitors, Darwin actually introduces into *human* mating the female choice that, for humans, he was to deny. The imagination in the form of a thought experiment confirms the overall theory. But at the same time it reaffirms the work of female choice, even among humans, against Darwin's own argument that inverts the pattern for the human because he could not believe that human females had a significant role to play in the development of the species.

Darwin defines the excess in nature that cannot be explained by natural selection as the 'pretty' or the beautiful in the eyes of the opposite sex. This is how to make sense of the astonishing adornments of peacocks and pheasants, for example. It builds on his recognition that there are 'pretty' things out there (and we know they are pretty because we as humans regard them as so). If human beings find them beautiful (one piece of evidence he gives is that women adorn themselves with bird feathers!), then birds must too. Victorian birds.

Darwin's standards of prettiness in the *Descent* are unabashedly human, though he often concedes that what is beautiful to certain animals is not beautiful to humans. In his section on bird display, he invokes artists to provide testimony to the amazing beauty of bird feathers and courting habits. After describing the designs on the tail of an Argus pheasant, for example, he notes, 'These feathers have been shewn to several artists, and all have expressed their admiration at the perfect shading' (ii. 91). In making his case for female choice, he writes:

> Many will declare that it is utterly incredible that a female bird should be able to appreciate fine shading and exquisite patterns. It is undoubtedly a marvellous fact that she should possess this almost human degree of taste, though perhaps she admires the general effect rather than each separate detail. He who thinks that he can safely gauge the discrimination and taste of the lower animals, may deny that the female Argus pheasant can appreciate such refined beauty; but he will then be compelled to admit that the extraordinary attitudes assumed by the male during the act of courtship, by which the wonderful beauty of his plumage is fully displayed, are purposeless; and this is a conclusion which I for one will never admit. (ii. 93)

Here again anthropomorphism is at work with a vengeance, but of course anthropomorphism is not for Darwin a mere sentimental lapse (though sometimes it may be that). It is rather a quite seriously worked out way of regarding a world in which there is an absolute continuity between human and animals. It is not so much anthropomorphism, then, as zoomorphism: that is, humans are animals, and therefore one can—as an animal oneself—in fact greatly understand non-human behaviour simply by imagining one's way into the animal's mind.

This passage is only one of many that register Darwin's remarkable power to break through provincial prejudices by thinking through them. He can do this in part because he is entirely convinced, beyond theory and into the depths of his own imagination, that human feeling and thought are grounded in animal consciousness. Even worms, in Darwin's imagination, are intelligent beings, thinking through how to get leaves into their holes or responding sensitively to the keys of E and G on the piano. The cultural prejudices of the Victorians may be the cultural prejudices of worms as well, but engaging those prejudices and thinking with and through them allowed Darwin to think beyond them, as he did in the matter of female choice.

But there is a striking aspect to this argumentation that places Darwin even deeper inside the ideology of his own moment at the same time as it opens up new possibilities. The most remarkable thing to me about this remarkable passage is that it participates in the same rhetorical methods as did the natural theology that Darwin had spent the best part of his argumentation trying to dismiss. Dov Ospovat, some years ago, showed that the development of Darwin's theory depended in great measure on the influence of the natural theology that he ultimately rejected. Only after reading Malthus did Darwin come to the view (and this, too, gradually) that there was no such thing as 'perfect' adaptation, and that the world, instead of being harmonious everywhere, was full of 'discord'.[20] But through this entire movement, the fact of 'adaptation' continued to play a major role, and adaptation carried with it the old rational force of natural theology.

At one point elsewhere in the *Descent* Darwin says about a courting display, 'We cannot believe such display is useless' (1882, 260). And in the passage I have just quoted he doggedly asserts that he will never admit that the courtship habits and style of the Argus pheasant are useless. Or consider this passage about monkeys: 'It is scarcely conceivable that these crests of hair and the strongly-contrasted colours of the fur and skin can be the result of mere variability without the aid of selection; and it is inconceivable that they can be of any ordinary use to these animals. If so, they have probably been gained through sexual selection, though transmitted equally, or almost equally, to both sexes' (ii. 308).

For a true Darwinian like me, such argument, entirely representative of the passages on sexual selection, takes the breath away. William Paley, whose *Natural Theology* Darwin read and admired, when he confronted the extraordinary contrivances of nature, expressed incredulity that anyone could see these things and not recognize in them design and, necessarily then, the evidence of a Designer. Paley's designer is, of course, God. In discussing sexual selection, precisely as he is asserting the intention of the participating animals, Darwin expresses the same consternation, disbelief, and refusal to accept the possibility that there is no intention at work. The difference is that Darwin's designer is not God but females. Darwin argues this way despite the fact that the *Origin* is largely given over to demonstrating the error of that kind of response. The deep Victorian need for order and meaning, which found expression in late-century attacks on Darwin's theory of natural selection, is alive and well and useful in *The Descent of Man*. Whether in conscious imitation of Paley's methods or not, Darwin employs here the very strategy of natural theology.

[20] Dov Ospovat, *The Development of Darwin's Theory: Natural History, Natural Theology & Natural Selection, 1838–1859* (Cambridge: Cambridge University Press, 1981). See in particular, 60–1.

The question, 'what is such and such a contrivance for?' is inescapable in evolutionary biology. In current controversy there is often debate about whether the ultimate benefit is the gene's, the organism's, or the group's (species'). But whichever answer one determines on, the answers themselves imply an objective, at least a quasi telos. The contrivance may be 'for' any one of these units, but it seems to be 'for' *something*, and it is taken as worthwhile, or rather necessary. There may be no adequate answer; it may be, as in many instances Darwin described it to be, simply the result of a correlation of growth of another adaptive part of the organism, a 'spandrel', as Stephen Jay Gould called it. But whatever position one takes in the end, the preliminary, thoroughly commonsensical, and therefore culturally loaded response is necessary. Something strikes us. We ask questions about it. It can only strike us if it fits into (or challenges) our culturally developed assumptions. Through sexual selection, Darwin introduces aesthetic taste and female intention as driving forces in the evolutionary process, and he could do that only because he had those attitudes and those assumptions.

Of course, this leaves out a great deal of the story. I have wanted for the most part simply to point to some things that look like paradoxes in the story of sexual selection and in Darwin's relation to that story in order to show that Darwin's theory is enabled by the culture that is often taken to have led him astray. The exposure of that enabling, many good critics have claimed, demonstrates Darwin's ideological complicity with Victorian sexism and undercuts his 'scientific' arguments. But this tells less than half the story. Some of the rest of that story would reveal that those very ideological assumptions are not merely wicked transmitters of vicious ideologies but a condition of the really valuable and original work Darwin did. Darwin's views, however they are locally hooked into ideologically repugnant aspects of Victorian culture, have no necessary connection to them nor to any other specific ideological positions. When other Victorian writers, particularly novelists such as Hardy or George Eliot, use or play variations on Darwin's ideas about sexual selection, they arrive with their own ideological baggage and they produce other interpretations that might well generate other ideological positions.

Sexual selection is an amazingly inventive and productive idea. Deriving from a very Victorian notion of what is 'pretty' and from a very Victorian sense of what is striking, it produces a scientifically interesting and probably correct understanding of evolutionary development. On the strength of it, one might make a case for the sexist Darwin as a kind of ideological hero. Certainly he believed in the inferiority of women. But female choice was about as revolutionary a concept as natural selection, and even now it is resisted more pervasively if not more intensely. That the theory of sexual selection is a product of Victorian culture is both unsurprising and remarkable since, if followed out to its fullest possibilities in directions Darwin established but did not follow, it might very well imply the intellectual superiority of women.

CHAPTER 3

Ordering Creation, or Maybe Not

HARRIET RITVO

By the middle of the nineteenth century systematics no longer constituted the cutting edge of natural history. As William Jardine, the editor of the Naturalist's Library, an influential series of zoology guides, put it, 'the age of superstitious reverence for categories ... has long passed away'.[1] Black-boxed and metonymized in the person of Linnaeus, classification had, in fact, come to occupy a somewhat unenviable position in the view of the two major audiences for zoology and botany. What in 1845 a *Westminster* reviewer of popular natural history books called its 'hard names and crabbed systems' lent an air of formidably arcane authority that discouraged, or even dismayed, amateur admirers of 'the *living* tenants of the fields, the woods, the waters . . . the birds that fill the air with their gladsome songs, and the ... [flowers] which enamel the earth with their exquisite tints, and impart their fragrance to the pure breezes of heaven'.[2]

And although the community of experts was not so easily awed by Latin and Greek, its members similarly found systematics to be boring and even repellent. Thus several decades later William Henry Flower, whose distinguished career included the Hunterian Professorship of Comparative Anatomy and Physiology at the Royal College of Surgeons, the presidency of the Zoological Society, and the directorship of what was then called the British Museum (Natural History), lamented that classification and its attendant practice of nomenclature were considered 'by many biologists of the present day . . . as the least attractive and least profitable branches of the subject'.[3] In order to oppose the widespread notion that taxonomy was an essentially clerical pursuit—a kind of scientific bookkeeping—Flower urged that 'in reality, *classification* is one of the, if not the, most important aims and ends of the study of morphology'.[4] Associating it with

[1] William Jardine, 'Memoir of Aristotle', in *The Natural History of Gallinacious Birds, Part I* (Edinburgh: W. H. Lizars, 1834), 19.

[2] [George Luxford], 'Popular Works on Natural History', *Westminster Review* 44 (1845), 208, 204.

[3] William Henry Flower, 'A Century's Progress in Zoological Knowledge', *Essays on Museums and Other Subjects connected with Natural History* (1898; Freeport, NY: Books for Libraries Press, 1972), 166. (Originally delivered as the Presidential address in the Department of Zoology and Botany, BAAS, Dublin 1878.)

[4] William Henry Flower, 'Introductory Lecture to the Course of Comparative Anatomy', *Essays on Museums*, 115. (Originally delivered at the Royal College of Surgeons, 1870.)

Darwinian theory, which was, of course, much closer to the cutting edge, he claimed that, 'a true classification . . . would be a revelation of the whole secret of the evolution of animal life'.[5]

This Victorian consensus that the main work of zoological and botanical classification had been accomplished by the heroic systematizers of the eighteenth century, leaving their successors to perform a laborious and uninspired mop-up operation, has shown a lot of staying power. The latinate nomenclature in which it has traditionally been expressed has retained its authoritative air, indeed becoming ever less accessible as knowledge of classical languages has become rarer. (The quality of the Latin and Greek embodied in new binomials has declined radically over time; elite Victorian nomenclators are doubtless spinning in their graves.) Modern biology textbooks continue to suggest that, after Linnaeus had sorted things out, taxonomy had few substantial intellectual contributions to make. This durable disparagement has in recent decades been reinforced from a very different quarter, one from which scientific bromides do not ordinarily receive intellectual support. The Foucauldian understanding of the Enlightenment episteme accords the project of universal classification a negative sign rather than a positive one, but it similarly assumes it to have been successful, even triumphant.[6]

As is often the case, however, the actual practices of Victorian zoologists and botanists suggest different conclusions than do their abstract theoretical formulations. Far from being routine and rigid—a puzzle that had been solved or a question answered—systematics continued to engage the most difficult challenges that naturalists faced, from the most technical to the most abstract. These challenges were numerous as well as formidable. The flood of new plants and animals from the remote ends of the world that had inspired the eighteenth-century efflorescence of taxonomy only increased as time went on. By the late nineteenth century an average of more than one thousand new genera of animals were being described each year (genera, not species; only animals, not plants), which was characterized in *Nature* as 'a simply appalling number'.[7] Nor were exotic lands the only source of organisms that required identification and naming. The boom in local natural history meant that novel forms were constantly being discovered much closer to home. As Charles Darwin noted in *On the Origin of Species*, the discovery of new organisms was directly correlated to the effort expended in searching for them: therefore 'it is in the best known countries that we find the greatest number of forms of doubtful value [that is, intermediate forms]'.[8]

The essential first step in dealing with any unrecognizable form, whether

[5] Flower, 'A Century's Progress', 167.

[6] See Harriet Ritvo, *The Platypus and the Mermaid, and Other Figments of the Classifying Imagination* (Cambridge, Mass.: Harvard University Press, 1997), 219 n. 58.

[7] 'The Zoological Record', *Nature* 27 (1883), 310.

[8] Charles Darwin, *On the Origin of Species* (1859; Cambridge, Mass.: Harvard University Press, 1964), 50.

clearly new or merely 'doubtful', was to give it a name that established its taxonomic location. Without classification and nomenclature, therefore, the practice of zoology and botany would have been not only impossible, but inconceivable. Naturalists would have been swamped by incomprehensible and meaningless data. And, indeed, sometimes they were swamped—although more often by raw material than by the information it could yield. Capable taxonomists were not always available when specimens in need of identification, description, and curation arrived in the metropolis, and potentially valuable collections, such as those that returned from the south seas with Captain James Cook, could be left to moulder and disintegrate.[9] Darwin's well-known difficulties in finding specialists willing to work on his *Beagle* specimens were far from unusual; what was less ordinary was his ultimate success.

Nor did difficulties end once a qualified specialist (ornithologist, malacologist, and so forth, as the case might be) had been persuaded to take a collection in hand. Received opinion to the contrary notwithstanding, classification was not merely a matter of recognizing into which (metaphorical) pigeonhole each specimen should be popped. Even if an appropriate pigeonhole existed, a range of practical problems might make it hard to recognize. The extent to which a dead plant or animal resembled its living former self varied widely, according to the vicissitudes of preservation, among other things. Both shape and colour, for example, might have been totally altered. In addition, important information about a given specimen—such as its age, its sex, its stage of the life cycle or the reproductive cycle—was often missing. A taxonomist working with a single specimen had no way of knowing how similar it was to others of its kind—that is, was it representative or was it anomalous, even a sport? And for many plants and animals in need of classification, there was no pre-existing pigeonhole. Despite the claims of almost all Enlightenment and post-Enlightenment systems to universality, previously unknown organisms frequently required the creation of novel categories.

All these difficulties had theoretical as well as practical dimensions. That is to say, they presented problems that could not be solved simply by the application of technical knowledge or skill. The connection between theory and practice was most obvious when the difficulties occurred at a relatively high taxonomic level. For example the discovery of the platypus and the echidna, Australian animals that were perceived to combine characteristics exclusive to mammals with characteristics that distinguished birds and reptiles, forced reconsideration of a zoological category (warm-blooded quadrupeds—which Linnaeus had rechristened as mammals) that had remained relatively stable for millennia, since the time of Aristotle.[10] Further, since they seemed intermediate between three cate-

[9] P. J. P. Whitehead, 'Zoological Specimens from Captain Cook's Voyages', *Journal of the Society for the Bibliography of Natural History* 5 (1989), 181–3.

[10] For further discussion, see Ritvo, *Platypus and Mermaid*, 1–15.

gories, rather than merely two, their oddness implicitly challenged the chain of being, then the dominant organizational metaphor for the organic world. One indication of the intractability of such problems is that their solutions have often proved unstable, and shifts in opinion have inevitably required further taxonomic revisions. For example, it is interesting that despite the late nineteenth-century consensus that the relationship between monotremes and marsupials and placental mammals had been definitively established, with monotremes and marsupials sharing one mammalian branch (called Marsupionta, at least by some taxonomists) and placentals evolving separately on another (predictably superior) one, biologists have recently produced fresh evidence about this issue, in support of the claim that monotremes evolved separately from marsupials as well as placentals.[11] So now we will have to cosy up—at least in a phylogenetic sense—to kangaroos and opossums.

But even the most subtle or apparently insignificant discriminations between forms could broach or expose theoretical problems. Perhaps the best-known classificatory story of the nineteenth century features what are now known as Darwin's finches. Of course, they were grouped under no such rubric during the Victorian period, and, as Frank Sulloway has persuasively demonstrated, this vernacular mode of referring to them has worked and continues to work to entrench what he calls 'a considerable legend in the history of science'.[12] This legend explains Darwin's theory of evolution by natural selection as a kind of 'Eureka' experience, inspired by his observation of what we would now call the adaptive radiation of various kinds of organisms when he visited the Galapagos archipelago in 1835.

This account has been repeatedly debunked by more sober historians, who have pointed out that the organisms whose diversification struck Darwin on the spot were tortoises and mockingbirds, that Darwin did not begin to take in the significance of even these relatively striking radiations until later in the voyage, and that he realized that the finches offered a still better example only after he had returned home and John Gould, the ornithologist whom he persuaded to classify the birds collected on the *Beagle* voyage, pointed it out. Darwin then realized that he had not even labelled his finch specimens according to local provenance—that is, which of the Galapagos islands they came from—and he had to consult specimens in the possession of several more meticulous collectors among his former shipmates, including Captain Fitzroy. (Darwin may have felt sheepish about this—he confused later historians by retroactively claiming to have noted the specific islands on which he had collected some of his specimens. But memory apparently failed him, and the resulting inaccuracies ultimately produced extra work for overly respectful subsequent taxonomists, as well as the retroactive alteration of

[11] Zhe-xi Luo, Richard L. Cifelli, and Zofia Kielan-Jaworowska, 'Dual Origin of Tribosphenic Mammals', *Nature* 409 (4 January 2001), 53–7.

[12] Frank Sulloway, 'Darwin and His Finches: The Evolution of a Legend', *Journal of the History of Biology* 15 (1982), 5.

Fitzroy's correct provenances (by museum curators) to bring them into accord with those of his scientific superior.)[13]

In fact, Darwin's initial response to the finches was not inspiration but the reverse. His attention was drawn to them by their numbers—they constituted the most numerous land birds in the Galapagos, both in terms of individuals and of kinds—rather than by any other quality. They were nondescript in several senses. When presenting his conclusions about Darwin's specimens to the Zoological Society in 1837, Gould declared that they were 'so peculiar in form that he was induced to regard them as constituting an entirely new group'.[14] But on the spot, the birds were less arresting, as scientifically inclined visitors both before and after Darwin have testified. (Of course aesthetic appeal is not supposed to constitute a criterion of zoological or botanical interest, any more than magnificence or ferocity or similarity to people are—but scientists are people too, after all.) For example, Captain James Colnett, who surveyed the islands in 1793, described the finches as 'not remarkable for their novelty or beauty', and David Lack, in his classic work of 1947, entitled simply *Darwin's Finches*, confessed that the birds were 'dull to look at, not only in their orderly ranks in museum trays, but also when they hop about the ground or perch in the trees . . . making dull unmusical noises'.[15]

Darwin himself found their plumage to be 'extremely plain'; he felt that 'like the Flora' of the Galapagos, they possessed 'little beauty'.[16] Far from sudden insight, his first scientific reactions emphasized 'inexplicable confusion' about 'the species of this family'. They all looked more or less similar. Although they varied in size, in colour (from brown to black), and 'in form of the bill', these variations seemed graduated rather than discontinuous, and therefore of no use in distinguishing kinds. To make things worse, he observed that 'there is no possibility of distinguishing the species by their habits, as they are all similar, & they feed together . . . in large irregular flocks'.[17]

Not until he returned to Britain did Darwin realize that he had been misled by difference as well as by similarity. When Gould finished his classification not only had he differentiated finch species where Darwin had seen only varieties, but he had recognized as ground finches several birds that Darwin had tentatively identified as belonging to whole different families (for example he had classified what became known as the cactus finch (Fig. 3.1) as an icterid—a member of the family that includes blackbirds and orioles—and he had classified what became known

[13] Frank Sulloway, 'The *Beagle* Collections of Darwin's Finches (Geospizinae)', *Bulletin of the British Museum (Natural History), Zoology Series* 43 (1982), 58–62.

[14] John Gould, 'Remarks on a Group of Ground Finches from Mr. Darwin's Collection', *Proceedings of the Zoological Society of London* 5 (1837), 4.

[15] David Lack, *Darwin's Finches* (1947; Cambridge: Cambridge University Press, 1983), 9, 11.

[16] Charles Darwin, *Journal of Researches into the Geology and Natural History of the Various Countries visited by H.M.S. Beagle* (London: Henry Colburn, 1846), 461.

[17] Nora Barlow (ed.), 'Darwin's Ornithological Notes', *Bulletin of the British Museum (Natural History)* 2 (1963), 261.

FIGURE 3.1 The cactus finch (*Cactornis scandens*), as illustrated by John Gould in Darwin's *Zoology of the Voyage of the Beagle*, 5 vols. (1839–43).

as the warbler finch as a kind of wren).[18] This was not so much telltale evidence of his inexperience—after all, although he was a very young naturalist, on the *Beagle* voyage he collected, and therefore had to classify, at least tentatively, specimens of every possible kind (geological as well as zoological and botanical)—as an indication of one of the reasons that the Galapagos finches have continued to fascinate ornithologists. The fact that they varied so little in colour and so much in shape and size of beak contradicted a commonplace of bird taxonomy: the predominance of beak over colour as an index of relationship. That is, as locomotion has tended to trump other considerations in the classification of vertebrates (so that many people have a hard time thinking of bats and whales as mammals, even though they may know better), so beak shape has tended to trump colour in the classification of birds.

By the end of the *Beagle* voyage Darwin had begun to reflect on the possible significance of the variability of the Galapagos fauna: 'such facts' would, he speculated, 'undermine the stability of Species'.[19] As they say, the rest is history (at least, sort of). Because the Galapagos played such a large part in Darwin's legend (although the finches are not specifically mentioned in the *Origin*, the

[18] Sulloway, '*Beagle* Collections', 57.
[19] Barlow, 'Darwin's Ornithological Notes', 262.

distinctiveness of the indigenous fauna and flora of oceanic islands such as the Galapagos, as well as their affinities with the biota of the closest continental land mass, formed an important piece of Darwin's argument) they became a magnet for subsequent researchers. Darwin and Gould were only the first Victorian naturalists to examine and classify the finches. Several major collecting expeditions were organized after the Galapagos became known as the laboratory of evolution.[20] As Osbert Salvin put it in 1875, when he presented his own taxonomic account to the Zoological Society: 'The ground is classic ground, and the natural productions of the Galapagos Islands will ever be appealed to by those occupied in investigating the complicated problems involved in the doctrine of the derivative origin of species.'[21]

Even the number of species of Darwin's finches was subject to argument for more than a century before settling down at fourteen (one of which does not live in the Galapagos, but on Cocos Island, which lies about 600 miles to the northeast); the number of genera remained open to discussion. Because the lines between species were often established statistically, on the basis of elaborate measurements, finch classification provided a field day for combatants in the most persistent taxonomic war—between splitters and lumpers. (Splitters recognize new taxa on the basis of relatively small differences, which tends to maximize the number of species, while lumpers follow a contrasting theory and practice.) Thus at the end of the nineteenth century committed lumpers (and authors of still another account of the Galapagos birds) Walter Rothschild and Ernst Hartert criticized one of their rivals (Robert Ridgway, an ornithologist at the Smithsonian Institution in Washington) for creating 'two new generic names (*Cactospiza* and *Platyspiza*) without need and to no practical purpose'.[22]

And the taxonomic story is not yet over. Since the 1970s a research team led by Peter Grant of Princeton University has been intensively studying the finches of a single small island in the Galapagos (Daphne Major, which is not much more than a large rock), and their work continues to revise accepted notions of both the means of evolution and its timetable.[23] Recently they have been working on hybridization between the established (or at least recognized) species, which seems to be more frequent and successful than had previously been assumed. As always, when the possibility of hybridization rears its head, species lines are jeopardized. And on a more abstract level, the icon of Darwinian evolution—the branching tree—may also tremble a bit.

[20] For a list of expeditions and publications, see Lack, *Darwin's Finches*, 9–10.

[21] Osbert Salvin, 'On the Avifauna of the Galapagos Archipelago', *Transactions of the Zoological Society of London* 9 (1876), 462

[22] Walter Rothschild and Ernst Hartert, 'A Review of the Ornithology of the Galapagos Islands' *Novitates Zoologicae* 6 (1899), 152.

[23] For a comprehensive report see Peter R. Grant, *Ecology and Evolution of Darwin's Finches* (Princeton: Princeton University Press, 1999). Jonathan Wiener has written a popular account of this research: *The Beak of the Finch* (New York: Vintage, 1995). Edward J. Larsen's *Evolution's Workshop: God and Science on the Galapagos Islands* (New York: Basic Books, 2001) offers a historical overview of scientific attention to the Galapagos.

Thus systematics remained heavily and significantly contested territory throughout the Victorian period and beyond. Far from exhibiting the frozen correctness of an earlier age—providing an inert scaffold upon which successors could build—latinate nomenclature masked disagreement and uncertainty. And if taxonomic practice suggested not the triumphant, monolithic, and smoothly progressive enterprise celebrated in biology textbooks and historical overviews, but one whose authority remained fragile and precarious, instability on the level of system or theory had more unsettling implications. The Grants' recent speculations about hybridity were far from the first gusts to shake the biological tree.

Indeed, ubiquitous though the tree seemed by the late nineteenth century, it was never without rivals as a metaphor for or graphic representation of the taxonomic system. It had displaced the chain of being, or the *scala naturae*, an ancient figure that organized nature as a linked, one-dimensional progression from the meanest animal (or vegetable or mineral, depending on the perspective of the systematist) all the way to humans (or superhumans, also depending on the perspective of the systematist), as the dominant visual metaphor of system.[24] By the first part of the nineteenth century the limitations of the chain—especially its Procrustean linearity—had become obvious to zoologists and botanists. In 1841, giving voice to an emerging scientific consensus, Hugh Strickland asserted that 'no *linear* arrangement . . . *can* express the true succession of affinities'. Using the newly dominant arboreal metaphor of system, Strickland urged his fellow naturalists 'to study Nature simply as she exists—to follow her through the wild luxuriance of her ramifications, instead of pruning and distorting the tree of organic affinities into the formal symmetry of a clipped yew-tree'.[25]

Nevertheless, the chain turned out to enjoy a vigorous afterlife. Although fewer and fewer naturalists explicitly endorsed it as a systematic model, it continued to shape the language of almost everyone who discussed the relationships among zoological and botanical groups, especially since the notion of linkage also reflected late-Victorian evolutionism. Thus if, from one perspective, a new orthodoxy had been established, from another, that of usage or practice, this changing of the metaphorical guard was far less clear. For example, the anatomist and racial scientist Robert Knox explained human distinctiveness in terms of the ostensibly discarded metaphor: 'The human family stands profoundly apart from all others, implying that in the great chain of being constituting nature's plan, some natural family filling up the link has disappeared.'[26] Writing near the end of the nineteenth

[24] The chain of being derived ultimately from the work of Aristotle (Hull, *Science as Process*, 82). On the chain generally, see Arthur O. Lovejoy, *The Great Chain of Being: A Study of the History of an Idea* (Cambridge, Mass.: Harvard University Press, 1936), esp. ch. 8.

[25] Hugh Strickland, 'On the True Method of Discovering the Natural System in Zoology and Botany', *Annals and Magazine of Natural History* 6 (1841), 192.

[26] Robert Knox, *The Races of Men: A Philosophical Enquiry into the Influence of Race over the Destinies of Nations* (London: Henry Renshaw, 1862), 503. It is possible that this adoption of an outmoded metaphor was semi-intentional, since the rejection of the chain was associated with the late eighteenth-century defence of monogenism (the specific unity of the human race) by British scientists, a position that Knox wished to counter. Nancy Stepan, *The Idea of Race in Science: Great Britain*

century, the sophisticated William Henry Flower could mix the rival metaphors without any apparent awareness of inconsistency: 'it is proposed . . . to treat of the horse . . . as one link in a great chain, one term in a vast series, one twig of a mighty tree'.[27]

Perhaps the most striking example of the multivocality of systematic discourse—or the freedom of the intellectual marketplace in which taxonomic metaphors competed—was the debate about the quinary system, which flared and sputtered into the third quarter of the nineteenth century. The brainchild of William MacLeay, who proposed it in his *Horae Entomologicae* of 1819, the quinary system was an elaborate and eccentric attempt to represent the complex, overlapping sets of resemblances among organisms. From the quinary perspective, the compounded linearity of the taxonomic tree was as unsatisfactory as the simple linearity of the taxonomic chain, because it similarly constrained the number of formal connections among animals.

Quinary nature was much more intricately organized than nature in the shape of a tree; it was also, in its fussy way, neater and more predictable. Quinarians arranged animals within a set of embedded circles, each of which consisted of five subsidiary circles, in turn subdivided into five smaller ones, and so on (Fig. 3.2). Thus, MacLeay divided the circle of the class 'Mammalia' into five orders: primates, ferae (carnivores), glires (rodents), ungulata (hoofed animals), and cetacea (marine mammals)—along with, awkwardly, a sixth 'transultant' order containing the ever-troublesome marsupials. An additional principle of organization helped to structure each group of five: one of its members was identified as 'typical', one as 'sub-typical' (these first two united under the rubric of 'normal'), and the remaining three as 'aberrant'. In a manuscript that detailed a scheme of mammalian classification, MacLeay identified the cats as the typical ferae, the hyenas as subtypical, and the weasels, dogs, and civets as aberrant, and therefore, among other things, less carnivorous; he similarly subdivided the bimana (in his view, although not in the view of every quinarian, the human group within primates) into the typical Europeans, the subtypical Asiatics (both 'normal' because 'civilized'), and the aberrant or 'savage' Americans, Africans, and Malays.

Several other kinds of connections between remoter animal groups were most effectively indicated by graphic representation or mapping. Every circle was drawn touching other circles of the same taxonomic rank at each of its five significant points. These abutments indicated analogic bonds, so that, according to MacLeay, mammals were connected to birds through the conjunction of the glires with gallinaceous fowl (chickens and their kin) and to fish through

1800–1960 (Hamden, Conn.: Archon Books, 1982), 6. A derivative of the metaphor of the chain—that of the missing link—retained more explicit intellectual power through the nineteenth century. See Gillian Beer, 'Forging the Missing Link', in *Open Fields: Science in Cultural Encounter* (Oxford: Oxford University Press, 1996), 113–45.

[27] William Henry Flower, *The Horse: A Study in Natural History* (London: Kegan, Paul, Trench, Trubner, 1891), 9.

(*a*)

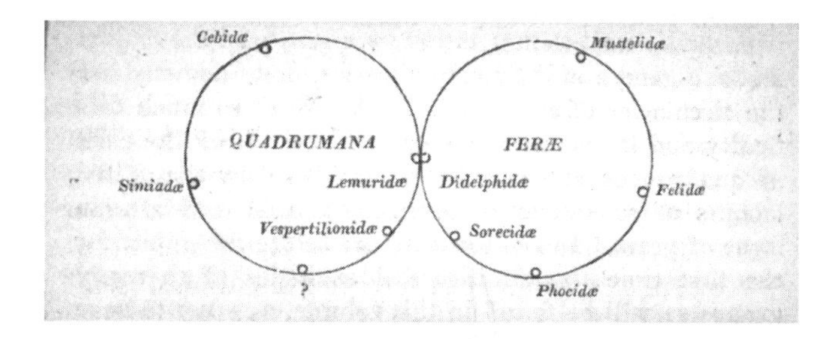

(*b*)

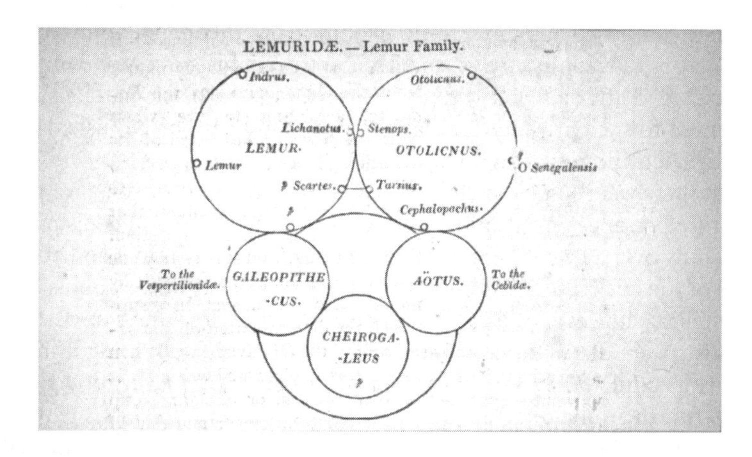

FIGURE 3.2 Quinarian circles from William Swainson's *On the Natural History and Classification of Quadrupeds* (1835). The circle of the Lemuridae represents an expansion of a single point on the circle of Quadrumana. Unlike William MacLeay, Swainson did not group humans with the other primates.

the conjunction of the cetacea with sharks. Finally, each circle repeated the same abstract pattern, so that theoretically any two circles, whether at the same taxonomic level or not, could be superimposed to yield a set of equivalences. Thus MacLeay related the primates to the ferae as follows: humans were analogous to the plantigrade carnivora (those that walked on the soles of their feet), apes to the digitigrade carnivora (those that walked on their toes), lemurs to bats, and tarsiers to insectivores. He noted that 'these analogies explain how some years ago, a Bear was exhibited to the London public as a wild Indian'.[28] There was also an empty or null category in primates corresponding to the amphibious carnivora (seals). The postulated existence of such place markers thus gave the system some predictive as well as analytic force, suggesting that

[28] William S. MacLeay, 'Draft Classification of Mammals on Quinary Principles', n.d., n.p., MS in Linnean Society Library, Case 5A, MS 241.

those categories would eventually be filled by creatures that must exist, but had yet to be discovered (or that had been seen but not properly recognized, like mermaids).

At first, discussion of MacLeay's ideas was primarily limited to a small circle of sympathetic spirits who belonged to the Zoological Club of the Linnean Society.[29] Gradually, however, quinarianism attracted wider attention; its dissemination and influence peaked in the 1830s, after it was taken up by William Swainson, a preternaturally productive author of semi-popular works about natural history.[30] Not surprisingly, much of this attention was negative; many naturalists disparaged MacLeay's system as a crackpot exercise in myth-making. For example, the young Charles Darwin, perhaps recognizing the threat it posed to his embryonic evolutionary theory, denounced quinarian systematics as 'vicious circles' and 'rigmaroles' in his correspondence and notebooks.[31] And after a few decades of skirmishing, the battle had been won, or so it seemed to such critics; MacLeay's system had been safely pushed beyond the pale of scientific respectability.

But there were some indications that quinarianism retained a more substantial scientific presence. Anti-quinarians felt a need to repeat their dismissals rather frequently over a period of several decades; the relatively impassioned language of some of the critics, such as Darwin, also suggested a seriously troublesome target. Most telling, however, was the fact that, at the same time that some establishment scientists were disparaging quinarianism, others were describing it in terms that were respectful and even admiring. In his progress report on zoology, presented to the fourth annual meeting of the British Association, Leonard Jenyns devoted more space to MacLeay's taxonomic system than to that of the magisterial Cuvier, particularly noting the 'influence which this theory has had over our own naturalists'.[32] J. E. Gray, the Keeper of Natural History at the British Museum, published a quinary classification of the mammalia, along the lines indicated in MacLeay's draft.[33] The pre-eminent anatomist Richard Owen characterized the ideas 'which have emanated from the naturalists of the English Quinary school' as

[29] Zoological Club of the Linnean Society, Minute Book 1823–1829. MA Drawer 36, Linnean Society Library.

[30] On Swainson, see Paul Lawrence Farber, 'Aspiring Naturalists and their Frustrations: The Case of William Swainson (1789–1855)', in Alwyne Wheeler and James H. Price (eds.), *From Linnaeus to Darwin: Commentaries on the History of Biology and Geology* (London: Society for the History of Natural History, 1985), 51–9; and David Knight, 'William Swainson: Naturalist, Author and Illustrator', *Archives of Natural History* 13 (1986), 275–90.

[31] *The Correspondence of Charles Darwin*, ed. Frederick Burkhardt *et al.*, in progress (Cambridge: Cambridge University Press, 1985–), iii. 109 n. and ii. 416; Charles Darwin, *Charles Darwin's Notebooks, 1836–1844: Geology, Transmutation of Species, Metaphysical Inquiries*, transcribed and ed. Paul H. Barrett, Peter J. Gautrey, Sandra Herbert, David Kohn, and Sydney Smith (Ithaca, NY: Cornell University Press, 1981), 354.

[32] Leonard Jenyns, 'Report on the Recent Progress and Present State of Zoology', *Report of the Fourth Meeting of the British Association* (London: John Murray, 1835), 155–6.

[33] J. E. Gray, 'An Outline of an Attempt at the Disposition of Mammalia into Tribes and Families', *Annals of Philosophy* 26 (1825), 344.

'remarkable for their novelty and boldness'.[34] The young Thomas Henry Huxley similarly found it suggestive, if not entirely convincing.[35]

Both theoretical praise of the quinarian system and practical application of its principles found their way into many of the manuals and other sources of zoological information that targeted the general public. For example, *Vestiges of the Natural History of Creation*, which appeared anonymously in 1844, became notorious for its evolutionary or transformist argument; less controversially, it devoted an entire appreciative chapter to MacLeay's system.[36] And even long after quinary ideas had become the primary property of zoological eccentrics and interlopers—people such as John Ruskin, who, near the end of a life mostly devoted to cultural struggles, made the subversive suggestion that 'we may be content with the pentagonal group of our dabchicks'—they retained the power to annoy and even infuriate canonical scientific authorities.[37] Thus in 1875 *Nature* editorially blistered an unspecified author for writing 'unmistakable, undeniable nonsense' about 'that silly "circular system" with its mystical numbers, its fives or its seven—the will-o'-the wisp of fancy that once led men's minds astray from the path where only they could find the truth they were earnestly seeking'.[38]

'Truth' was a rather uncompromising term to apply to so speculative an enterprise as systematics. Its use exemplified the process by which volatility and conflict were rewritten as consensus and progress. This process tended to deemphasize or ignore the pervasiveness of metaphor in scientific discourse and scientific thought, as it more generally worked to isolate scientific work from the surrounding culture in which that work took place. Such a narrow focus offered clear benefits to experts eager to establish their claim to authority over their chosen fields, but it also came at an intellectual cost (although its payment was long deferred). Both the story of finch classification and the larger taxonomic questions suggested by that story demonstrate the necessity of acknowledging the figurative dimension of science, the blurredness and permeability of the border between science and other Victorian cultural practices. As Gillian Beer has so generatively argued, context gives science greater significance, not less rigour or prestige.[39]

[34] Richard Owen, 'Mammalia', in Robert Bentley Todd (ed.), *The Cyclopedia of Anatomy and Physiology*, 5 vols. (London: Sherwood, Gilbert, & Piper, 1835–59), iii. 242. He followed this statement with several pages of direct quotation from MacLeay.

[35] Adrian Desmond, *Huxley: The Devil's Disciple* (London: Michael Joseph, 1994), 89–90.

[36] Robert Chambers, *Vestiges of the Natural History of Creation and Other Evolutionary Writings*, ed. James A. Secord (Chicago: University of Chicago Press, 1994), 236–73.

[37] John Ruskin, *Love's Meinie. Lectures on Greek and English Birds* (1881), in E. T. Cook and Alexander Wedderburn (eds.), *The Works of John Ruskin*, 39 vols. (London: George Allen, 1903–12), xxv. 112.

[38] 'Zoological Nonsense', *Nature* 12 (1875), 128.

[39] For Gillian Beer's earliest elaboration of this theme see *Darwin's Plots: Evolutionary Narrative in Darwin, George Eliot and Nineteenth-Century Fiction* (London: Routledge, 1983).

CHAPTER 4

Chances Are: Henry Buckle, Thomas Hardy, and the Individual at Risk

Helen Small

> *Brian, mistaken for the Messiah, attempting to dissuade a vast crowd of followers*: You are all individuals.
> *Crowd [in unison]*: Yes, we are all individuals!
> *Brian*: You are all different.
> *Crowd*: Yes, we are all different!
> *Lone voice of protest from front of crowd*: I'm not.
>
> (*Monty Python's Life of Brian*, 1979)

This essay is about that old philosophical conundrum, neatly fingered here by Monty Python, of how one can plausibly assert individualism in a context where being one of a crowd is of the essence. Strictly speaking, there should be no paradox: there is nothing logically inconsistent about the claim that societies are constituted of individuals. Nevertheless, to *announce* individuality as the grounds for commonality is to find oneself teetering on the edge of absurdity. At the level of enunciation, 'We are all individuals' is as impossibly self-defeating as the effort to dissent from individualism: 'I'm not.' Recent work on the nineteenth century has helped to isolate the force this logical and expressive difficulty had for those many Victorian writers who sought to give value to the concept of a general culture or society or public without ignoring or negating the constitutory differences of individuals.[1] These are also questions that have presented renewed difficulties for literary, cultural, and political theory of late—not least for those critics interested in different scales of collectivism within a global politics.[2]

I am grateful to Melanie Bayley, whose M.Phil. dissertation on 'Victorian Theories of Chance and the Late Novels of Thomas Hardy' (University of Oxford, 2001) I supervised while researching and writing the essay that became this chapter, and with whom I discussed some of these ideas in detail as they took shape.

[1] Particularly Alison Winter's work on consensus in *Mesmerised: Powers of Mind in Victorian Britain* (Chicago: University of Chicago Press, 1998), Mary Poovey's *Making a Social Body: British Cultural Formation, 1830–1864* (Chicago: University of Chicago Press, 1995), and George Levine's discussion of scientific epistemology and the constraints on the expression of individualism, *Dying to Know: Science, Epistemology and Narrative in Victorian England* (Chicago: University of Chicago Press, 2002).

[2] See, for example, Neil Smith, 'Contours of a Spatialized Politics: Homeless Vehicles and the Production of Geographical Scale', *Social Text* 33 (1992), 54–81.

Here I want specifically to probe the nature of the individual's relationship to the group as it presented itself in the mid-nineteenth century in the light of powerful moves within British (and, indeed, European and American) culture to subordinate individual experiences to larger patterns of behaviour discernible in societies, races, and—in the widest view—species.[3] Since my subject is likeness observed within—or imposed upon—difference, I am going to bring together, for the sake of argument, a group of writers not all of whom directly knew each other, and (holding intertextuality at bay for the moment) whose influence on each other was sometimes only one-way, but who together serve to sharpen the question of what might be done with the oddity of individual experience in a context where individualism no longer seemed, in the relative order of things, to matter. Henry Buckle was one of the most problematically famous figures on the European intellectual scene in the mid- to late nineteenth century. Even his detractors, who were many, acknowledged that his colossally ambitious *History of Civilisation in England* did more than any other book of its day to popularize the concept that history might be rethought on scientific principles, and that all human actions were governed by ineluctable, mathematically expressible laws. John Venn, the Cambridge logician, and Leslie Stephen, editor of the *Cornhill Magazine*, were among the most forceful of his opponents, finding (from quite different perspectives) serious problems with the definitions of individualism presented by Buckle's work and life. Thomas Hardy was one of the many writers who puzzled over the legacy of these debates, especially in the domain of ethics. In reading Buckle's *History* and Venn's critique of it against Hardy's *The Return of the Native*, I want to ask what concept of the individual and of individual moral agency might prove rescuable from a philosophical context so perplexed by individualism, and so often avowedly hostile to its mere 'irregularity'. The concluding section of the essay brings the discussion back to the present day by thinking about more recent attempts from within moral philosophy to find answers to these same questions.

Individualism, it should be acknowledged from the outset, is a flag over disputed ground. A more comprehensive account of its vicissitudes in the nineteenth century would have to take into account its capacity to embrace such distinct notions as the dignity of man, autonomy, privacy, self-development, and political, economic, and religious as well as ethical individualism.[4] As Steven Lukes notes in his admirably succinct history of the term, 'individualism' played a smaller and less readily characterizable role in British intellectual and political history than in France (where there was a strong tradition of seeing *individualisme*

[3] It will be clear from what follows that I have chosen not to pursue here the Foucaultian view that those generalities that 'subordinate' the individual actually individualize, or create the individual, in the first place. That approach is not incompatible with much of what is said in this essay, but I have prefered a vocabulary that allows a different kind of ethical debate about exceptionalism and ordinariness, and about moral agency in relation to chance.

[4] These are some of the distinct categories under which Steven Lukes discusses the term in *Individualism* (New York: Harper & Row, 1973).

as opposed to the interests and well-being of society), or in Germany (where the Romantic tradition permitted individualism to be absorbed into the concerns of the group or nation without necessary contradiction), or in America (where it was, and remains, one of the rhetorical touchstones of American democracy). Britain reflected something of all those responses, but the Millite liberal tradition to which Buckle, Venn, and Hardy were all, in significant degree, indebted saw individualism as a persistent dilemma: both virtue and vice, progressive and regressive in its social, intellectual, and ethical tendencies. The particular texts and histories explored in this essay played only a small part in these much larger and more complex histories of debate about the value of individual concerns and interests in relation to those of society. But then the perceptible problem of these texts' particularity or otherwise is, of course, a telling indication that such debates are far from over.

*

The valuing of the general or collective above the individual or particular underpins mid-Victorian intellectual thought in numerous ways. To isolate just three: the logical extrapolation from particular facts to general truths is the central tenet of inductive philosophy and thus of nineteenth-century scientific rationalism. Not that induction was uncontested, of course. There were major debates—particularly within Kantian philosophy—over its definition and its viability. Nevertheless, the operative assumption that knowledge could proceed only by moving from specific empirical observations to general claims was a *sine qua non* of the logical approach to truth in all fields of knowledge. A second, related mode of attention to the general came through the new science of statistics and probability, lucidly explored in Ian Hacking's 1990 book *The Taming of Chance*.[5] As Hacking shows, the concept of the social norm, and of standard laws of deviation and dispersal from that norm, was central to modern sociology and rapidly spread its influence through other disciplines. A third mode of reasoning from individual cases to generalities, arguably the most foundational of all, was—and remains—classification. Without it we have no possibility of making sense of the world: no genres, no narratives, no language.

But the identification of likenesses at the expense of differences is also, we know, a necessary fiction. Herbert Spencer put the philosophical case against its validity very clearly in his *Principles of Biology*:

> [H]abitual use of [classificatory terms of division] needful for purposes of convenience, has led to the tacit assumption that they answer to actualities in Nature. . . . [But it] is a wholly gratuitous assumption that organisms admit of being placed in groups of equivalent values . . . The endeavour to thrust plants and animals into these definite partitions, is of the same nature as the endeavour to thrust them into a linear series. . . . it does violence to the facts. Doubtless the making of divisions and sub-divisions is extremely useful; or

[5] Cambridge: Cambridge University Press.

rather it is absolutely necessary. Doubtless, too, in reducing the facts to something like order, they must be partially distorted. So long as the distorted form is not mistaken for the actual form, no harm results. But it is needful for us to remember, that while our successively subordinate groups have a certain general correspondence with the realities, they inevitably give to the realities a regularity which does not exist.[6]

But Spencer was writing in 1861, and working hard to counter a newly popular style of thought *not* strongly disposed to remember that fact.[7] As recently as October 1857 the *Westminster Review* had dedicated a lengthy review article to a volume that it had no hesitation in identifying as 'the most important work of the season; and . . . perhaps the most comprehensive contribution to philosophical history that has ever been attempted in the English language'.[8] Henry Buckle's *History of Civilization in England* was an attempt to rescue the study of history from a lamentable absence of scientific principle by applying classification, statistics, and inductive logic:

The unfortunate peculiarity of the history of man is, that although its separate parts have been examined with considerable ability, hardly any one has attempted to combine them into a whole, and ascertain the way in which they are connected with each other. In all the other great fields of inquiry, the necessity of generalization is universally admitted, and noble efforts are being made to rise from particular facts in order to discover the laws by which those facts are governed. . . . among [historians] a strange idea prevails, that their business is merely to relate events, which they occasionally enliven by such moral and political reflections as seem likely to be useful.[9]

Buckle proposed to draw together the findings of political economy, ecclesiastical scholarship, physical science, and statistical sociology, in order to discover the regular and predictable patterns that he believed underlay the apparent irregularity and unpredictability of human history. He was not out to win friends among fellow historians. Early in the General Introduction, he observes bruisingly that faith in an underlying law of progress is standard among men of science; that it is not generally found among historians must be a reflection not just of the greater complexity of social phenomena but of the investigators' 'inferior ability'. There is, he asserts, no historian 'who in point of intellect is at all to be compared with Kepler, Newton, or many others that might be named' (i. 6–7).

Enter Buckle as the Newton of History, ready to expel the demon Chance from the field. Human actions might wear the appearance of arbitrariness, individuals

[6] Herbert Spencer, *Principles of Biology* (London, 1864), i. 304–5.

[7] Spencer does not cite Buckle directly in the *Principles of Biology*. His *Autobiography* records that he knew Buckle, and introduced him to Huxley, and that, though he never read the *History of Civilisation*, he had 'dip[ped] into it' sufficiently to form the opinion that Buckle was 'top-heavy'—not just physically, as Huxley had observed, but in his writing. 'Buckle had taken in a much larger quantity of matter than he could organize; and he staggered under the mass of it.' *An Autobiography*, 2 vols. (New York: D. Appleton, 1904), ii. 4.

[8] [Mark Pattison], 'Henry Buckle, *History of Civilization in England*, Vol. I', *Westminster Review*, NS 12 (October 1857), 375–99.

[9] Henry Thomas Buckle, *History of Civilization in England*, 2 vols (London: John W. Parker & Son, 1857–61), i. 3.

might assert their free will, but in the aggregate, and across time, they were subject to fixed mathematical laws. In one of the most notorious passages of the *History of Civilization*, he drew on the work of the French statistician and sociologist Adolphe Quetelet, to adduce the example of suicide:

In a given state of society, a certain number of persons must put an end to their own life. This is the general law; and the special question as to who shall commit the crime depends of course upon special laws; which, however, in their total action, must obey the large social law to which they are all subordinate. And the power of the larger law is so irresistible, that neither the love of life nor the fear of another world can avail any thing towards even checking its operation. (i. 25–6)

The sheer absolutism of Buckle's determinism made him a key figure in the nineteenth century's 'taming of chance'. As Hacking puts it, the *History of Civilization* initiated a debate that 'raged in England for more than a decade. No topic was more intensely discussed before it faded into oblivion.'[10] The first edition of 1,500 copies sold out within a few weeks; numerous editions and translations (into French, German, Dutch, and—four times—into Russian) were called for. Dostoevsky 'read and reread' it,[11] as did George Eliot, George Henry Lewes, John Stuart Mill, Thomas Macaulay, Charles Darwin, Walter Bagehot—virtually everyone on the intellectual scene.

What Hacking omits to mention is that almost none of them liked it. Buckle was the talk of the town, but almost all the talk was against him. Arnold characterized him, memorably, in *Culture and Anarchy* as 'a fanatical partisan of light and the instincts which push us to it',[12] a hopelessly misdirected enthusiast for progress. Darwin was more impressed, finding it '*wonderfully* clever and original', until he met 'the great Buckle' and found him insufferably overbearing and glib. Most of the leading scientific naturalists—Huxley, Tyndall, Spencer—welcomed the book in so far as it spoke to a newly popular endorsement of the aims of physical science, but regretted its intellectual bluntness and, above all, the overweeningness of the man. When George Eliot's friend Sara Hennell penned some satirical verses in response to the *History of Civilization*, Eliot confessed, 'I am afraid I feel a malicious delight in them, for he is a writer who inspires me with a personal dislike.'[13] Eliot was particularly scandalized by Buckle's denial of heredity and of race: 'he holds that there is no such thing . . . !'—a point taken up by G. H. Lewes in a scathing dissection of 'Mr Buckle's Scientific Errors' for *Blackwood's Edinburgh Magazine*.[14] That was in 1861, shortly after the appearance of the long-awaited second edition, and by then Lewes could justifiably open his review with the observation:

[10] Ian Hacking, *The Taming of Chance* (Cambridge: Cambridge University Press, 1990), 126.

[11] *Prison Notebook*, quoted in Hacking, *Taming of Chance*, 125.

[12] Matthew Arnold, *Culture and Anarchy and Other Writings*, ed. Stefan Collini (Cambridge: Cambridge University Press, 1993), 140.

[13] 6 October 1858, *The George Eliot Letters*, ed. Gordon S. Haight, 9 vols. (New Haven: Yale University Press, 1954–78), ii. 485–6.

[14] 40 (November 1861), 582–96.

It must be owned that Mr Buckle has given great and often just offence, both by his matter and his manner. On the one hand, a large class of thinking men is offended by certain speculative conclusions which he advocates; and, on the other hand, many of those who are inclined to agree with his conclusions are irritated by the arrogance of his tone. Mr. Buckle is not a modest man; . . . Even the admiration excited by his erudition is qualified by the bad taste of his ostentation.[15]

Buckle's talent for giving offence to other writers did him no harm in publishing terms. Adrian Desmond and James Moore, in their biography of Darwin, liken the huge street sales of the *History of Civilization* to the reception earlier in the century of that other classic of popular secularization, the *Vestiges of Creation*.[16] But in important ways the response to Buckle was the exact opposite of the *Vestiges* phenomenon as analysed in Jim Secord's *Victorian Sensation* (2000). The remarkable success of the *Vestiges* was substantially a result of its anonymity (Robert Chambers's authorship was not formally revealed, and not generally known, until 1884, forty years after its first publication). The book's 'unparented' status permitted it to be interpreted with a much larger measure of philosophical and social licence than would otherwise have been the case, though it could also pose 'acute moral dilemmas': 'what if the book was not lofty aristocratic philosophy, but a piece of infidel propaganda?'[17] Buckle's book was, by contrast, thoroughly 'authored', and Buckle himself 'on display' in London. He moved in the higher social circles, and had been presented at court. Shortly after the publication of the *History*, he was elected to the Athenaeum and to the Political Economy Club and invited to lecture at the Royal Institution. He addressed an 'overflowing and enthusiastic audience' on 'The Influence of Women on the Progress of Knowledge', speaking for an hour and forty minutes 'in a "beautifully modulated voice," and without once referring to a few notes which he had set down'.[18]

To that extent, Buckle might look a comfortably accommodated member of the fraternity of gentleman scholars—only no such fraternity can be assumed on the basis of social status alone. It says something about his ambiguous position within the London world of letters that his election to the Athenaeum, sponsored by Hooker and Huxley, was a notoriously contested affair, with 9 black balls cast

[15] [G. H. Lewes], 'Mr Buckle's Scientific Errors', *Blackwood's Edinburgh Magazine* 40 (November 1861), 582–96.

[16] Adrian Desmond and James Moore, *Darwin* (1991; London: Penguin Books, 1992), 463.

[17] James A. Secord, *Victorian Sensation: The Extraordinary Publication, Reception, and Secret Authorship of 'Vestiges of the Natural History of Creation'* (Chicago: University of Chicago Press, 2000), 23.

[18] Even before the publication of the book, he was in some measure a public figure, having established a name for himself as a brilliant chess player. (He beat Andersson, Loewenthal, and Kieseritzki, three of the leading chess players in Europe, at a public contest during the Great Exhibition.) Leslie Stephen, 'Buckle, Henry Thomas', *Dictionary of National Biography*, ed. Sir Leslie Stephen and Sidney Lee, 22 vols. (Oxford: Oxford University Press, 1921–2), iii. 208–11 (209–10). Also Alfred Henry Huth, *The Life and Writings of Henry Thomas Buckle*, 2 vols. (London: S. Low, Marston, Searle, & Rivington, 1880), i. 56–62.

against 264 white, and more decisive opposition only narrowly averted.[19] The reluctance with which Buckle was accepted as an influence in British intellectual circles became the more evident after his sudden death in 1862 at the age of just 40. A bizarre print war began in the obituaries, and continued for more than three decades, between his friends and, not exactly his enemies, but those like John Venn, Leslie Stephen, and Grant Allen who, while admiring Buckle's learning and ambition, wished, in Allen's phrase, to see the intellectual claims of the man 'exploded like an inflated wind-bag'.[20]

Among the many rebuttals of Buckle, *The Logic of Chance*, by the Cambridge mathematician John Venn, was arguably the most efficient. Published in 1866, it included, in passing, a serious attempt to counter the misconceptions the *History of Civilization* had popularized. Venn found Buckle's claims about suicide absurd, but he believed their influence to be widespread and increasing. Venn set about clarifying the terms on which probability claims were valid, and explaining their implications for the concept of individual free will. 'Every agent, whether or not his conduct form part of any table of statistics, finds himself the centre of a sphere of action,' Venn acknowledged. The business of probability theory is not with whether or not individuals will try to alter their conduct (in this case, to avoid suicide). It merely describes the results of individual efforts or failures of effort in the aggregate.

> All that any person could then mean by talking about the inutility of efforts would [be] . . . that the difference, according as they were made or not, would be little or nothing.
>
> It will scarcely be maintained, in this sense, that motives are feeble or efforts at suppression ineffective. Any considerable alteration in the belief of people as to a future world, or in their comfort in this world, would unquestionably have a great influence upon the number of murders or suicides. . . . it will not I apprehend be denied that a great deal might be done towards *increasing* the annual number of those who destroy themselves,—by removing the police, for instance, from the neighbourhood of the Serpentine and Waterloo Bridge.[21]

It is a defence of individual freedom, but only indirectly and only in a strictly limited sense of the words. Unless a general and structural change occurs in the population's conditions of living or philosophical outlook, 'Little or no difference' will be observed in the aggregate behaviour of a population. An individual may exert his or her moral agency and refuse the temptation to suicide, but that exertion can make no difference to the behavioural patterns of the group.

Although Venn's interest was not primarily in psychology, *The Logic of Chance*

[19] J. Hooker letter to T. H. Huxley, 26 January 1858, quoted in Desmond and Moore, *Darwin*, 464. Also Stephen, 'Buckle', 209, and Huth, *Life and Writings*, i. 251–2.

[20] Grant Allen, *Charles Darwin*, English Worthies Series (New York: D. Appleton, 1893), 198. Quoted in John Mackinnon Robertson, *Buckle and His Critics: A Study in Sociology* (London: Swan Sonnenschein, 1895), 23.

[21] John Venn, *The Logic of Chance: An Essay on the Foundations and Province of the Theory of Probability, with Especial Reference to Its Application to Moral and Social Science* (London: Macmillan, 1866), 359–60.

also offered a necessary warning against the psychological misreadings Buckle's use of statistics invited. Venn insisted on the neutrality of the statistical observer. Two conditions had to be obeyed:

> (1) That the observer should leave the things which he observes to work out their courses undisturbed by any interference on his own part. (2) That he should adhere consistently to the position of an observer, and not in imagination step down and take a place amongst the things which he observes. (344)

He drew a further, and important, distinction between actual and perceived chances:

> A man, for example, is out with a friend, whose rifle goes off by accident, and the bullet passes through his hat. He trembles with anxiety at thinking what might have happened, and perhaps remarks, 'How very near I was to being killed!' . . . If the reader will try to analyse his feelings just after he has had a narrow escape himself, or witnessed one in others, I think he will find that this fallacy is generally to some extent involved in them. The mere proximity to danger cannot be the cause of the anxiety, for in other cases where the danger was equally near, but from which the notion of chance is excluded, no such anxiety is felt. . . . [If], as is very often the case when all the chances are contemplated, even that amount of proximity to danger which was actually experienced was extremely improbable, then no justification can be offered for the subsequent anxiety. (268–71)

For Venn, perceptions of risk too frequently ran counter to the actualities of risk. Indeed, 'risk' in its guise as chance psychologized, appeared to him the primary problem with the public's understanding of mathematical logic. For a pragmatist, Venn's logic too easily leaves psychology behind. To tell a man that he is unjustified in feeling anxiety when a bullet has just sailed within an inch of his skull is asking for excessive neutrality. Moreover, the scenarios Venn offers are not ones a philosopher or a literary critic would find sufficiently complex or testing in moral terms—this despite the subtitle of his book: *An Essay on the Foundations and Province of the Theory of Probability, with Especial Reference to Its Application to Moral and Social Science.* There is no consideration here of individual agency, or perceived agency, in relation to the probability of events in a given series. Venn's insistence on objectivity precludes investigation of such questions except under the title of 'fallacies'.

The Logic of Chance has been described as 'the last word' on Buckle's determinism, so far as probability theory went, but it was far from the last or most damaging word more generally. In 1886 Leslie Stephen wrote the entry for the *Dictionary of National Biography* which has remained until now the first port of call for any researcher wanting information about Buckle. It brings into a different kind of relief the questions Buckle's work and life raised about the nature and value of individualism in a wider historical perspective. Stephen is crushingly direct in assessing his subject's intellectual shortcomings. After a brief paragraph reviewing Buckle's lineage,[22] he gets personal:

[22] His descent from Sir Cuthbert Buckle, once lord mayor of London; the family firm of shipowners with the improbably Dickensian name of Buckle, Bagster, & Buckle.

Buckle was a very delicate child, unfit for the usual games. By Dr. Birkbeck's advice his parents were careful not to over-stimulate his brain. His early education was conducted by a most devoted mother, who would read the Bible to him for hours. He scarcely knew his letters at eight, and till eighteen had read little but 'Shakespeare,' the 'Pilgrim's Progress,' and the 'Arabian Nights;' . . . For a time he was sent to the school of Dr. Holloway in Kentish Town, on the condition that he should learn nothing but what he chose. He won a prize in mathematics, to which his attention had been accidentally drawn. His father offered him any additional reward he pleased, whereupon he chose the reward of being taken away from school. (208)

He was 13.

Stephen draws a portrait, a caricature perhaps, of a mollycoddled child growing into a wayward and helplessly neurotic adult. Extensive European travel, with a view to mending his health, was the making of Buckle intellectually. He became a freethinker and a radical, and acquired a working knowledge of the language of every country he stayed in. By the age of 28 he 'could read nineteen languages with facility and converse fluently in seven, though [Stephen can't resist the barb] he was incapable of acquiring a tolerable accent even in French' (208). The personal portrait is sad enough—repeated mental and physical collapses, hopeless romantic attachments to cousins, banned on principle by his parents[23]—but it is in his account of Buckle's intellectual achievements that Stephen does most harm:

The 'History of Civilisation in England' won for its author a reputation which has hardly been sustained. The reasons are obvious. Buckle's solitary education deprived him of the main advantage of schools and universities—the frequent clashing with independent minds—which tests most searchingly the thoroughness and solidity of a man's acquirements. Specialists in every department of inquiry will regard him as a brilliant amateur rather than a thorough student. (211)

Having earlier registered Buckle's commitment to the empiricism of Mill, Stephen here implicitly accuses him of mistaking the essential criterion for Millite rationalism: the understanding that truth will emerge only through the 'free expression and interplay of as many points of view as possible'.[24] In pursuing a wilfully idiosyncratic and largely solitary process of self-education and research (Stephen notes that Buckle was in the habit of purchasing books he needed, and selling them when done with, rather than use libraries and clubs) he had cut himself off from the very social and institutional mechanisms by which truth claims establish their value. Stephen also reiterated a criticism he had made in an

[23] Stephen depicts an incurable mummy's-boy, who never married. Buckle eschewed a career in the law, fearing the strain on his health. He guarded his energies anxiously for his literary endeavours, eating 'only bread and fruit' at luncheon in order to keep his brain clear and gave up public chess matches after his Great Exhibition performance because he 'grudged' the distraction. Stephen attributes his death to the combined effects of overwork and inability to withstand the deaths of his mother, and soon afterwards, of a favourite nephew. The combined shocks left him too weak to combat a fever contracted at Nazareth, where he had travelled in search of health.

[24] D. A. Hamer, *John Morley: Liberal Intellectual in Politics* (Oxford: Clarendon Press, 1968), 73–4.

earlier review of Buckle for the *Fortnightly Review*:[25] 'Buckle spoke cordially of the early writings of Darwin and Mr. Herbert Spencer, but he came too early to assimilate their teaching or to divine its importance. His speculations are already antiquated, because he was without the method which has come to be regarded as all-important by thinkers of his own school' (211). In short, he had missed the essential scientific insight of the mid-nineteenth century onward that there is no such thing as a 'given state of society'—societies, like species, are constantly evolving, and the law of variation precludes deterministic laws of the fixed and regularizing kind he had sought to describe. As Edward Clodd put it, succinctly, from the vantage point of 1885, had Buckle understood Darwin and Spencer 'his book must have been rewritten. . . . The notion of a constant relation between man and his surroundings is . . . untenable.'[26]

That criticism was unanswerable, even by Buckle's most devoted adherents, but it left Leslie Stephen with the awkward question of why Buckle's reputation among the wider reading public should have remained, up to the 1890s, so much in excess of his achievements—or, to put it another way, why the leading intellectuals of the day needed to work so hard *against* Buckle. Stephen had no answer, but one is suggested by John Mackinnon Robertson's 1895 book *Buckle and His Critics*. Robertson, a freethinker who, like Buckle, had left school at the age of 13 and thereafter been 'almost entirely self-educated',[27] homed in immediately on what he saw as the establishment's hostility to the concept of the autodidact, researching and writing independently of the prevailing intellectual culture. He objected roundly to Stephen's implied sneer against the 'brilliant amateur'[28] and frankly questioned the 'propriety' of a dictionary entry which supplies not just biographical facts but prejudicial opinions.[29]

A further reason for the unpalatability of Buckle's style of individualism to many of his contemporaries emerges from the final chapter of *Henry Buckle and His Critics*. Buckle held, very much against the grain of current liberal thinking, and of the ostensible message of his own *History*, that induction worked best not by collective efforts but by force of individual genius in the old-fashioned Romantic mould: the insight of a Newton or a Kepler (or a Buckle). Robertson sympathetically endorses that Romanticism, depicting his subject finally as an exceptional individual in whom, as in Goethe, the poetic and scientific powers of mind were united: 'a man not at all typically English, inasmuch as he combined the feelings of an emotional type with the reflectiveness of the most studious type. The combination recalls the sombre theory of recent years that all genius is pathological.'[30]

For most of its duration, then, from the early 1860s through to the 1890s, the

[25] 'An Attempted Philosophy of History', *Fortnightly Review*, NS 27 (1880).
[26] *Myths and Dreams* (London: Chatto & Windus, 1885), 3–4.
[27] As described by his biographer for the *DNB*, Harold Laski.
[28] Robertson, *Buckle and His Critics*, 40 and 41; quoting Stephen, 'Buckle', 211.
[29] Robertson, *Buckle and His Critics*, ch. 2, esp. 40.
[30] Robertson, *Henry Buckle*, 526.

argument over Buckle's *History of Civilization* and its endorsement of general laws over individual actions, doubled as an argument about the nature of access to knowledge. There are a number of ironies, here, one being that the reaction against Buckle's general laws was itself a powerful example of a social model of knowledge in which the collective and evolving enterprise is valued above the merely individual contribution (increasingly seen as a sorry hangover of Romanticism).[31] Another, and simpler, is that the great exponent of the irrelevance of individual choices to the deterministic laws governing the life of the group should have become so strikingly the odd man out in the history of the taming of chance. That Buckle nevertheless attracted such a popular, and pan-European, following might be read in several ways—and I'll come back to some of them later. But to the large extent that his popular following *was* a following, and not a refutation, it might be seen to indicate a desire for clearer and cleaner alternatives than the evolutionary narrative was offering—for a choice between absolute determinism, on the one hand, or Romantic self-determination on the other. That was Stephen's best guess. I want to suggest another, I hope less patronizing to the 'uncultivated reader' (Stephen's phrase): namely that Buckle's mistaken valuation of the general at the expense of the individual, by its very outmodedness, provided a means to articulating questions about the nature of the individual's relationship to the group and about the ethics of individualism.

*

Between late 1876 and early 1878, when Hardy was researching and writing *The Return of the Native*, he was also assiduously taking notes—and having his wife Emma take notes for him—on the problem of the individual's place in the general scheme of things. Critics interested in Hardy's meliorist leanings have focused particularly on the transcriptions of passages from Auguste Comte's *Social Dynamics, or the General Theory of Human Progress* (vol. iii, 1876).[32] But Comte was only one of the writers whom Hardy was plundering for claims that there are irresistible laws governing human experience. Representative quotations include:

[from Comte] 'The normal type of Human existence is one of complete unity.' (724)
'All progress, whether of the individual or of the race, consists in developing & consolidating that unity'. (725)
The Individual Mind—represented in the Mind of the Race. (747)

[31] T. W. Heyck, *The Transformation of Intellectual Life in Victorian England* (London: Croom Helm, 1982) discusses the growing professionalism of that culture. Heyck alludes to Buckle's ideas (133–7), but does not discuss his place within the social structures of Victorian intellectual life.

[32] Even his late claim to 'evolutionary meliorism' is not easily supported by his other public writings and pronouncements. This is, of course, a perennial topic of debate among Hardy critics, but for a short summary see *The Literary Notes of Thomas Hardy*, ed. Lennart A. Björk, i. *Notes* (Göteborg: Acta Universitatis Göthoburgensis, 1974), 287, 306–7. For a representative opposition of views, see J. O. Bailey's broadly pro-meliorist *The Poetry of Thomas Hardy: A Handbook and Commentary* (1970) *v.* Martin Seymour-Smith, *Hardy* (London: Bloomsbury, 1994), 325–6, 625, 634.

[from Spencer] Progress of classification—Classif^{n}. reaches its complete form by slow steps. The classif^{n}. eventually arrived at is one in which the segregation has been carried so far that the objects integrated in each group have more attributes in common with one another than they have in common with any excluded objects. (887)

And perhaps most tellingly of all, the following extract from John Tulloch's review of Caro's *Problèmes de morale sociale* for the *Edinburgh Review* in October 1876:

According to M. Caro [the French statistician and sociologist] . . . a slow but irresistible change is proceeding . . . Facts are elevated to the height of principles. . . . Numbers are reckoned the final reason of things & only organ of justice. Moral responsibility is denied, & the right of punishment esteemed a social usurpation. . . . Human destiny is only the amelioration of the species. No other or higher prospect awaits man . . . (872)[33]

In collecting these testimonies to the primacy of the collective pattern over individual experience, Hardy would unquestionably have come across Buckle's reputation and his work. He would have known the argument of the *History of Civilization*, if not at first hand then through his reading of Spencer and Huxley and his perusal of the periodical press (especially *Blackwood's*, the *Fortnightly Review*, the *Edinburgh Review*).[34] He was also well acquainted with many of those who were sceptical about Buckle's achievement—most significantly Leslie Stephen, who was Hardy's editor at the *Cornhill*, and whose philosophy Hardy acknowledged as having influenced him for many years 'more than that of any other contemporary'.[35]

Jude the Obscure might seem the obvious Hardy text to read alongside the debate over Buckle's statistical determinism and his personal reputation, given that novel's interest in the hostility of the intellectual establishment to the autodidact. But Jude is not really a Bucklean individual: his version of self-improvement is too far removed from Buckle's education through the benefits of money and travel; unlike Buckle's, Jude's model of learning is underpinned by participation in a class effort at self-advancement (the Artizan's Mutual Improvement Society of Aldbrickham); and Jude's efforts to study for the priesthood are at a far remove from Buckle's scientific agnosticism and his desire to work out from the history of religion, *inter alia*, to general laws. More simply, *Jude* belongs to a much later period in which Hardy had become interested in the darker prospects of civilization as expressed in the philosophy of Schopenhauer and von Hartmann. In *The Return of the Native* he was still pondering more immediately over Huxley, Spencer, Comte, Mill, and, of course, Darwin, all of whom offered a comparatively progressive view of history, even if optimism was, notoriously, not Hardy's response.

33 Björk (ed.), *Literary Notes*. References in the text are to item numbers. The passage from Spencer on pp. 66-7 above is also cited.

34 Michael Millgate confirms that there is no record that Hardy possessed copies of either *The History of Civilization* or Venn's *Logic*. See his forthcoming listing of Hardy's library.

35 Florence Emily Hardy, *The Life of Thomas Hardy, 1840–1928* (London, 1962), 100.

Gillian Beer has written compellingly of Hardy's deep, but conflictual, indebtedness to Darwin—his sense of chance and change being 'the permanent medium of experience and thus of language' so that plot, for example, must work to frustrate determinism and to confront the individual human being with variation, with multiple possible futures 'only one of which can occur and thus be verified in time, space, and actuality'.[36] The particular problem for much of Hardy's writing, she observes, is finding a scale for the human, 'a scale that will neither be unrealistically grandiose, nor debilitatingly reductive, which will accept evanescence and the autonomy of systems not serving the human, but which will still call upon Darwin's often repeated assertion: "the relation of organism to organism is the most important of all relations" '.[37] Thoroughly Darwinian as *The Return of the Native* is, there remains one surprisingly unremarked aspect to the deployment of evolutionary perspectives in the novel—namely, the disjunction between the Darwinianness of Hardy's view of the world, and the distinctly non-Darwinian philosophy of the men and women in the novel. *The Return of the Native* is set in the two decades immediately preceding Darwin, the 1840s and 1850s, when a student of sociology such as Clym Yeobright could not have available to him the knowledge that governs Hardy's own attentiveness to variation and evolution. In approaching Clym through Buckle's perversely popular outdatedness, I want to bring to the fore the question of anachronism within Hardy's narrative, which seems to me absolutely germane to the ethical dilemmas *The Return of the Native* identifies and the notion of individual moral agency it ends up with. For those who want an argument more by direct influence than by conjunction there is one tantalizing possibility—though it remains to my mind no more (nor less) than a coincidence. Diggory Venn, Hardy's reddleman, may or may not take his name from John Venn, Leslie Stephen's first cousin.

*

Egdon is a place in which the sheer weight of history presses both the landscape and its inhabitants towards the appearance of uniformity. Its rate of progress is so slow as to look, deceptively, like unchangingness, and its indifference to the individual human form so strong as to seem like absorption: (famously) 'There was something in its oppressive horizontality which too much reminded [Clym] of the arena of life; it gave him a sense of bare equality with, and no superiority to, a single living thing'.[38] The individual can appear, in this context, not even as a unit but as a mere part of a much larger whole: 'The scene was strangely homogenous, in that the vale, the upland, the barrow, and the figure above it amounted only to unity. Looking at this or that member of the group was not observing a

[36] Gillian Beer, *Darwin's Plots: Evolutionary Narrative in Darwin, George Eliot and Nineteenth-Century Fiction* (London: Routledge & Kegan Paul, 1983), 239.

[37] Beer, *Darwin's Plots*, 249.

[38] Thomas Hardy, *The Return of the Native* (1878), ed. Amanda Hodgson (London: J. M. Dent, 1995), 207–8.

complete thing, but a fraction of a thing' (21). Yet the 'sense' and 'seeming'ness of unity are persistently contested by a narrative counterdrive to discern particularity within homogeneity. Hardy's prose strives to develop, in his terms, an eye and an ear and even a foot for the participant units in the general picture. The tread of the night-walker on Egdon Heath finds out the irregularities and textural differences that masquerade as sameness to the eye during the day: 'To a walker practised in such places a difference between impact on maiden herbage, and on the crippled stalks of a slight footway, is perceptible through the thickest boot or shoe' (62). Hardy's most refined attentiveness is reserved, however, for sound, registered so closely that the sense of hearing becomes as if tactile:

> the linguistic peculiarity of the heath . . . was a worn whisper, dry and papery, and it brushed so distinctly across the ear that, by the accustomed, the material minutiae in which it originated could be realised as by touch. It was the united products of infinitesimal vegetable causes, and these were neither stems, leaves, fruit, blades, prickles, lichen, nor moss.
>
> They were the mummified heath-bells of the past summer, originally tender and purple, now washed colourless by Michaelmas rains, and dried to dead skins by October suns. So low was an individual sound from these that a combination of hundreds only just emerged from silence, and the myriads of the whole declivity reached the woman's ear but as a shrivelled and intermittent recitative. Yet scarcely a single accent among the many afloat . . . could have such power to impress a listener with thoughts of its origin. One inwardly saw the infinity of those combined multitudes; and perceived that each of the tiny trumpets was seized on, entered, scoured and emerged from by the wind as thoroughly as if it were as vast as a crater. (60)

The narrative gaze might seem akin to that of a botanist or biologist in the field, seeking out the particulars from which larger truths will emerge in due course, but the direction of thought is rather that of an artist attentive to the formal properties which, by a reverse of the classifier's logic,[39] emerge only with proximity. The ear of the narrative strains after 'original' distinctions in the natural world, lost to the sense but retrievable to the mind. To locate those individual sounds is to give form and meaning to an otherwise uncanny ghost of sound, as if the listener is intent on learning the fundamental particularities of a foreign language.

Similarly, Eustacia and the mummers, approaching the Yeobrights' house, hear music first as dissociated sounds which only assume meaning with proximity:

> Every now and then a long low note from the serpent, which was the chief wind instrument played at these times, advanced further into the heath than the thin treble part, and reached their ears alone; and next a more than usually loud tread from a dancer would come the same way. With nearer approach these fragmentary sounds became pieced together, and were found to be the salient points of the tune called 'Nancy's Fancy'. (134)

[39] At least the systematic 'lumper's' logic described by Harriet Ritvo in this volume, Ch. 3.

Such careful and, in *The Return of the Native*, constant redirection of the senses from the general impression to the proximate detail, happens also at the level of characterization and moral thought in the novel. When Clym gives up a career in the Paris jewellery industry to throw in his lot with the inhabitants of his native Egdon Heath, he does so in the grip of a philosophical conviction usually attributed to Hardy's reading of Comte but in tune with that much wider belief, urged on by Buckle, in the insignificance of individual lots seen against the larger picture of things:[40] 'Yeobright loved his kind. He had a conviction that the want of most men was knowledge of a sort which brings wisdom rather than affluence. He wished to raise the class at the expense of individuals rather than individuals at the expense of the class. What was more, he was ready at once to be the first unit sacrificed' (174). Almost immediately, the narrative puts this self-sacrificial ethos under sceptical scrutiny. The reader is given to understand that Clym is impossibly misplaced to achieve his goal of educating Egdon Heath out of bucolic superstition and into modern Enlightenment. One can't, Hardy asserts, with an apparent gesture toward Eustacia Vye, bypass the necessary developmental stage of 'luxury'.

Eustacia's Romantic individualism is the main counterforce in the narrative to Clym's self-abnegating ethic. An egoist who can remark without emotion that 'I have not much love for my fellow-creatures,' she is not immoral (as Hardy's first manuscript version would have had it) so much as anachronistic. As Bert G. Hornback puts it, 'her supposed size is from the beginning but romantic or rhetorical inflation'[41]—the inflation, one might add, of an older literary idiom. A Delphic Oracle, a Cleopatra, a Heloïse, she belongs to the past or, rhetorically, to the timelessness of a future heaven in which the divisions of history will have dissolved. On Egdon she is impossibly without a context, uncomprehended by anyone around her except, notionally, by Clym who warns her not to suppose, in her inexperience, 'that I cannot rebel, in high Promethean fashion, against the gods and fate as well as you. I have felt more steam and smoke of that sort than you have ever heard of' (251). Unlike Clym, she has not 'matured' into that frame of mind which sees 'nothing particularly great . . . and nothing particularly small' in life. Nor has she assumed the apparent carelessness of what happens that goes with such greater (though still inadequate) philosophical maturity.

When her aspirations are rebuffed, Eustacia is left morally and emotionally with no possible sense of agency 'tak[ing] a standing-point outside herself, observ[ing] herself as a disinterested spectator' (331). That description comes at the moment when, having left Clym and taken refuge in her grandfather's house, she discovers that her attendant, Charlie, has lit on her behalf a fire that will be

[40] Hardy himself is characteristically evasive: 'Mentally he was in a provincial future, that is, he was in many points abreast with the central town thinkers of his date. Much of this development he may have owed to his studious life in Paris, where he had become acquainted with ethical systems popular at the time' (174).

[41] *The Metaphor of Chance: Vision and Technique in the Works of Thomas Hardy* (Athens, Ohio: Ohio University Press, 1971), 29.

interpreted by Wildeve as her summons. The characteristically Hardyan predicament of not owning one's own individuality, of watching it neutrally as an object not connected with oneself, is explicitly linked to the condition of unresolved action: 'To have lost', the preceding sentence reads, 'is not as disturbing as to wonder if we may possibly have won.' That 'may possibly have' hedges its bets. Does Hardy mean that 'we have lost, but under other circumstances might have won' or does he mean 'we do not yet know whether we have lost or won'? Eustacia's version of individualism belongs very much to that domain of uncertainty about how one's own actions are to be judged in relation to the unfolding of time.

Hardy's presentation of her is, famously, riven with ambivalence. Anachronistic though her egoism is—a hangover, like Buckle's version of Romantic intellectualism—it is also something commanding the narrative's fascination and sympathy. At the nadir of Eustacia's fortunes, the narrative voice for the first time explicitly resists the estranging force of her egoism by invoking an imaginary spectator who, seeing closely, could not but sympathize: 'Any one who had stood by now would have pitied her' (345). In that claim, Eustacia's selfish individualism, her failure to connect with her fellow-creatures, becomes the ground for invoking a general response in which 'any one' would feel the same.

To put it another way, the rejection of Eustacia's outmoded individualism becomes the means by which other individuals are imagined coming together as a community of sympathy elsewhere, most obviously as readers, though Hardy also, I think, means to invoke a wider moral and imaginative engagement than the reading scenario. But her story does not offer a model for thinking either more directly about how one acquires a vision of community, or about individual moral agency within a community. The first of those problems has its clearest expression in Hardy's representation of Diggory Venn. If there is a Vennian (that is, John Vennian) statistical observer at work in this novel, it is the reddleman, who, in a suggestive image, knows the secrets of Egdon's irregularities well enough to be able to 'ascen[d] the valley in a mathematically direct line' (379). At several points Venn seems to attract the narrative's endorsement as a 'coolly' but 'generously' detached observer of the life around him. He charts the movements of the population, an uncanny attendant on their individual actions. As Melanie Bayley has pointed out, his adopted trade of selling reddle, 'a pigment put around the necks of rams and transferred to ewes on mating so that the number of impregnations can be recorded', is 'appropriate to his role in the plot of observing, marking, who is mating with whom'.[42] At several points his is the adopted point of view on a crucial scene, although he does not himself contribute to the action;[43] at others, Hardy introduces his viewpoint as a purely imaginary way into a description ('Had the reddleman been watching he might have recognized [Eustacia] as the woman who had first stood there so singularly' (59)). Venn prompts the narrative to its most explicitly scientific posture of neutrality: 'That

[42] Bayley, 'Victorian Theories of Chance'.

[43] See esp. ch. 9.

[his] keen eye had discerned what Yeobright's feeble vision had not—a man in the act of withdrawing from Eustacia's side—was within the limits of the probable' (262).

But Venn's status as objective quasi-statistical observer of the group is of course as specious as Clym's claim to be an integrated member of that group. Unlike John Venn's ideal observer, the reddleman does not permit the objects of his study to 'work out their proper courses undisturbed by any interference'. He 'step[s] down and take[s] a place among the things he observes' not just imaginatively but actually, attempting to clear a path for Thomasin to be happy with Wildeve. He is not just an attendant on plot, but an instigator of plot. ('The reddleman, stung with suspicion of wrong to Thomasin, was aroused to strategy' (87)). The famous scene in which he dices with Wildeve on the heath to win back what he thinks is Thomasin's money is surely Hardy making deliberate play with the logic of probability.

> Down they sat again, and recommenced with single guinea stakes; and the play went on smartly. But Fortune had unmistakably fallen in love with the reddleman tonight. He won steadily, till he was the owner of fourteen more of the gold pieces. Seventy-nine of the hundred guineas were his, Wildeve possessing only twenty-one. The aspect of the two opponents was now singular. Apart from motions, a complete diorama of the fluctuations of the game went on in their eyes. A diminutive candle-flame was mirrored in each pupil, and it would have been possible to distinguish therein between the moods of hope and the moods of abandonment, even as regards the reddleman, though his facial muscles betrayed nothing at all. Wildeve played on with the recklessness of despair. (230)

The tension between image and interpretation provides a perfect illustration of the conflict between scientific neutrality and psychology scrutinized in John Venn's *Logic*. The diminutive reflection of the scene in each player's pupil is a record of emotion, driving and distorting the calmness of action: near stillness is, palpably, tension.

And what of the reader? There is nothing in the laws of chance to say that Wildeve's win against the hapless Christian Cantle and then Venn's against Wildeve are impossible, but the inevitable non-neutrality of plot puts the scene outside the reader's sense of what is probable. For the sake of the narrative, Venn needs to win. As Gillian Beer put it, in her 1990 Bateson lecture, 'The Reader's Wager':

> Venn must win. That's how such stories end. The whole is determined. The reddleman does win. He does so by virtue of his virtue: the cards connive, under the ordering hand of the writer. . . . A moral drama is enacted, a popular story. Gambling is under the sway of some larger law that orders the chances.

But the chapter and the book 'The Fascination' end with a vertiginous loosing of hazard again.[44]

[44] 'The Reader's Wager: Lots, Sorts, and Futures', The F. W. Bateson Memorial Lecture, *Essays in Criticism*, 30 (1990), 99–123 (121).

Venn promptly goes on to give all the money he has won to Thomasin, not knowing that half of it was meant for Clym, and thereby puts in motion the events that lead ('predictively', as Beer says) to Mrs Yeobright's death and all the misunderstandings and further disaster that follow from it.

Once the impurity of hazard and the possibility of error are admitted back into Hardy's novel so too are those urgent questions of individual moral agency that have been held in abeyance through the characters' assumption of a mistakenly remote or wilfully 'self-sacrificial' relationship to the group. Clym's moral failing towards his mother is explicitly associated with a failure to recognize the marks of individuality in a person to whom he bears an ethical responsibility by virtue of *nearness*—a responsibility revealed to him only with the emotional consequences of error. That failure is registered as a literal failure of recognition. So Clym, approaching the body of his dying mother on the heath, can hear a moan and make out a feminine form but remains 'not absolutely certain that the woman was his mother till he stooped and beheld her face, pallid, and with closed eyes' (287). A few pages earlier that non-recognition had been the other way around: Mrs Yeobright follows a labourer whom she sees as 'a furze-cutter and nothing more—wearing the regulation dress of the craft, and thinking the regulation thoughts, to judge by his motions'. Not until she observes 'peculiarities in his walk' is she, in Hardy's pointed phrase, 'attracted to his individuality' (273).[45]

That it should be her death that precipitates a radical rethinking of the nature of individuality in Clym and in the novel is fitting in the light of one of the most substantial passages offering elucidation of her character earlier in the novel:

> She had a singular insight into life, considering that she had never mixed with it. There are instances of persons who, without clear ideas of the things they criticize, yet have had clear ideas of the relations of those things. Blacklock, a poet blind from his birth, could describe visual objects with accuracy; Professor Sanderson, who was also blind, gave excellent lectures on colour, and taught others the theory of ideas which they had and he had not. In the social sphere these gifted ones are mostly women; they can watch a world which they never saw, and estimate forces of which they have only heard. We call it intuition.
>
> What was the great world to Mrs Yeobright? A multitude whose tendencies could be perceived, though not its essences. Communities were seen by her as from a distance; she saw them as we see the throngs which cover the canvases of Sallaert, Van Alsloot, and others of that school—vast masses of beings, jostling, zigzagging, and processioning in definite directions, but whose features are indistinguishable by the very comprehensiveness of the view. (190)

It is a strange passage, not least because the insight is so much in excess of any specific description of her that the novel provides, so that the capacity identified in Mrs Yeobright for seeing the world intuitively in its larger outlines seems to be replicated here by the narrative itself. Less charitably, it is possible to suspect Hardy of using her as a vehicle for a statement about ways of seeing the world

[45] Similarly, Eustacia is so intent 'upon Wildeve's fortunes that she forg[ets] how much closer to her own course were those of Clym' (294).

which he required for the philosophical development of his story but which comes, ironically, at the cost of her particularity. The passage is doubly curious, in that it describes positively a mode of perception that the rest of the novel predominantly strives to counteract in the sphere of human activity. To see human life in its collective outlines is here to be gifted with 'intuition'. The paragraph that follows retracts, however, insisting that the 'completeness' of Mrs Yeobright's 'reflective life' is not matched by her lived experience. Her philosophical 'assurance' has never been built upon; her 'once elastic walk' has been 'deadened by time', and her 'pride of life . . . hindered in its blooming by her necessities'.

With her death, the moral cost of Clym's seeing life too much in its 'tendencies' and not its 'essences' comes to the fore, and requires in the process a significant recasting of the concept of individuality. In his grief that his mother died believing herself cast off by him, Clym rebukes himself with what looks to those around him like irrational extravagance:

> 'I cannot help feeling that I did my best to kill her.'
>
> 'No, Clym.'
>
> 'Yes, it was so; it is useless to excuse me! My conduct to her was too hideous—I made no advances; and she could not bring herself to forgive me. Now she is dead! . . . I dare say she said . . . a hundred times in her sorrow, "What a return he makes for all the sacrifices I have made for him!" ' . . .
>
> 'You give yourself up too much to this wearying despair,' said Eustacia. 'Other men's mothers have died.'
>
> 'That doesn't make the loss of mine less. Yet it is less the loss than the circumstances of the loss. I sinned against her, and on that account there is no light for me.' (302–3)

Having surrendered too readily the value of his individuality (he can't, of course, sacrifice his individuality) he now has to try to wrest back something of individual moral agency through remorse, but can only do so by—valuably—complicating the notion of the individual.

That complication finds a helpful elucidation in late twentieth-century philosophical writing about the place of chance, or luck, within our conception of the moral life. In his *Lectures on the History of Moral Philosophy* (2000), John Rawls provides a lucid account of how, in Kant's view of morality, we decide whether or not to identify with the society around us: in his terms, how we decide whether or not we are prepared to '*will a social world*'. As the first limit on such a decision, Kant reduced the notion of individualism in play: 'The first limit is that we are to ignore the more particular features of persons, including ourselves, as well as the specific content of their and our final ends and desires.'[46] For Kant, the apprehension of moral law required a necessary abstraction from the particular. It also, relatedly, required that moral agency was seen not to apply to circumstances in

[46] John Rawls, *Lectures on the History of Moral Philosophy* (Cambridge, Mass.: Harvard University Press, 2000), 175. For Bernard Williams's response to earlier, less Kantian Rawls, see *Ethics and the Limits of Philosophy* (London: Fontana Press, 1985).

which luck plays a hand, and which are, therefore, both non-representative and outside the sphere of choice. In his now classic essay, 'Moral Luck', Bernard Williams resists these conditions strongly, asking why post-Kantian thought should have continued so resolutely to set its face against a fuller notion of individualism, and against luck as a component of our ideas about morality. There are, as he points out, different kinds of luck, which may affect in different ways our success or failure to achieve the projects we choose for ourselves. When Clym suffers the serious damage to his eyesight that prohibits him from pursuing his plan of starting a school, he becomes subject to what Williams would call 'external' luck. (This is not, of course, the case if one considers—as one may well do—that his loss of sight is symbolic of a deeper failure.) But when he realizes that his mother has died believing that he cast her off, and when—still later—he realizes that his wife has died without receiving a gesture of reconciliation from him, these are *intrinsic* forms of luck. They affect his ability to justify his actions, as the loss of sight could not. Such cases, Williams suggests, generate a state of mind that he describes as 'agent-regret'—a response that goes beyond the spectator's 'how much better if it had been otherwise' and becomes of the order of Clym's assumption of a logically tenuous, but deeply felt, responsibility: 'I made no advances; and she could not bring herself to forgive me. Now she is dead!'

Williams's own example of agent-regret is as follows:

> The lorry driver who, through no fault of his, runs over a child, will feel differently from any spectator, even a spectator next to him in the cab, except perhaps to the extent that the spectator takes on the thought that he himself might have prevented it, an agent's thought. Doubtless, and rightly, people will try, in comforting him, to move the driver from this state of feeling, move him indeed from where he is to something more like the place of a spectator, but it is important that this is seen as something that should need to be done, and indeed some doubt would be felt about a driver who too blandly or readily moved to that position. We feel sorry for the driver, but that sentiment co-exists with, indeed presupposes, that there is something special about his relation to this happening, something which cannot merely be eliminated by the consideration that it was not his fault.

As far as sympathy goes,

> the discussion of agent-regret about the involuntary also helps us to get away from a dichotomy which is often relied on in these matters, expressed in such terms as *regret* and *remorse*, where 'regret' is identified in effect as the regret of the spectator, while 'remorse' is what we have called 'agent-regret', but under the restriction that it applies only to the voluntary. The fact that we have agent-regret about the involuntary, and would not readily recognize a life without it (though we may think we might), shows already that there is something wrong with this dichotomy: such regret is neither mere spectator's regret, nor (by this definition) remorse.[47]

Hardy's novel leaves the reader in the awkward position of the spectator at a fatal accident who feels bound to say to the truck driver/Clym, with Diggory

[47] Bernard Williams, *Moral Luck* (Cambridge: Cambridge University Press, 1981), 28, 30.

Venn, that 'you can't charge yourself with crimes in that way'. The distance between Venn's spectatorship and Clym's own recognition of involuntary but morally unavoidable agency leaves Yeobright's individuality at a level of complexity that finds no answering voice anywhere else in the novel. By the end of the novel, Clym is no longer an adherent of 'creeds and systems of philosophy'. His chosen life as an itinerant preacher who takes as his subject 'the opinions and actions common to all good men' marks him out as still desiring to assist the general life around him, but also as irretrievably set apart from those he addresses. Their combination of intolerance and a much deeper tolerance for him is also, by extension, the reader's and is expressed in the one term for moral good that Hardy held to tenaciously throughout his life, even in his most pessimist moments: kindness. ('Loving kindness' was his usual, biblical formulation, the use of kindness by itself here suggesting more readily 'being of human kind', part of the human collectivity.)

> He left alone creeds and systems of philosophy, finding enough and more than enough to occupy his tongue in the opinions and actions common to all good men. Some believed him, and some believed not; some said that his words were commonplace, and others complained of his want of theological doctrine; while others again remarked that it was well enough for a man to take to preaching who could not see to do anything else. But everywhere he was kindly received, for the story of his life had become generally known. (396–7)

This famous closing paragraph of *The Return of the Native* suggests that one of the things 'common to all good men' is the recognition that the supposedly exceptional individual among them—Clym Yeobright, Henry Buckle— generally none the less lags behind where other good men think he should be. His insights will seem merely 'commonplace' or 'wanting in doctrine', or will be merely the best that he can do given that he cannot do anything else. But Hardy also suggests that the recognition that another's philosophy falls short of the point at which we could all pitch our belief is what, paradoxically, sustains our sense of commonality with him. In other words, it is in falling short of the crowd that he establishes himself as one of their number—and allows them to recognize themselves as a crowd: 'We are', after all, 'all individuals.'

*

No more than proximate as their relationship was within mid-Victorian culture, Buckle's *History of Civilization in England* and Hardy's *Return of the Native* serve to raise the question of what the value of proximity may be in modifying our sense of the individual's relation to the wider group. Their ability to take that question beyond the merely structural stems not a little from their both foregrounding the presence of the mistaken, or mistimed, within a progressive history. Why should the mistaken Buckle have continued to attract readers even decades after his death? Why should the mistaken Clym still earn, without irony, the title of a hero?

('The nicest of all my heroes', Hardy called him, 'and *not a bit* like me.') The answer includes both sympathy and its opposite. Other people's mistakes and excesses are, as every scholar knows, precisely what we need to sharpen and define our own, more nuanced understanding by. This was true for Leslie Stephen and John Venn and all those other nineteenth-century intellectuals who attempted to rectify Buckle's mistaken philosophy or expose the outmodedness of his personal style of living. It was also true for the many first readers and reviewers of Hardy's novel, perplexed by their *liking* for a hero whose misery is accidentally incurred and has nowhere to go beyond self-reproach.

Hardy was particularly drawn to Comte's notion that 'human progress is a "looped orbit", sometimes going backward by way of gathering strength to spring forward again'.[48] What I hope this chapter has shown is that such loops are not only mathematical metaphors. They are intrinsic to our interest in those mistaken ideas that do not fit a neatly linear idea of intellectual history, but that help to define what the 'proper' line (or lines) of intellectual history might be. They also lie at the heart of those backward-looking motions of regret for actions in which we had no voluntary agency, but which are nevertheless constitutive of our moral relationship to others, and without which, as Williams puts it 'we would not readily recognize a life'.[49]

[47] Robert Schweik, 'The Influence of Religion, Science, and Philosophy on Hardy's Writings', in Dale Kramer (ed.), *The Cambridge Companion to Thomas Hardy* (Cambridge: Cambridge University Press, 1999), 67. And see Björk (ed.), *Literary Notes*, 307.

[49] Williams, 'Moral Luck', 30.

CHAPTER 5

The Psychology of Childhood in Victorian Literature and Medicine

SALLY SHUTTLEWORTH

In 1838, as Darwin painstakingly filled his notebooks, working and reworking the ideas that were to yield his theories of biological evolution, he took time to compose his 'Autobiographical Fragment' where he attempted to chart his earliest memories.[1] The following year, with the birth of his first son William, he started a notebook that he would maintain for four years, recording the development of his children. The three forms of writing complement one another: Darwin's reflections on the origins of species give birth to his personal speculations on the processes by which the infant and child transmute into adult forms. His study of infant life, however, was to wait even longer than his theory of species for publication: it fed into *The Expression of Emotions in Man and Animals* (1872), but was only published in its own right in 1877 as 'A Biographical Sketch of an Infant'.[2]

Darwin's seeming reluctance to publish his thoughts on child development was matched by a similar reticence in the emerging fields of psychology and psychiatry. Child development studies were not established in England until the 1880s with the work of James Sully,[3] whilst in psychiatric literature there was no systematic discussion of childhood mental disorders until around the 1870s.[4] The numerous tomes on childhood diseases, or manuals of domestic medicine, dealt

[1] Gavin de Beer (ed.), *Charles Darwin, T. H. Huxley: Autobiographies* (Oxford: Oxford University Press, 1983).

[2] Charles Darwin, 'A Biographical Sketch of an Infant', *Mind: A Quarterly Review of Psychology and Philosophy* 2 (1877), 285–94. The full text of the original notebook is published in *The Correspondence of Charles Darwin*, ed. F. Burkhardt *et al.*, in progress (Cambridge: Cambridge University Press, 1985–), iv. 410–33.

[3] Sully's major work, *Studies of Childhood*, was published in 1895, but he published articles on the area from the 1880s and his *Outlines of Psychology with Special Reference to the Theory of Education* (1884) offered an explicit focus on the child mind. Parallel work was proceeding in Germany with the work of W. Preyer whose *Die Seele des Kindes* (*The Mind of the Child*) was published in 1881; and in France with the work of Bernard Perez, whose text, *The First Three Years of Childhood*, was published in England with a preface by Sully in 1885.

[4] There are individual treatments of the subject before this time, but it was not really established as a field until the 1870s. One of the earliest discussions is J. Crichton Browne, 'Psychical Diseases of Early Life', *Asylum Journal of Mental Science* 6 (1860), 284–320.

almost entirely with physical ailments, with the small but interesting exception of discussions of 'night-terrors'. Such absences are all the more surprising when set alongside the literature of the period where one can trace a clear line from Rousseau and the Romantic cult of the child. The Wordsworthian notion that the 'Child is father of the man' shapes the narratives of childhood development to be found in the works of Dickens, Eliot, and Charlotte Brontë.[5] Dickens's *Oliver Twist* (published in the same year that Darwin wrote his autobiographical fragment), Brontë's *Jane Eyre* (1847), or Eliot's *The Mill on the Floss* (1860), all chart in great detail the growth of a child's mind. This seeming asymmetry between disciplinary fields is intriguing. As Gillian Beer has so consummately shown, the relations between science and literature are not to be reduced to a simple hierarchical influence model, but can be pursued through the imaginative reworkings of plot, analogy, and metaphor.[6] In this case the workings of silence, occlusion, and contradiction are also crucial. The comparatively slow emergence of child psychology should be attributed not to a lack of interest amongst the medical and scientific professions, but rather to a fundamental uncertainty as to how to define and demarcate the sphere of childhood.

Within the emerging field of anthropology, women, children, and savages, or the lower classes, were repeatedly linked together as figures who stood outside the unstated norms of white, middle-class masculinity. The figure of the child, I would suggest, lies at the heart of nineteenth-century discourses of gender, race, and selfhood: a figure who is by turns animal, savage, or female, but is located not in the distant colonies, nor in the mists of evolutionary time, but at the very centre of English domestic life. Although there has been considerable work recently on Victorian medical constructions of the female mind and body, there has been comparatively little attention paid to medical discourses of childhood. In this chapter I will look at various literary texts in the light of these debates which, as is customary in medicine, focus not on the normal but on the abnormal, as medics circled round and round the question of whether a child could become insane. The question lies at the heart of changing interpretations of childhood in the nineteenth century.

When Jane Eyre in childhood breaks out into a fit of rage the Gateshead servants see her as a 'mad cat' and gaze on her 'as incredulous of my sanity', concluding that 'it was always in her'.[7] Underneath the cover of a docile demeanour she had lived a double life, nursing the germs of this passionate fury which finally breaks out, shattering social decorum and expectations of 'natural'

[5] William Wordsworth, 'My Heart Leaps Up'.

[6] See Gillian Beer, *Darwin's Plots: Evolutionary Narrative in Darwin, George Eliot and Nineteenth-Century Fiction*, 2nd edn. (Cambridge: Cambridge University Press, 2000). These ideas are further developed in *Open Fields: Science in Cultural Encounter* (Oxford: Clarendon Press, 1996). For further interesting explorations of the imaginative territory cohabited by Victorian literature and science see George Levine, *Darwin and the Novelists: Patterns of Science in Victorian Fiction* (Cambridge, Mass.: Harvard University Press, 1988).

[7] Charlotte Brontë, *Jane Eyre*, ed. Margaret Smith (Oxford: Oxford University Press, 1980), 12. All references to this work will be cited hereafter in the text.

social behaviour. From the very opening of the novel we learn that Jane Eyre is already an outcast. She is not to be allowed into the family group until she has obeyed the paradoxical injunction to acquire a 'more sociable and child-like disposition . . . something lighter, franker, more natural as it were' (*Jane Eyre*, 7). How does a child learn to be 'child-like'? The contradictions in Mrs Reed's position mirror those of the wider culture, as medical writers and social commentators sought to define the boundaries of that threateningly unknowable species of humanity, the child.

Although the servants readily classify Jane as insane, psychiatric writers of the first half of the nineteenth century were less willing to commit themselves, despite the fact that theories of partial insanity, emerging from the 1830s, gave new possibilities of thinking through this question. Sufferers from moral insanity could be of perfectly sound intellect, but merely disordered in their moral judgement; sufferers from monomania could be insane in one aspect of their lives, but otherwise perfectly normal. Perhaps most alarmingly, monomaniacs and the morally insane could frequently live amongst the general populace undetected, their insanity lying latent, or not recognized as such.[8] Like Jane Eyre, they finally break cover, confirming to all around that 'it was always in her' (12).

Ideas of partial insanity gave rise to crucial debates on the issue of individual responsibility. When is an individual legally responsible for his or her actions? Women, who were held to be more liable to insanity due to the ways in which their minds were subjected to the unstable tyrannies of their reproductive systems, were an important focus of these debates.[9] There was confusion, however, around the subject of the child.

Questions of childhood responsibility, and of possible insanity, are of course pressing contemporary issues in the aftermath of the Jamie Bulger case[10] and in the light of the Blair government's tough stance on juvenile crime. Newspaper headlines denounce child savages, whilst images of childhood innocence (often problematically sexualized) are used to sell virtually every consumer product conceivable. Similar contradictions and confusions, I will suggest, were present in Victorian attempts to interpret and control the boundaries between the adult and child state.

Can a child be insane? If, as was increasingly the case in the nineteenth century, insanity comes to define the space occupied by those who stand outside the curve of social normality, where should the child, who also stands outside social normality, be placed? And at what point can the child be deemed to pass into the normative spectrum of adulthood? The answer to at least the first of these questions was

[8] For further details see Roy Porter, *Mind-Forg'd Manacles: A History of Madness in England from the Restoration to the Regency* (Cambridge, Mass.: Harvard University Press, 1987) and Sally Shuttleworth, *Charlotte Brontë and Victorian Psychology* (Cambridge: Cambridge University Press, 1996), ch. 3.

[9] For a discussion of these issues see Roger Smith, *Trial by Medicine: Insanity and Responsibility in Victorian Trials* (Edinburgh: Edinburgh University Press, 1981).

[10] In this case, two 10-year-old boys took the toddler Jamie Bulger, who had become separated from his mother in a crowded Liverpool shopping mall, and tortured him to death.

fairly clear by the last decades of the century. The pessimistic evolutionary psychology of the post-Darwinian era highlighted the ways in which children could inherit the germs of insanity from their parents, although there was still some debate as to whether this insanity would actually manifest itself as such until adulthood.[11] Following Haeckel's formulation in 1866 of the theory that 'ontogeny recapitulates phylogeny', evolutionary psychologists increasingly mapped the ways in which the stages of childhood re-enacted the forms of our animal or 'savage' ancestors' lives. As the anthropologist John Lubbock noted in *Prehistoric Times*, 'the life of each individual is an epitome of the history of the race, and the gradual development of the child illustrates that of the species'.[12] Lombroso took these theories one stage further to suggest a fundamental association between the mind of the criminal, who is arrested at an early stage of development, and that of the child.[13]

At the mid-century, however, the question of whether a child could be insane was one that could not easily be answered. Commentators repeatedly stated that a child could be imbecile (as in Brontë's creation of her heroine's alter ego, the cretin, in *Villette*), but not insane. J. E. D. Esquirol, for example, argued in *Mental Maladies* that 'It is only at puberty, during the earliest menstrual efforts, or during, and after a too rapid growth, that we begin to notice certain cases of mental alienation. . . . Mental alienation might, therefore, be divided, relative to ages, into imbecility for childhood, mania and monomania for youth, lypemania or melancholy for consistent age, and into dementia for advanced life.'[14] Behind this attempt to map mental illness onto a truncated version of the seven ages of man, one can trace both a desire to preserve the innocence of childhood, and also an unwillingness to allow the child access to the (often doubtful) privileges of a fully formed intelligence.[15]

One of the problems in this literature concerns the demarcation of childhood, and youth, from adulthood.[16] Texts give widely variant definitions of the age range they deem to be covered by the term 'child', or 'youth'. Since the concept of adolescence was also in the process of formation there are no accepted divisions between childhood and puberty, for example. One of the first attempts to formalize these distinctions, T. S. Clouston's 'Puberty and Adolescence Medico-Psychologically

[11] For details on the psychology of 'degeneration' see, Daniel Pick, *Faces of Degeneration: A European Disorder c. 1849–c. 1918* (Cambridge: Cambridge University Press, 1989) and William Greenslade, *Culture, Degeneration and the Novel* (Cambridge: Cambridge University Press, 1994).

[12] John Lubbock, *Pre-Historic Times, as Illustrated by Ancient Remains, and the Manners and Customs of Modern Savages*, 2nd edn. (London: Williams & Norgate, 1869), 558.

[13] For a discussion of the impact of Cesare Lombroso's theories on conceptions of childhood in England, see Hugh Cunningham, *Children of the Poor: Representations of Childhood since the Seventeenth Century* (Oxford: Blackwell, 1991), ch. 5.

[14] J. E. D. Esquirol, *Mental Maladies. A Treatise on Insanity*, trans. E. K. Hunt (Philadelphia: Lea & Blanchard, 1845), 34. Esquirol was a pupil of Pinel, and one of the strongest influences on the development of early Victorian psychiatry.

[15] The association of innocence and imbecility is a recurrent theme in the literature of the time, from Wordsworth's 'The Idiot Boy' to George Eliot's story 'Brother Jacob' (1864).

[16] For further discussion of the problems of demarcation see Carolyn Steedman, *Strange Dislocations: Childhood and the Idea of Human Interiority, 1780–1930* (London: Virago Press, 1995), 7.

Considered' (1880), situates the onset of puberty between 11 and 14, but defines adolescence as the period from 18 to 25.[17] Other texts extend the period of youth to 30.[18] Whilst literary texts, particularly Shakespeare, were routinely invoked by nineteenth-century psychological texts as forms of case studies, Clouston actually draws his definition of female adolescence from George Eliot's depiction of Gwendolen Harleth in *Daniel Deronda*: 'This authoress is by far the most acute and subtile psychologist of her time, and certainly the character I have mentioned is most worthy of study by all physicians who look on mind as being in their field of study or sphere of action.'[19] Since Gwendolen is actually in her early twenties, Clouston's judgement suggests we should take literally Eliot's labelling of her as a 'Spoiled Child'.[20] The difficulties we face disentangling the precise ages suggested by nineteenth-century terms for non-adult states mirror the Victorians' own confusion and lack of certainty in this area.

Despite the reiterated claims in the first half of the century that children could not become insane, psychiatric texts none the less offered examples of childhood insanity, often with no reference at all to the apparent inconsistency so displayed. Esquirol, for example, offers three cases of homicidal monomania in childhood. All are girls; the first, aged 7, wishes to kill her mother; the second, also aged 7, wishes to kill her stepmother. Esquirol gives an account of his interview with this second child:

> Her replies were made without bitterness or anger; and with composure and indifference. Why do you wish to kill your mother? Because I do not love her. Why do you not love her? I do not know. Has she treated you ill? No. Is she kind to you? Yes. Why do you beat her? In order to kill her. How! In order to kill her? Yes, I desire that she may die. Your blows cannot kill her; you are too young for that. I know it. One must suffer, to die. I wish to make her sick, so that she may suffer and die, as I am too small to kill her at a blow. When she is dead who will take care of you? I do not know. You will be poorly taken care of, and poorly clothed, unhappy child! That is all one with me; I will kill her; I wish her dead.[21]

The dialogue, designed to show the unreason of the child, actually highlights the impercipience of Esquirol, whose monotonal questioning is utterly defeated by the child's quiet persistence.[22] The third case, also a girl, is of an 11-year-old, who (with shades of the recent Jamie Bulger case) lures younger children to a well and then pushes them down.[23] Esquirol's commentary on these cases is interesting; in

[17] T. S. Clouston, 'Puberty and Adolescence Medico-Psychologically Considered', *Edinburgh Medical Journal* 26 (1880), 5–17.

[18] See Esquirol, *Mental Maladies*, e.g. 379.

[19] Clouston, 'Puberty and Adolescence', 14.

[20] The first book of *Daniel Deronda* (1876), which introduces Gwendolen Harleth, is entitled 'The Spoiled Child'.

[21] Esquirol, *Mental Maladies*, 371.

[22] In effect, although not design, the passage is similar to Wordsworth's poem, 'We Are Seven', where the logic of the child finally forces the condescending questioner to acknowledge the limitations of his own frame of thought.

[23] The phenomenon of a child murdering another child is not, as the papers now might have us believe, a new occurrence; the psychiatric texts in the nineteenth century offer numerous examples.

the first case the girl had learnt onanism (masturbation) from older children and hence been corrupted; the second child had had her mind poisoned against her stepmother by careless talk from her grandmother. In the third case the girl had simply been brought up without the requisite moral training. We are not in the domain of Freud here; there is no attempt to analyse the family dynamics that could lead a child to hate her mother or stepmother. Equally, there is no attempt to speak of inherited traits, or essential sinfulness: the children themselves are not the target of blame. Immorality and crime have proximate but external causes.

Running parallel, and in some respects counter, to this emerging psychiatric discourse in nineteenth-century England was a whole tradition of morally improving literature, often with an evangelical bent, focused on the child. Against the model of childhood innocence proclaimed by Wordsworth and Coleridge, Evangelicals continued to insist on the doctrine of original sin, following Hannah More's argument that 'it was a fundamental error in Christians, to consider children as innocent beings, whose little weaknesses may perhaps want some correction, rather than as beings who bring into the world a corrupt nature and evil dispositions'.[24] Children's literature at this period was overwhelmingly concerned with correction and moral example. In *Ellen, or The Naughty Girl Reclaimed* (1811), for example, Ellen initially refuses to do her lessons, but is 'reclaimed' for studious docility by the end.[25] (Fig 5.1) Readers are offered their own do-it-yourself humiliation kit: a dunce's cap to cut out and place on Ellen's head, to ensure her redemption.

Two of the primary concerns in this literature were with childhood passion and with lying; both were to be resolutely controlled. These were also, however, precisely the areas that childhood psychiatry was to target: in each case the child is threatening to disrupt adult constructions of the human and the 'real'. When Jane Eyre answers back in fury to her aunt, Mrs Reed, she is a familiar figure from instructional, and to a lesser extent, medical texts: a passionate, deceitful child. What is radically new, however, is the tone of voice and stance of the speaker:

> 'I am not deceitful; if I were, I should say I loved *you*; but I declare, I do not love you: I dislike you the worst of anybody in the world except John Reed; and this book about the liar, you may give it to your girl, Georgiana, for it is she who tells lies, and not I.' . . .
>
> Shaking from head to foot, thrilled with ungovernable excitement, I continued:—
>
> 'I am glad you are no relation of mine: I will never call you aunt again as long as I live. I will never come to see you when I am grown up; and if anyone asks me how I liked you, and how you treated me, I will say the very thought of you makes me sick, and that you treated me with miserable cruelty.'
>
> 'How dare you affirm that, Jane Eyre?'
>
> 'How dare I, Mrs Reed? How dare I? Because it is the *truth*.' (36–7)

[24] Hannah More, 'Strictures on Female Education', in *The Works of Hannah More*, 18 vols. (London: 1818), vii. 67; quoted in Hugh Cunningham, 'The Rights of the Child from the Mid-Eighteenth to the Early Twentieth Century', in *Children at Risk. Aspects of Education, Journal of the Institute of Education, The University of Hull* 50 (1994), 2–16 (5).

[25] Anon., *Ellen, or, The Naughty Girl Reclaimed, A Story, Exemplified in a Series of Figures* (London, Printed for S. and J. Fuller, 1811).

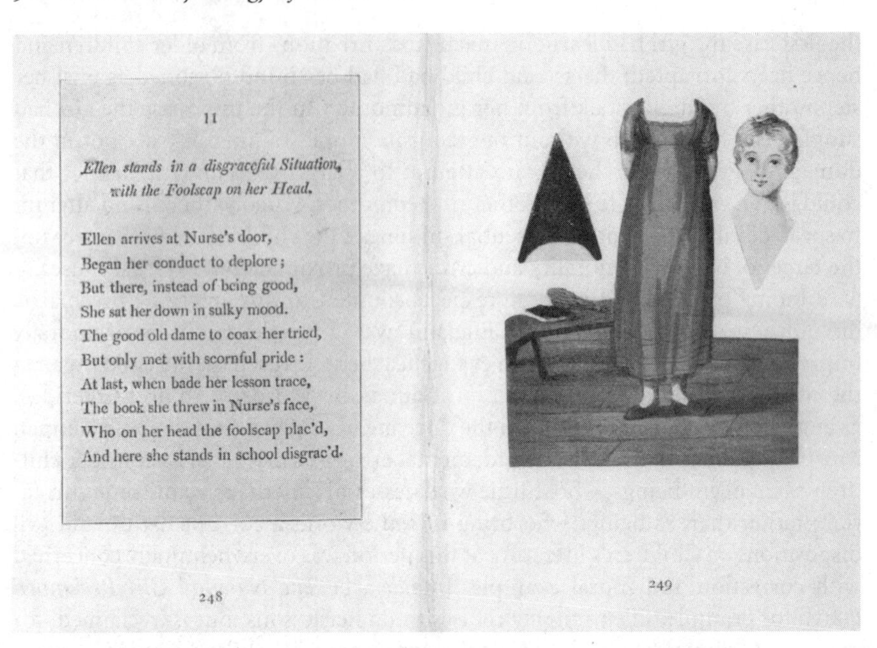
11

Ellen stands in a disgraceful Situation, with the Foolscap on her Head.

Ellen arrives at Nurse's door,
Began her conduct to deplore;
But there, instead of being good,
She sat her down in sulky mood.
The good old dame to coax her tried,
But only met with scornful pride:
At last, when bade her lesson trace,
The book she threw in Nurse's face,
Who on her head the foolscap plac'd,
And here she stands in school disgrac'd.

248

249

FIGURE 5.1 *Ellen, or, The Naughty Girl Reclaimed* (1811).

In place of the bafflingly errant child in educational or medical texts, presented for readerly consumption almost in disbelief, we are offered the voice of a child strong in her sense of anger and injustice. Although campaigns for industrial reform had made the notion of the child as victim a familiar one, anger had been reserved for the adult observer or commentator. Children were usually rendered as passive sufferers, thus retaining the moral value invested in their 'innocence'.[26] Jane Eyre, however, speaks out in passionate defiance, with a vehement sense of her rights.

Sully, in *Studies of Childhood* (1895), noted his agreement with Preyer and Darwin that a child was a creature of animal appetites and passions, but, he added, the child has more than 'the brute's sensual fury': 'Self-feeling, a germ of the feeling of "my worth" enters into this early passionateness and differentiates it from a mere animal rage.'[27] Brontë, writing nearly fifty years earlier, clearly endows her angry child-heroine with an equivalent sense of self-worth, yet in doing so she is running counter not only to the Evangelical tradition represented by Mr Brocklehurst, but also to psychological and social discourses of the time. As Hugh Cunningham has argued, discussions of child rights in the nineteenth

[26] Blake's *Songs of Innocence and Experience* is of course a partial exception. For an account of one of the first factory novels describing the sufferings of children, Frances Trollope's, *Michael Armstrong, the Factory Boy* (1839), see Inga-Stina Ewbank, *Their Proper Sphere. A Study of the Brontë Sisters as Early-Victorian Female Novelists* (Göteborg: Scandinavian University Books, 1966), 25–6.

[27] James Sully, *Studies of Childhood* (1895; 2nd edn., London: Longmans, Green & Co., 1896), 235.

century invariably focused on the child's rights to protection from the social spaces and occupations of adulthood; both Evangelical and Romantic visions of childhood were grounded on divisions between the childhood and adulthood state, and hence actually worked to undermine any sense of equal rights.[28] In its passionate sense of injustice and a child's rights to equal treatment with adults, *Jane Eyre* is unusual, although it possesses a direct contemporary parallel in Harriet Martineau's *Household Education* (1849), a book designed, Martineau commented, as 'for the Secularist order of parents'.[29] The similarities between *Household Education* and *Jane Eyre* are remarkable, not only in their descriptions of childhood states of terror, but in their coupling of these perceptions with a firm statement of a child's rights.[30] For Martineau, 'Every member of the household—children, servants, apprentices—every inmate of the dwelling must have a share in the family plan; or those who make it are despots, and those who are excluded are slaves.'[31] Like Brontë, whose heroine feels like a 'revolted slave' as she struggles to escape from 'insupportable oppression' (14–15), Martineau brings a sharp political edge to her analysis of the ways in which adult tyranny creates the seeming insanities of childhood terrors and passions.

THE LIAR

Jane's anger against her aunt is initially focused on the accusation that she was a liar. Mr Brocklehurst, on leaving, had presented her with a book 'entitled the "Child's Guide;" read it with prayer, especially that part containing "an account of the awfully sudden death of Martha G—, a naughty child addicted to falsehood and deceit" ' (35). At Lowood she is subsequently forced to stand on a stool, exhibited to the gaze of all as that most shameful of all beings, '—a liar!' (67). Horror of the lie was not confined, however, to the Brocklehursts of Victorian culture, but spread across social, domestic, and medical literature.[32] The disruptive figure of the liar clearly threatened the fragile boundaries of the Victorian

[28] Cunningham, 'The Rights of the Child', 12–13.

[29] *Harriet Martineau's Autobiography, with Memorials by Maria Weston Chapman*, 3 vols. (London: Smith, Elder & Co, 1877), ii. 293; Harriet Martineau, *Household Education* (London: Edward Moxon, 1849). A large part of *Household Education* had been previously published in the *People's Journal* (1846–7), so it would have been possible for Brontë to have read sections whilst she was writing *Jane Eyre*. Martineau later recalled Brontë's claim that 'she had read with astonishment those parts of "Household Education" which relate my own experience. It was like meeting her own fetch,—so precisely were the fears and miseries there described the same as her own, told or not told in "Jane Eyre." ' *Harriet Martineau's Autobiography*, ii. 324.

[30] Valerie Sanders in *Reason over Passion: Harriet Martineau and the Victorian Novel* (Brighton: Harvester Press, 1986) discusses some of the similarities between the two texts, 138–44.

[31] Martineau, *Household Education*, 2.

[32] For an excellent analysis of the preoccupation with lying in Victorian culture see John Kucich, *The Power of Lies: Transgression in Victorian Fiction* (Ithaca, NY: Cornell University Press, 1994). Kucich does not consider, however, the question of lying in relation to childhood.

world, suggesting the possibility of alternative truths. Domestic childcare manuals focused firmly on the need to extirpate all forms of untruth. Thus in *How do You Manage the Young Ones?*, Old Chatty Cheerful, F. H. H. S. (Fellow of the Happy Home Society) advises mothers to 'Watch for a lie . . . as you would watch for some destructive viper'.[33]

It would appear that such household manuals were on the side of Gradgrinding truth, in opposition to the redemptive forces of fancy, so strongly advocated by Dickens in *Hard Times* (1854). The division is not so clear cut, however. Language similar to that of Old Chatty Cheerful is to be found even in Martineau's *Household Education*. Martineau, herself a successful novelist, laments the fact that education tends to stifle imagination, and advocates strongly the development of the faculty of Wonder, but none the less she preaches the stern duty of absolute truthfulness. Thus a poor child who, from inattention, has told a slight untruth is held before us as an example of moral turpitude: 'When a moral disease so fearful as this appears, parents should never rest till they have found the seat of it, and convinced the perilled child of the deadly nature of its malady.'[34] Although Martineau recognizes that children will often recite wonderful dreams, or wonderful things they have seen in their walks, such as 'giants, castles, beautiful ladies riding in the forests', she classifies such accounts as lies, and maintains that it is up to the parents to correct even the most trifling of misstatements, for 'All peace is broken up when once it appears there is a liar in the house.'[35]

CHILDHOOD INSANITY

In his 1838 autobiographical fragment, and much later autobiography, 'Recollections of the Development of my Mind and Character' (1876), Darwin accorded high prominence to his childhood practice of lying which gave rise both to shame and exhilaration, in Darwin's own memorable phrase, to 'pleasure like a tragedy'.[36] As Gillian Beer has argued, 'It may be that the length of the account of his story-telling or lying, compared with his other memories, registers an elation and a creative disturbance newly felt again in 1838 by the young Darwin, akin to that which he had experienced as a ten-year-old.'[37] Paradoxically, however, the very impulse that spurred the creation of that magnificent story of species origin became, under the subsequent development of evolutionary psychology, evidence of human degeneration.

The language of disease, so frequently employed with reference to childhood lying, fed through into nineteenth-century medical discourse on childhood insanity. If insanity has become a partial state, to be detected perhaps only in occasional

[33] Old Chatty Cheerful, *How do You Manage the Young Ones?*, Household Tracts for the People, 9th edn., 46th thousand (London: Jarrold & Sons, n.d.), 17.

[34] Martineau, *Household Education*, 166.

[35] Ibid. 163, 170.

[36] de Beer, *Autobiographies*, 5.

[37] Beer, *Darwin's Plots*, 26.

deviations from customary behaviour, how then is the child who builds his own fantasies to be judged? James Crichton Browne, writing in 1860, was one of the first voices to reverse the verdict that a child could not suffer insanity. Childhood, with all its imaginative projections, becomes instead the very definition of insanity: 'Monomania, or delusional insanity, we believe to be more common during infancy and childhood than at any other period of life.' Imaginative 'castle building' is denounced as a 'pernicious practice': 'much mental derangement in mature life, we believe, is attributable to these reveries indulged in during childhood'.[38] Literary and biographical accounts of childhood are then ransacked, including Anna Jameson's descriptions of her intense internal life in her early years, to demonstrate the prevalence of mental disorders in childhood. Although Jane Eyre's terror in the red room is not cited, it is directly parallel to many of the descriptions offered of childhood 'delusional insanity'.

The protected status of childhood rapidly disappears in post-1860 psychiatry: children are not only capable of experiencing forms of insanity previously reserved for adulthood, they are also often more prone. According to George Savage, writing in 1881, children between 5 and 10 have shown signs of moral insanity, or 'moral perversion', which could also be accompanied by intellectual or sexual precocity.[39] Clearly, under this definition, Jane Eyre would join Bertha in the ranks of the 'moral maniacs'.[40] Savage, like Crichton Browne, singles out lying as the first symptom of insanity: the morally insane child lies 'with such wonderful power that he lies like truth'. Moral insanity and 'the power of romancing as a genius' are 'scarcely to be distinguished'.[41] Sexuality is also linked up with moral insanity, since the liar often becomes a masturbator. Savage is here drawing on a century of anti-masturbation literature, such as Ritchie's *A Frequent Cause of Insanity in a Young Man* (1861) which painted the dire consequences of masturbation, from the moral degradation of lying through to insanity and death (this was one of the few forms of recognized childhood insanity before this point).[42] Although this literature placed masturbation, and not lying, as the primary causal activity, the connections are clear. Both involve a form of hidden life, outside the domain of adult control. Esquirol's first case of homicidal mania, it should be remembered, was attributed to the practice of onanism. This is not to

[38] Crichton Browne, 'Psychical Diseases of Early Life', 303.

[39] George H. Savage, 'Moral Insanity', *Journal of Mental Science* 27 (1881), 150.

[40] In a letter to her editor, W. S. Williams, Charlotte Brontë had defended her creation of Bertha in the light of medical theories of moral insanity: 'I agree . . . that the character is shocking, but I know that it is but too natural. There is a phase of insanity which may be called moral madness, in which all that is good or even human seems to disappear from the mind and a fiend-nature replaces it.' T. M. Wise and J. A. Symington, *The Brontës: Their Lives, Friendships and Correspondence*, 4 vols. (Oxford: Basil Blackwell, 1933), ii. 173.

[41] Savage, 'Moral Insanity', 151, 150.

[42] Robert P. Ritchie, *An Inquiry into a Frequent Cause of Insanity in Young Men* (London: Henry Renshaw, 1861). The discourse in England on the dangers of masturbation that followed the translation of Tissot's treatise, *Onanism*, in 1781, was given renewed vitality in the 1840s with the publication of Lallemand's work on spermatorrhoea (a 'disease' whose primary symptom was the involuntary loss of semen).

suggest that the preoccupation with childhood lying and insanity can all be traced to a sexual cause; rather that lying and insanity, together with sexuality, were all forms of behaviour that suggested a secret form of internal life. Childhood could no longer be seen as 'empty', a space to be traversed before the acquisition of adult powers, or innocent.

Crichton Browne was not backward in proclaiming the disturbing propensities now to be discerned in infant life. Nymphomania, he announced, could be found in children as young as 3. His readers are asked to redraw entirely their map of childhood: 'The mind of childhood, that which we are accustomed to look upon as emblematic of all that is simple, and pure, and innocent, may be assailed by the most loathsome of psychical disorders, viz, satyriasis, or nymphomania; the monomania affecting the sexual instinct.'[43] This view is confirmed in William Acton's text, *The Functions and Disorders of the Reproductive Organs in Childhood, Youth, Adult Age, and Advanced Life* (1857) which opens with the statement that 'In a state of health no sexual idea should ever enter a child's mind', but then goes on to show that children seem rarely to be in this requisite state of health.[44] The ground is being prepared for Freud's later, more benign, observations on the polymorphous perversity of child sexuality.

THE EVOLUTIONARY PERSPECTIVE

The tendency to treat children as sexual beings was further intensified by the development of evolutionary psychology from the 1860s onwards (although it should be noted that medical and anthropological works throughout the century had commonly grouped 'women, children and savages' together as a sub-category of humanity).[45] Crichton Browne warns parents in 1883 that they 'should remember that children are not little nineteenth-century men and women, but diamond editions of very remote ancestors, full of savage whims and impulses, and savage rudiments of virtue'.[46] Evolution was transferred from the forests and plains, and unimaginable aeons of geological time, to the interstices of the domestic space. Each parent is invited to witness the developmental processes of millions of years enacted before his or her own eyes. Projections of childhood now stress the animalistic, primitive forms of emotion exhibited by the child, yet there is also an attempt to claw back possibility, to alleviate the strong bestial projections, as in Crichton Browne's reference to the 'savage rudiments of virtue'.

[43] Crichton Browne, 'Psychical Diseases of Early Life', 308.

[44] William Acton, *The Functions and Disorders of the Reproductive Organs in Childhood, Youth, Adult Age, and Advanced Life Considered in their Physiological, Social and Moral Relations* (1857), 4th edn. (London: John Churchill, 1865), 1.

[45] See Cynthia Eagle Russett, *Sexual Science: The Victorian Construction of Womanhood* (Cambridge, Mass.: Harvard University Press, 1989).

[46] J. Crichton Browne, 'Education and the Nervous System', in Malcolm Morris (ed.), *The Book of Health* (London: Cassell, 1883), 379.

The effects of the evolutionary perspective on child psychology are complex, and not simply to be linked to the downgrading of the child to a lower form of being. Darwin, for example, in his 'A Biographical Sketch of an Infant', takes comfort in the fact that the night-time terrors that so disturbed his son could be given an understandable cause: 'May we not suspect that the vague but very real fears of children, which are quite independent of experience, are the inherited effects of real dangers and savage superstitions during ancient savage times.'[47] Paradoxically, the inexplicable is controlled and tamed by the attribution of a savage history.

Whilst Lombroso firmly identified the criminal mind with that of the child (as a form of arrested development), other theorists sought to glory in the child's visible heritage. Alexander Chamberlain concludes his 1900 text, *The Child: A Study in the Evolution of Man*, with an impassioned vision:

> The child, in all the helpless infancy of his early years, in his later activity of play, in his naiveté and genius, in his repetitions and recapitulations of the race's history, in his wonderful variety and manifoldness, in his atavisms and his prophecies, in his brutish and in his divine characteristics, is the evolutionary being of our species, he in whom the useless past tends to be suppressed and the beneficial future to be foretold. In a sense, he is all.[48]

At once atavism and prophecy, the child, in this fin-de-siècle projection, becomes the bearer of all future hope.

Such optimism is not matched, however, by other writers of the period. T. S. Clouston, in his 1883 *Clinical Lectures on Mental Diseases*, asks gloomily, 'What child is born in a civilised country without inherited brain weaknesses of some sort or in some degree?'[49] The child, in the pessimistic psychology of the late century, is doubly burdened: the carrier of both primitive, animalistic passions, and also the attenuated nerves of an overdeveloped civilization. In Henry Maudsley's phrase, he is ruled by 'the tyranny of his organisation'.[50]

In 1895 two very different texts dealing with the child mind were published: James Sully's *Studies of Childhood*, the first major English study of child development, and Crichton Browne's *On Dreamy Mental States*. Whilst Sully offered a Wordsworthian, celebratory response to the primitive, imaginative qualities of the child mind, Crichton Browne placed his earlier theories in an even stronger degenerative framework.[51] Dreamy mental states, which emerged in girls at 8 or 9, and boys at 10, were 'revivals of hereditarily transmitted or acquired states in new

[47] Charles Darwin, 'A Biographical Sketch of an Infant', 287.

[48] Alexander Chamberlain, *The Child: A Study in the Evolution of Man* (London: Walter Scott, 1900), 464.

[49] T. S. Clouston, *Clinical Lectures on Mental Diseases* (London: J. & A. Churchill, 1883), 528.

[50] Henry Maudsley, *Body and Mind: An Inquiry into their Connection and Mutual Influence, Specially in Reference to Mental Disorders* (London: Macmillan, 1870), 75.

[51] Sully followed the now standard line of seeing the child as the key to a race's history, but celebrated the qualities which the child shared with the primitive mind: fancy, myth, and story-telling (see ch. 2, 'The Age of Imagination'). He was concerned to rescue the child from the domain of the lie, however, suggesting that, 'The idea of perpetrating a knowing untruth, so far as I can judge, is simply awful to a child who has been thoroughly habituated to the practice of truthful statement' (264–5).

or special combinations'.[52] If left unattended they would lead to progressive degeneration of the mind over the following generations. Hardy's *Jude the Obscure*, also published in 1895, follows the latter, more pessimistic model of childhood, offering in Father Time a terrifying example of a child burdened by history.[53] The seeds of disorder are clearly sown within the Fawley family, and are seen in the child Jude, who 'had held his outer being for some long tideless time' in the surroundings of his aunt's shop, but had inwardly lived within his gigantic dreams.[54] (Dreamy mental states, Crichton Browne had argued, involve 'an exaltation of subject consciousness, and a degradation of the power of attention', and create in the second generation problems with ideas of space, and in the third and fourth, the loss of a sense of personal identity.)[55]

Jude is a problematic child, who is 'an ancient man in some phases of thought' (22). In Father Time, the offspring of Jude and Arabella, we find an intensification of this problem: he is less a child than 'Age masquerading as Juvenility' (290). He becomes a literal enactment of the anthropological vision of the child as the embodied history of his race. More specifically, he becomes the expression of all the mistakes enacted in the relations of Jude, Arabella, and Sue: 'He was their nodal point, their focus, their expression in a single term' (356). Father Time, however, is not only the concentrated focus of past history, he is also a form of prophecy. All sentimental versions of childhood are delivered a mortal blow by Father Time's deliberate act of hanging himself and his two siblings. His actions are explained according to the psychology of the time: he possesses a 'morbid temperament' and is thrown into a 'fit of aggravated despondency' by the news of their destitution, and yet another child on the way (355). Jude reports the verdict of the doctor: ' "It was in his nature to do it. . . . He says it is the beginning of the coming universal wish not to live" ' (355).[56] The assessment is pure Maudsley, who argued not only that children were ruled by their inheritance, an inheritance that expressed all the mistakes of their parents and immediate forebears, but that suicide would be an inevitable response to such unremitting degeneration. Maudsley maintains that Goethe was quite right to make Werter commit suicide, rather than developing a clearer, calmer insight into his sorrows:

[52] James Crichton Browne, *The Cavendish Lecture. On Dreamy Mental States. Delivered before the West London Medico-Chirurgical Society, On Thursday, June 20, 1895* (London: Baillière, Tindall & Cox, 1895), 6.

[53] Hardy was at this time a friend of Crichton Browne. Crichton Browne had sent a letter to Hardy in 1892, inviting him to lecture at the Royal Institution, and praising *Tess*, which 'examines the psychologic tissues with a powerful lens free from chromatic aberration' (MS letter, 16 November 1892; Dorset County Museum). Hardy subsequently records discussing scientific issues with him at the Royal Society. See Florence Emily Hardy, *The Life of Thomas Hardy, 1840–1928* (London: Macmillan, 1972), 254.

[54] Thomas Hardy, *Jude the Obscure* (1897), ed. P. Ingham (Oxford: Oxford University Press, 1985), 17–18. All references to this work will be cited hereafter in the text.

[55] Crichton Browne, *On Dreamy Mental States*, 20.

[56] For a more detailed discussion of this topic, see Sally Shuttleworth, ' "Done because We Are Too Menny": Little Father Time and Child Suicide in Late-Victorian Culture', in Phillip Mallett (ed.), *Thomas Hardy: Texts and Contexts* (Basingstoke: Palgrave, 2003), 133–55.

suicide was the natural and inevitable termination of the morbid sorrows of such a nature. It was the final explosion of a train of antecedent preparations, an event which was as certain to come as the death of a flower with a canker at its heart. Suicide or madness is the natural end of a morbidly sensitive nature, with a feeble will, unable to contend with the hard experiences of life.[57]

As an expression of atavism and prophecy, Father Time is far removed from Chamberlain's optimistic vision of the potentiality contained in the child. The scene of the hanging, with its violation of all novelistic decorum, could be seen as the end of the Victorian novel. As Maudsley's comments suggest, the novel of development, the *Bildungsroman*, has nowhere to go: *The Sorrows of Werter* is now a more appropriate literary model than *Wilhelm Meister*. In the light of degenerationist theory, marriage, that customary form of ending, with its achieved or implied complement of children, becomes a signal not of achieved fruition but rather of impending doom.

Jude the Obscure was published shortly before another novel that also sounds the deathknell for childhood: Henry James's *What Maisie Knew* (1897). Lunatics and criminals, Maudsley had argued, are 'as much manufactured articles as are steam-engines and calico-printing machines'.[58] The child similarly had an inescapable legacy stamped upon it, the product of its parents' and ancestors' weaknesses and profligacy. James is not concerned with evolutionary psychology, but his child is equally a manufactured product, a child denied the possibilities of innocence. In the face of the appalling machinations of her parents and step-parents, innocence becomes for Maisie a commodity to be produced on demand. James takes up the questions at the centre of the developing science of childhood: what does a child know? How does it think, and how develop? Whilst his answer is free from the evolutionary negativity of Hardy, it is equally dispiriting. Childhood for Maisie, that disturbingly absent presence at the heart of the book, is a matter of performance; a sequence of roles quickly learnt and enacted in answer to adult pressure. In her 'doom of a peculiar passivity', Maisie absents herself from her own history, and instead seizes upon the roles she is required to play that map the negative aspects of childhood or 'primitive' innocence.[59] Thus when 'she began to be called a little idiot, she tasted a pleasure new and keen' (20), and in practising 'the pacific art of stupidity' she quickly achieves, in James's biting phrase, 'a hollowness beyond her years' (58), as she strives to destroy any appearance of psychological depth. Lying, that other supposedly primitive and childhood flaw, is an activity she would willing embrace, if only the adults would

[57] Maudsley, *Responsibility in Mental Disease* (London: Henry S. King, 1874), 272. The reference is to *The Sorrows of Werter* (1774). For discussions of Hardy's debt to Maudsley see Patricia Gallivan, 'Science and Art in *Jude the Obscure*', in Anne Smith (ed.), *The Novels of Thomas Hardy* (London: Vision Press, 1977), 126–44, and Peter A. Dale, 'Thomas Hardy and the Best Consummation Possible', in John Christie and Sally Shuttleworth (eds.), *Nature Transfigured: Science and Literature, 1700–1900* (Manchester: Manchester University Press, 1989), 201–21.

[58] Maudsley, *Responsibility in Mental Disease*, 28.

[59] Henry James, *What Maisie Knew*, ed. Douglas Jefferson (Oxford: Oxford University Press, 1980), 85. All further references to this work will be given hereafter in the text.

give her a lead: 'She would have pretended with ecstasy if he [her father] could only have given her the cue' (138). Even the kindly Mrs Wix is similarly demanding, as she too imposes her own evolutionary model on Maisie, wanting her to progress from primitive amorality to the first stages of the moral sense that she believes will be evidenced by the emergence of jealousy: 'Maisie met her expression as if it were a game of forfeits for winking. "I'd *kill* her!" That at least, she hoped as she looked away, would guarantee her moral sense' (214). In this inverse evolutionary fable, Maisie learns to be a child, to adopt the poses of stupidity, pretence, and amorality customarily associated with the child and primitive races, but in reality shown working at their highest power amongst the corrupt adults of her acquaintance. In her eagerness to instil in Maisie some moral sense, the well-intentioned Mrs Wix only compounds the errors, as Maisie is stimulated to utter her most immoral sentiments in the text.

Maisie offers another version of 'Age masquerading as Juvenility', although her 'age' is acquired, not inherited. Like Father Time, Maisie becomes the 'nodal point' of her parents' lives. In their explorations of the death of childhood, these literary texts pinpoint some of the difficulties inherent in defining its boundaries. At the very point in time when childhood became of crucial interest in psychological theory it seems, paradoxically, to disappear from sight, to become a mere iteration of parental or species history, rather than an entity in itself.[60] Whether following the traces of parental influence, or the unfolding of humanity's evolutionary history, psychologists were looking at child history as a map to another, alternative domain. As Sully noted, the child is a 'monument of his race, and in a manner a key to its history'.[61] Although the child, and its development, was receiving unprecedented levels of attention across medical and historical sciences at this period, the question of what it meant to be a child became increasingly unclear.[62]

Mrs Reed's instruction to Jane Eyre to acquire a more 'natural', 'child-like disposition' brilliantly encapsulates the conflicts and contradictions that surround the social category of 'the child'. Jane Eyre and Father Time are both children who refuse, or indeed are unable, to be children, to fit social expectations of the childlike, whilst Maisie, through the very fact of possibly 'knowing' ceases to be a child (234). The half-century separating *Jane Eyre* (1847), *Jude the Obscure* (1895), and *What Maisie Knew* (1897), witnessed radical shifts in medical and psychological constructions of childhood. No longer was the question whether a child could become insane, but rather whether hereditary taints would leave a

[60] This was, of course, also the 'golden age' of children's literature (see Humphrey Carpenter, *Secret Gardens: A Study of the Golden Age of Children's Literature* (London: Allen & Unwin, 1985)). The relations between this literature written for children, and the science devoted to uncovering the development of the mind of the child, are complex, and still little discussed.

[61] Sully, *Studies of Childhood*, 9.

[62] Following Darwin's 'Biographical Sketch of an Infant', the pages of *Mind* saw innumerable discussions of child development, whilst the general periodical press printed frequent articles on babies and their relation to evolutionary development. The British Association for Child Study, of which Sully was a member, was formed in the early 1890s.

child with any option of full sanity. The pessimism of the *fin de siècle* found expression in the evolutionary psychology of the era. *Jane Eyre*, *Jude* and *Maisie* position themselves very differently, however, in relation to the psychology of their respective times. Charlotte Brontë writes in awareness of debates about possible child insanity, but challenges their underlying assumptions. Hardy, on the other hand, absorbs, and intensifies, the most pessimistic projections of Maudsley. James's vision is less bleak, but childhood in *Maisie* is similarly a masquerade or performance, a projection of the adult world.

The 1890s witnessed the coming of age of the science of child development, but also, concomitantly, a dissolution of the boundaries of childhood. In his detailed individual study of his son's early years, Darwin both opened up the area of infant development for future research, and laid the foundations for a science that would abolish the category of childhood as previously understood. Innocence was replaced with experience, as the child came into the world bearing the marks of familial and racial history, offering itself up as a 'key' to lost worlds. An individualized study of one infant gave birth to a science that would subsume the particularity of childhood within a larger narrative of human development.

The story of child psychology in the nineteenth century cannot be told, however, without exploring the interconnections between the literary and scientific fields. Questions of child development, opened up so tentatively by Darwin in his notebook jottings of the 1830s, lay at the heart of the far more expansive novelistic projections of the era that so effectively focused cultural attention on the mind of the child. Literary texts opened up the silences of science, initially leading the way and then exploring, commenting upon, and often challenging the formulations of the newly emerging scientific domain. In place of the general pronouncements, and brief anonymous case histories of the psychological textbook, where the child was rarely given its own voice, the novel offered detailed, albeit fictional, analyses of development, placing the mind of the child at their heart. Whilst Charlotte Brontë sought, quite radically, to validate the voice of a passionately angry child, challenging early Victorian medical and religious models of childhood, Hardy and James offer more muted, but equally devastating reflections on late-Victorian constructions of childhood. In this formative decade of child psychology, they chronicle not the development of childhood, but rather its extinction.

CHAPTER 6

A Freudian Curiosity

RACHEL BOWLBY

'There can be no doubt about Little Hans's sexual curiosity.'[1]

In the course of an essay on islands, Gillian Beer dwells with relish on Robinson Crusoe's simple, repeated declaration: 'I was curious.'[2] Defoe's words suggest a whole distinctive attitude to individuality and knowledge, with curiosity poised between the purposeful and the pleasurable of possible discovery. We can see the beginning, here, of how later, in the nineteenth century, curiosity could figure as a distinctly modern disposition of openness to the world.[3] We are born curious; happy the man, be he poet or *flaneur*, who knows how to retain or how to return to the bright curiosity now confidently attributed to the child.

In his theories of childhood development, Freud gives curiosity a very grand narrative, linking it in complex ways to what a well-known article of 1909 boldly calls 'The Sexual Enlightenment of Children'. Enlightenment, *Aufklärung*: the term seems to have its full historical and philosophical weight, suggesting the proud achievement of unbiased knowledge in the face of the prejudices and superstitions of a previous age. It is a word he uses often, especially in writings of the early years of psychoanalysis. But whose enlightenment is it and what and how do they learn? In Freud's writings, the sexual enlightenment 'of' children implicitly includes both what they find out (or try to) for themselves, and the question of sex education, what their carers do or should impart to them. Freudian children are passionately curious researchers, fighting for knowledge in the face of the shifty evasions of a prejudiced older generation.[4] So is Freud. For there is also, at

[1] Sigmund Freud, 'Analysis of a Phobia in a Five-year-old Boy' ('Die sexuelle Neugierde unseres Hans leidet wohl keinen Zweifel') ('Little Hans') (1909), *Standard Edition of the Complete Psychological Works of Sigmund Freud*, trans. and ed. James Strachey in collaboration with Anna Freud, 24 vols. (London: Hogarth Press, 1955–74), x. 9. Texts by Freud are all cited from this edition. Hereafter *SE*, followed by volume number.

[2] 'Island Bounds', in Rod Edmond and Vanessa Smith (eds.), *Islands in History and Representation* (London: Routledge, 2003), 32–42 (35).

[3] For a fascinating analysis of curiosity in French versions of modernity, see John House, 'Curiosité', in Richard Hobbs (ed.), *Impressions of French Modernity: Art and Literature in France 1850–1900* (Manchester: Manchester University Press, 1999), 33–57.

[4] See Adam Phillips, *The Beast in the Nursery* (London: Faber & Faber, 1998), for other reflections on children's curiosity in Freud.

another remove, an analogy, made more or less explicitly, between children's enlightenment and the knowledge claimed for psychoanalysis itself, which enlightens the world in relation to children's sexuality and the sexual aspect of neurotic illness in adults. To begin with, Freud thinks, all children attempt the achievement of personal enlightenment in matters of sexuality for themselves; at the same time, psychoanalysis is itself a form of enlightenment in an area that has been shrouded in mystery. Observation—the direct evidence of eyes and ears, in studying both the child and the adult patient—reveals and validates a knowledge that comes before the distorted forms of established philosophy and religion. Freud will often goad or cajole readers taken to be unconvinced, if not hostile to psychoanalysis, by insisting that its findings in this field are clear as day, if only they could divest themselves of the prejudices that prevent them from recognizing the truth.

Post-Enlightenment views of children's development tend to look upon them as natural researchers and explorers who come into the world ready to find out and learn and develop. In childcare manuals, the standard view today casts the child as a gifted scientist who engages as soon as possible in all kinds of experimentation with the world around it, adding all the time to its supply of usable knowledge. Children are not necessarily doomed to dumbing or darkening down; offered a stimulating environment, they may go on extending their spontaneous reach for the light. In the commonest, post-Romantic version of individual development, curiosity comes naturally; knowledge enters and illuminates an open mind. But stories of collective, cultural enlightenment may work in other ways too. The sun of knowledge shines in upon a darkness it must actively drive away; and if there is nothing new under the sun, enlightenment presents itself as knowledge never before seen as such. If the knowledge is being claimed as new, its success will require that it expose prior knowledge as clouded—wrong or preliminary or corrupt.

Freud's versions of enlightenment, which involve elements of both these, are much more complex and peculiar than may appear at first sight, both in relation to the new knowledge that psychoanalysis claims as its form of cultural enlightenment and in relation to the new world encountered by the child. Psychoanalytical enlightenment involves an impossible conversion, to beliefs which are bound to be resisted; it is a revolution in thought whose refusal is inseparable from its acceptance. Children's curiosity and striving for knowledge are not spontaneous but emerge, like cultural enlightenment, as a reaction to and a rejection of the claims of the older generation. And they end not so much in knowledge as in failure or further prejudice. At the end of the day, Freudian enlightenment is a depressing business.

In relation to psychoanalytic knowledge, Freud does not only employ enlightenment as a standard metaphor; he also refers directly to the history of sun-knowledge. In the 1910 text on Leonardo da Vinci, to which we shall return, Leonardo is credited with having anticipated Copernicus's hypothesis that it is the earth that orbits the sun, and not the other way around. At the end of 'The Resistances to Psychoanalysis' (1925), the Copernican discovery is linked both to

Darwin's theory of evolution and to psychoanalysis itself. All three met with resistance because they not only transformed the frame of reflection, but brought about what Freud calls 'a severe blow to human self-love', to 'men's narcissism'.[5] Psychoanalysis is elevated to the role of paradigm theory, since its own frame, here involving collective human 'narcissism', can be used in a general way for the interpretation of other blows to anthropocentrism. At the same time, a significant advance in knowledge goes together with a diminishment of human pretensions. Enlightenment thus begins to look less straightforward in its structure, as an apparently clear line of progressive illumination is muddied or muddled by the inevitable intervention of 'self-love'. Revolutionary progress in human knowledge directly concerns it; and self-love in the form of resistance necessarily impinges upon the acceptance of such new knowledge.

RELIGION AND BELIEF

In one of the lectures he delivered at Clark University in Worcester, Massachusetts in 1909—enlightening America or 'bringing them the plague', as he is supposed to have said—Freud's theme, as often, was the reluctance of those encountering it for the first time to believe what psychoanalysis has to say about sexuality. Freud puts his American audience in that position, and offers them a parable of conversion. Others have felt this way—they 'began by completely disbelieving' the claim about the significance of sexuality—until 'their own analytic experience compelled them to accept it'. The initial situation is not just indifference or neutrality or ignorance; it is disbelief, even *complete* disbelief. The change from one belief to another is brought about not by persuasion (by another) but by new knowledge based on the evidence of 'their own . . . experiences'. And this evidence is irresistible: belief is *compelled*. They have heard the truth with their own ears. And it turns out that there are cases of such hypothetical converts available in the present entourage, 'a few of my closest friends and followers' who have accompanied Freud on the journey to Worcester. They are now themselves the evidence: 'inquire from them, and you will hear that they all began by completely disbelieving . . .'[6]

There is a hesitation here between the language of science and truth, and the language of religion and belief. Scientifically, an experiment is validated by the subject's own senses, the neutral truth observed. Religiously, one belief is exchanged for another. Both aspects are included in the reference immediately afterwards to 'a conviction of the correctness of the thesis'. There is a personal assurance about the correctness, and there is the question of the correctness itself, which is not the same. Freud then goes on to elaborate a fascinating analogy in relation to the process of conversion, or learning, here drawing on the difficulties

[5] Freud, 'The Resistances to Psychoanalysis' (1925), *SE* xix. 221.

[6] 'Fourth Lecture', Freud, *Five Lectures on Psycho-analysis* (1910), *SE* xi. 40.

rather than the opportunities of the analytic situation. Patients, it seems, are as reluctant to reveal what is sexual as sceptics are to believe in its importance:

A conviction of the correctness of this thesis was not precisely made easier by the behaviour of patients. Instead of willingly presenting us with information about their sexual life, they try to conceal it by every means in their power. People are in general not candid over sexual matters. They do not show their sexuality freely, but to conceal it they wear a heavy overcoat woven of a tissue of lies, as though the weather were bad in the world of sexuality. It is a fact that sun and wind are not favourable to sexual activity in this civilized world of ours; none of us can reveal his [or her] erotism freely to others. But when your patients discover that they can feel quite easy about it while they are under your treatment, they discard this veil of lies, and only then are you in a position to form a judgement on this debatable question.[7]

Here we have—at least at first sight—a classic scenario of surface falsehood and underlying truth. There is a hint of the flasher in that heavy overcoat obscuring the sexuality beneath it, and also a suggestion of feminine seductiveness when the coat becomes a 'veil' of lies. Yet in one way there is nothing enigmatic about this coat of lies, which serves to protect sexuality against the bad weather of the civilized world it has to live in. Exposure would be inappropriate, it is implied; in this world the weather is bad, the sun is the wrong sun—'sun and wind are not favourable'. In these adverse circumstances, it would not be right to reveal what is underneath the coat. Only in the specific enabling situation of the treatment, where the weather is good, is it feasible to 'discard this veil of lies'.

The tensions revealed in this passage are indicative, I think, of a problematic that surfaces repeatedly in Freud's discussions of sexuality and enlightenment in relation to the new knowledge of psychoanalysis. Psychoanalysis is enlightenment, in an emancipatory view of the new science as illumination, as a demonstration and observation of unbiased truth against the darkness of other views, those associated with false civilization and religion. But Freud also insists that one object of knowledge for psychoanalysis is, precisely, the resistance to knowing or seeing the truth as it is. 'People . . . in general' do not want to know, in particular do not want to know about sexuality; and why, as well as what, they don't want to know is part of what psychoanalysis endeavours to unravel or uncloak.

Freud's references to children and sexual knowledge are often, in part, directed against those who refuse to acknowledge a truth that is really as clear as day. Children, he claims, are sexually active and sexually curious, not innocently presexual in relation to a truth that is a corrupting one, whose impact must be kept from them for as long as necessary. But children's own 'enlightenment' in sexuality is not straightforward: it is neither an innate knowledge nor a process of simply seeking or simply acquiring the facts.

Children's preliminary sexual speculations, as described in detail in the 'Little Hans' case of 1909, and summarized at the same time in 'The Sexual Enlightenment of Children', typically involve different theories of the origin of

[7] Ibid. 40–1.

babies and the difference of the sexes. Linked as they evidently are, these two questions are generally represented as separate issues, so that which of them comes first, in importance and in time, is itself, for Freud, a significant question, as we shall discover. The idea of children's sexual enlightenment makes the 'age' that of an individual rather than a culture: even though it may happen in comparable or even identical ways for all children, they each have to follow the path to sexual knowledge on their own, making mistakes, suffering setbacks, with luck (but it does not always happen) achieving a solution. As we trace the various versions of this path that Freud maps out, we shall see that to reach the state of enlightenment, if you do, is not, in practice, to find either happiness or the truth; at best you acquire a belief.

THE STORK GIVES BIRTH TO THE CYNIC

Curiosity—the first stage on the way to getting knowledge—is not a natural, primary disposition in Freudian children. It is, however, presented as a universal stage of development, which descends upon the child as a compelling response to a change, or potential change, in its situation. In most of Freud's descriptions, the first such change, chronologically and in other respects, is the birth of a sibling, or the observation and inference of the likelihood of such an event from the evidence of other families. A new child decentres the place of the older one; and children then asking the question 'where do babies come from?' express both the urgency of that particular need to know, and also a wish to master, to explain for themselves, the altered situation.

The question, Freud claims, persists, because it does not meet with an adequate answer. Children are told a story, but it is an absurd story, the fable of the stork that brought the new baby out of the water. And they readily cotton on to its fabrication, to its being no better than a cloak for the truth that is actually being covered up. This, says Freud, has two crucial results. First, there begins a mistrust of grown-ups as fundamentally deceitful, saying one thing but knowing another; and a corresponding segregation of children's own thoughts and discoveries as secret from grown-ups. Knowledge becomes something hidden, valuable, open to wilful distortion or dissimulation in the mode of its presentation. It is not simply or primarily a nugget of truth, possessed or not possessed, but a token of exchange to be traded and tricked over in a battle between those deprived of it and those who want to keep it to themselves. Second, Freud claims, children set off on a path of solitary research. In the logic of his own story, this doesn't follow directly since the self-imposed isolation is only from adults. So he assumes, without making it explicit, that the disappointment in adult communications of knowledge will have a knock-on effect on children's relations with their peers. Instead of banding together in a research group in competition with the grown-ups, they all go their own lonely ways.

Initially, the child endowed grown-ups, parents in particular, with omnipotence and knowledge. Parents, Freud says in 'Family Romances' (1909), are 'the source of all belief', and he is fond of making this structure the prototype of religion, in which God takes over the parental role.[8] The premiss is that such a strong, stable source of belief is both necessary and bound to be challenged. The stability of the child's world gets shaken up in one way by the intrusion, or feared intrusion, of another child, but the security of *belief* in the parents is a slightly different matter, even though it is linked to this earth-shatteringly unhappy event. The parents' failure to give a truthful answer to the baby question produces an indeterminacy in relation to knowledge: about who knows it, and what they do with it. It now becomes part of a power game, involving quest and concealment, and the child sets off on its own to pursue it independently of the adults. This is the religious secession, born of scepticism, that becomes for Freud the forerunner and prototype of all intellectual inquiry.

There is, however, a distinct sting in the tail of what partly develops as a classic tale of heroic resistance to superstition and triumph over obstacles. The young researcher never quite makes it. And in fact, it is not clear whether there might have been, or should be, an end to be reached: whether the quest for sexual knowledge or for knowledge in general has a conclusion. The child researcher comes up with various unsatisfactory hypotheses and encounters various impediments to knowledge, but his or her attempts to find out are also in some way thwarted by what Freud presents as an internal difficulty, the immaturity of the sexual apparatus. Inconclusiveness, the inability to get at the answer or solve the enigma, is inherent to children's investigations. And that is not the end of it. In 'On the Sexual Theories of Children' (1908), Freud witheringly declares that 'the first failure has a crippling effect on the child's whole future'.[9]

Adult research, it then comes as no surprise, is equally problematic, equally sad. Is it worth interminably seeking to know when what you thereby don't do is to live the rest of your life? This is the point that Freud makes, in striking language, in relation to Leonardo, the boy researcher who never gave it up: 'The postponement of loving until full knowledge is acquired ends in a substitution of the latter for the former. A man who has won his way to a state of knowledge cannot properly be said to love and hate; he remains beyond love and hatred.'[10] What is perhaps most remarkable about this is the endorsement of the childhood order of things. Loving (and hating) comes first; the urge to know is secondary, and should not, now, overlay and replace the primary importance of loving well. The 'state of knowledge' is represented as a kind of aloofness, away from the world and from the reality of life; it is not a noble goal. Nor is it applied to a follower of philosophy, that branch of knowledge-seeking that regularly incurs Freud's snide devaluation. Leonardo's researches were practical as much as theoretical, like those of psychoanalysis itself. Freud's identification with Leonardo the

[8] Freud, 'Family Romances' (1909), *SE* ix. 237.

[9] Freud, 'On the Sexual Theories of Children' (1908), *SE* ix. 219.

[10] Freud, *Leonardo da Vinci and a Memory of his Childhood* (1910), *SE* xi. 75.

artist-scientist, the great man who was ahead of his time in his thinking, is clear, leading him even to wonder why it was that this all-round Renaissance man who was involved in so many forms of investigation was interested only in external things, not also in psychology.[11]

In Leonardo, Freud constructs the peculiar case of a man whose urge to know is to be understood both as a dangerous instinct that has got out of hand, coming to dominate his existence in a debilitating way, and at the same time as the source of nothing less than a world-shaking intellectual revolution:

> And finally the instinct, which had become overwhelming, swept him away until the connection with the demands of his art was severed, so that he discovered the general laws of mechanics and divined the history of the stratification and fossilization of the Arno valley, and until he could enter in large letters in his book the discovery: *Il sole non si move* [The sun does not move].[12]

Leonardo may have shown the relations of sun and earth to be other than they appear, he may have enlightened the world with a finding that alters the focus of all human knowledge; but the impetus is pathological, a single instinct that has taken over. He discovers new connections in one field, but at the cost of losing the links and the balance in himself.

More than any other text of Freud's, the study of Leonardo treats the instinct for knowledge as though it is on a par with the two primary instincts with which he is concerned at this time, the sexual and the self-preserving; and this is so even though he also provides one of the most detailed of his descriptions of how this third instinct arises as a secondary formation, following the thwarting of the child's first sexual question. Because it is presented as a full-blown instinct, the instinct for knowledge has to suffer a subsequent fate that resembles that of the sexual instinct: at a certain point its childhood freedom has to be curbed because of adverse circumstances and turned in other directions, each of which is more a compromise than a continuation. Let us look at this process in more detail.

THE FIRST FIRST QUESTION

Freud enters into speculation about the peculiarities of Leonardo's situation as a child in order to answer the question of how he became the exceptional man that he was. In doing so, as happens so often in the case studies, he in fact comes to show that the abnormal example is only a strong instance of the normal. The question is where does Leonardo's 'overwhelming' instinct for knowledge come from, and Freud's preliminary answer is to set it alongside the sexual instinct, asserting that the force of both in children is underestimated. He has been describing the way in which what were formerly sexual instinctual forces will end up being directed towards professional activity in adult life:

[11] Ibid. 76–7. [12] Ibid. 76.

There seem to be special difficulties in applying these expectations to the case of an over-powerful instinct for investigation, since precisely in the case of children there is a reluctance to credit them with either this serious instinct or any noteworthy sexual interests. However, these difficulties are easily overcome. The curiosity of small children is manifested in their untiring love of asking questions; this is bewildering to the adult as long as he fails to understand that all these questions are merely circumlocutions and that they cannot come to an end because the child is only trying to make them take the place of a question which he does *not* ask. When he grows bigger and becomes better informed this expression of curiosity often comes to a sudden end.[13]

Seasoned readers of Freud may recognize something familiar in this combination of endlessness and sudden end in a childhood phenomenon: it is just what happens to the sexual instinct itself. Sexuality and 'this serious instinct' for investigation seem destined to meet; they in fact turn out to be inseparable. Freud continues:

Psycho-analytic investigation provides us with a full explanation by teaching us that many, perhaps most children, or at least the most gifted ones, pass through a period, beginning when they are about three, which may be called the period of *infantile sexual researches*. So far as we know, the curiosity of children of this age does not awaken spontaneously, but is aroused by the impression made by some important event—by the actual birth of a little brother or sister, or by a fear of it based on external experiences—in which the child perceives a threat to his selfish interests.[14]

Again, the 'threat to his selfish interests' recalls another, better known infantile scene, when the boy is confronted by the fact of girls' lack of a penis and infers that the castration with which he has been threatened is a real possibility. This scene is elaborated much later in Freud's own researches than the present passage; as we shall see, it forms a crucial element in what I will call the second first question.

It should by now be clear that Freudian curiosity is always, primarily, sexual curiosity—or at least, as with the question of new babies, a curiosity that finds a sexual answer. Freud may hesitate, as we shall see in more detail, about which sexual question comes first—whether it is the origin of babies, or the difference between the sexes, or (a distant third) the obscure problem of 'being married', what exactly it might be that couples do with each other. But ultimately, in a striking reversal of the Genesis story, it is sexuality that precedes and engenders curiosity and knowledge, rather than the other way around. 'Thirst for knowledge seems to be inseparable from sexual curiosity,' Freud says in the 'Little Hans' case.[15] In German the link is verbally more explicit, between the (general) *Wissbegierde*, 'thirst for knowledge', 'knowledge-longing', and the (specific) '*sexuelle Neugierde*', 'sexual curiosity', 'sexual longing-for-new'. Yet is not just that sexuality spurs the curiosity; ultimately sexuality and curiosity are bound together, emerging as the joint source of the stories and explanations that human beings

[13] Ibid. 78. [14] Ibid. 78. [15] 'Little Hans', 3.

require to live by.[16] They need to know, and the 'urge to know', as Freud sometimes calls it, matters more than the knowledge itself, its truth or falsehood.

So far, in the discussion of Leonardo, Freud has described the typical development of curiosity. There is a slight hesitation about its applicability. Although we have moved swiftly from the one great man to 'small children' in general, Freud is not sure whether to go to the opposite extreme and make this a universal. Many? Most? The most gifted? He asks the question but passes on to the sexual connection, which becomes the focus:

Researches are directed to the question of where babies come from, exactly as if the child were looking for ways and means to avert so undesired an event. In this way we have been astonished to learn that children refuse to believe the bits of information that are given them—for example, that they energetically reject the fable of the stork with its wealth of mythological meaning, that they date their intellectual independence from this act of disbelief, and that they often feel in serious opposition to adults and in fact never afterwards forgive them for having deceived them here about the true facts of the case.[17]

Here we have the same conjunction of tactics and reactions noted in relation to the Clark University lecture, the year before the text on Leonardo. In the initial stage, there are three points of interest. First, research is not spontaneous but triggered by an event. Second, it is not disinterested: the event impinges, or is felt to, upon the child, and that is the reason for starting investigations. Third, the aim of research is not knowledge for its own sake but applied knowledge: to find out how to prevent the occurrence or recurrence of the 'undesired event'. The aim is both negative and self-interested; and at this point the only reason for calling the research sexual has to do with the true answer, of which the child is ignorant.

The second stage comes with children's dissatisfaction at the responses they get. Where first they unquestioningly believed, now they 'refuse to believe' and 'energetically reject'. Disbelief is not just a state but a positive 'act'; it is like a secession from the parents, to whom they are henceforth in 'serious opposition'. This involves both 'intellectual independence' and the more personal motive of revenge for a wrong committed. Knowledge is now understood to be bound up with the possibility of withholding it from another, or dissimulating it by presenting a false version. Belief in the parents is withdrawn not because they are themselves lacking in the truth, but because they have kept it back. These elements are clearly brought together once again in 'On the Sexual Theories of Children' (1908):

[F]ar more children than their parents suspect are dissatisfied with this solution and meet it with energetic doubts, which, however, they do not always openly admit. . . . It seems to me to follow from a great deal of information I have received that children refuse to believe the stork theory and that from the time of this first deception and rebuff they nourish a distrust of adults and have a suspicion of there being something forbidden which is being

[16] This is one implication of Jean Laplanche and J.-B. Pontalis's enlightening essay, 'Fantasy and the Origins of Sexuality' (1964), trans. in Victor Burgin, James Donald, and Cora Kaplan (eds.), *Formations of Fantasy* (London: Routledge, 1986), 5–34.

[17] *Leonardo*, 78–9.

withheld from them by the 'grown-ups', and that they consequently hide their further researches under a cloak of secrecy.[18]

This time with a cloak rather than a coat, the passage includes the theme of secrecy that is underlined in the Clark lecture.

The final part of the passage from *Leonardo* describes the results of children's subsequent researches: 'They investigate along their own lines, divine the baby's presence inside its mother's body, and following the lead of the impulses of their own sexuality form theories of babies originating from eating, of their being born through the bowels, and of the obscure part played by the father.'[19] Here we can see another explanation of children's theories slipping in alongside the primary impetus to find the truth that the parents have not imparted. The young researchers are 'following the lead' of the oral and anal stages of their development, a lead which is both a forward direction and a limitation: it is not yet the whole story. These two aspects are elaborated further both in 'On the Sexual Theories of Children' and in the 'Little Hans' case, with a slightly different line-up of infantile theories. In the essay Freud makes two supplementary, closely related claims. The first is that each of the mistaken hypotheses does in fact contain an element of the truth; and the second, that there is a specific obstacle to getting at the whole truth, which is children's ignorance, at this stage, of the existence and role of the vagina: of there being a second sex. In the first category, Freud alludes to the fact that 'so many animals' do in fact give birth anally,[20] and describes children's understanding of sexual intercourse, if they observe it, as a violent act. The grain of truth here is the frequent survival in adult sexuality of the sadistic and masochistic trends that dominate during a particular period of childhood; more locally and anecdotally, the child may be witnessing aspects of an unhappy relationship—'the spectacle of an unceasing quarrel' carried over into the night, or a mother sexually reluctant because fearful of becoming pregnant again.[21] By far the lengthiest account, however, is devoted to the fragment of truth in the assumption attributed to all children that there is only one sex. When small boys—the paradigm here is Little Hans and his baby sister Hanna—see female genitals, they do not see the difference, or see it only as contingent: here is something small that will grow in time. And they are not completely wrong, says Freud, because in evolutionary terms the clitoris really is an undeveloped penis.

Clearly, there is a connection between the one-sex theory and the second of Freud's points, that ignorance of female difference is the block to complete knowledge at this stage. In 'Little Hans', Freud states: 'His conviction that his mother possessed a penis just as he did stood in the way of any solution.'[22] 'On the Sexual Theories of Children' makes the same point with more elaboration: '[A]t this juncture the enquiry is broken off in helpless perplexity. For standing in its way is his [the child's] theory that his mother possesses a penis just as a man

[18] 'On the Sexual Theories', 213–14.
[19] *Leonardo*, 79.
[20] 'On the Sexual Theories', 219.
[21] Ibid. 221–2.
[22] 'Little Hans', 135.

does, and the existence of the cavity which receives the penis remains undiscovered by him.'[23] One theory or conviction bars the way to seeing another, but this very theory is naturalized in the image of the all-pervasive penis shutting out other ways of imagining the female sex. For the vagina appears only as complementary, as that which 'receives the penis'. It is not, for instance, seen as the place from which babies first emerge.

THE SECOND FIRST QUESTION

Even when the issue of the origin of babies is ostensibly the starting point, it is apparent that more and more it is taken over, in Freud's account, by the related question of the difference between the sexes. Following the trajectory of his own theories, we can see that this parallels or resumes the pattern by which the castration complex comes to occupy the foreground, taking precedence over the Oedipal triangle. The son whose place is challenged by the arrival of a new sibling gives way to the boy whose masculinity is threatened by the existence of femininity.

In terms of children's theories, as opposed to Freud's own, the growing dominance of the sexual difference question is signalled expressly by Freud in a footnote to an essay of the mid-1920s, 'Some Psychical Consequences of the Anatomical Difference between the Sexes'.[24] Appearing when and where it does, the changeover evidently has something to do with the conscious turn of attention to girls' sexual understanding, and its possible differences from that of boys. In this area, the conditions of research turn out to be different from those that surrounded the first first question, that of the origin of babies. These differences can be divided into three categories, all of which are themselves closely connected to the question of sexual difference. There is first, the reason for undertaking the research; second, the outcome as it affects the disposition of the researcher; and third, the kind of knowledge that the research produces. Let us take these one by one.

With the question of where babies come from, as we have seen, there is an external reason for undertaking research, namely the arrival or possibility of another child. Where do they come from and, concomitantly, how can they be prevented from turning up in the first place? As we have seen, Freud believes that 'the curiosity of children of this age does not awaken spontaneously'.[25] With the second question, on the other hand, early research does come naturally, and in boys it is penis-led:

[23] 'On the Sexual Theories', 218.

[24] Freud, 'Some Psychical Consequences of the Anatomical Difference between the Sexes' (1925), *SE* xix. 252.

[25] *Leonardo*, 78.

> This part of the body, which is easily excitable, prone to changes and so rich in sensations, occupies the boy's interest to a high degree and is constantly setting new tasks to his instinct for research. . . . The driving force which this male portion of the body will develop later at puberty expresses itself at this period of life mainly as an urge to investigate, as sexual curiosity.[26]

This time we have a curiosity that is unrelated to external events and arises from a biological fact of sexuality.

The essay in which this passage occurs, 'The Infantile Genital Organization' (1923), is traditionally seen as a turning point in Freud's own theory, because it introduces the concept of the phallic phase, paving the way for the central significance of castration for both sexes. But the passage above, applying to boys, is preceded by an apology for only being able to describe the boy's situation, since 'the corresponding processes in the little girl are not known to us'.[27] For the first time, ignorance of girls is an issue. As a barrier for the researcher to full knowledge into children's sexuality it figures just like the child's ignorance of the vagina in the first account of sexual researches: knowledge of girls is marked in its absence, just like knowledge of the vagina was before. This recognized lacuna then opens up the way for the discussions of differences in girls' development that follow in the next few years, beginning in 1924 and 1925 with 'The Dissolution of the Oedipus Complex' and 'Some Psychical Consequences', through to 'Female Sexuality' and the lecture 'Femininity' in the early 1930s, where the girl's difference has become the principal subject of investigation.

In the first account, boys' (and possibly girls') researches into the origin of babies always end in failure from a combination of external and internal factors: lack of evidence (the unknown difference of the sexes) and lack of capacity (the boy's sexual organization is not yet ready for him to imagine the truth). This results, Freud says, in a debilitating attitude of mind that never goes away: 'The impression caused by this failure in the first attempt at intellectual independence appears to be of a lasting and deeply depressing kind.'[28] Or again, in a statement partly quoted already: 'This brooding and doubting . . . becomes the prototype of all later intellectual work directed towards the solution of problems, and the first failure has a crippling effect on the child's whole future.'[29] In this first scenario, the boy winds up depressed and unfit for life. In the second scenario, though, when the first question, that of sexual difference, is the very one that blocked the solution in the first, the results, if far from happy, are not quite so dark. Doubt and delay are two elements from the first scenario that maintain a role, but in different sequences.

Doubt, or disbelief, was initially a loss of trust in parental honesty, the breakaway independent research fostered by the 'energetic rejection' of the stork story as patently fabricated. The 'brooding and doubting' that remain to haunt the

[26] 'The Infantile Genital Organization: An Interpolation into the Theory of Sexuality' (1923), *SE* xix. 142–3.

[27] Ibid. 142.

[28] *Leonardo*, 79.

[29] 'On the Sexual Theories', 219.

unsuccessful researcher make him sound like a nineteenth-century cleric, afflicted with nebulous but insoluble 'doubts' about a faith he can neither take on trust any more, nor replace with a compelling confidence of his own.

In the later version, doubt derives from the possibility of there being fellow human creatures who lack a penis. It is once again, but now more directly, related to a threat to the self. The new baby threatened the child's special position *vis-à-vis* his parents; the threat of castration is directed to what now for the first time emerges as the boy's masculinity, in urgent distinction from the femininity against which it is defended. In every account Freud gives of boys' realization of the difference, the process is a slow one. Like a soundtrack that has to be linked to the appropriate footage, the verbal threat and the visual perception must be brought together before the new conviction clicks into place. Some of Freud's descriptions show doubt and disbelief arising from the first sight of the girl's different genitals, and clarified in the light of an adult's mock threat of castration as a punishment; most often, the threat is itself the first stage, unheeded until connection is made with the same sight. For instance:

> [T]o begin with the boy does not believe in the threat or obey it in the least. . . . The observation which finally breaks down his unbelief is the sight of the female genitals. Sooner or later the child, who is so proud of his possession of a penis, has a view of the genital region of a little girl, and cannot help being convinced of the absence of a penis in a creature who is so like himself.[30]

This breaking down of unbelief, leading to inescapable conviction, tells the reverse story from that of the first first question. Here we have the making of a reluctant convert; there it was a reluctant rebel with his (or her) faith shattered by a false story. The combination here of complete scepticism ('not . . . in the least') and forced acceptance ('cannot help being convinced') also repeats the Clark lecture account of the making of psychoanalytic believers: those who 'began by completely disbelieving' Freud's claim about the significance of sexuality were eventually 'compelled . . . to accept it' in the light of 'their own analytic experience'.

Less often, the delay is said to occur in the other sequence, between observation (not taken seriously) and threat (which gives it meaning):

> It is not until later, when some threat of castration has obtained a hold upon him, that the observation [of a girl's genital region] becomes important to him: if he then recollects or repeats it, it arouses a terrible storm of emotion in him and forces him to believe in the reality of the threat which he has hitherto laughed at.[31]

Earlier, as in 'The Infantile Genital Organization', there is a kind of slow-motion skittle effect in the realization. Gradually, more and more women are understood to lack penises after all until finally, when the mother has fallen along with all the rest, it has to be recognized that all women are without one, that to be a woman

[30] 'The Dissolution of the Oedipus Complex' (1924), *SE* xix. 175–6.
[31] 'Some Psychical Consequences', 252.

is to be without a penis, no more and no less.[32] In this passage the quest for the origin of babies is clearly subsequent to the investigation of the problem of having or not having a penis. After this, an accommodation is made to the new knowledge that enables the boy to pass definitively from childhood to manhood in a fairly clear-cut way:

> In his case it is the discovery of the possibility of castration, as proved by the sight of the female genitals, which forces on him the transformation of his Oedipus complex, and which leads to the creation of his super-ego and thus initiates all the processes that are designed to make the individual find a place in the cultural community.[33]

The 'proof' compels ('forces on him') certain changes, but those changes are smooth and sure, neatly resituating the Oedipal prince as one young man among others, out of the family and into the 'cultural community'. Failure and depression have gone from the account; instead, this is a modestly happy ending, a new beginning.

The girl's castration story, of course, is rather different. Now it is for her, not for the boy (or the child in general) that there is no settled conclusion, only a set of loose ends and surreal equivalences. Babies remain in her picture (while they have disappeared from the boy's), but strangely transformed. No longer potential rivals whose origins must be sought, they have turned into belated substitutes for unobtainable penises. Leaving aside for the moment the epistemological framework here, it is notable that the girl's pathway to this equivocal knowledge occurs according to a temporal pattern that sharply contrasts with the boy's.[34] The boy takes his time, doubting and delaying, but ultimately comes to terms with it; it provides the means for him to move out into the wider, post-familial world. The girl sees and understands instantly, without a doubt; but she never really moves on, remaining caught up in the nexus of futile reactions into which castration casts her. Essentially, where the boy (initially) doubts, the girl (subsequently) 'clings' desperately to the impossible hope of one day acquiring what she in fact can never get. She has a moment of illumination, but it is wholly negative: 'A little girl behaves differently. She makes her judgement and her decision in a flash [*im Nu*]. She has seen it and knows that she is without it and wants to have it.'[35] These short declarative clauses stand out in a single paragraph, further exposing the peculiarity of this moment of development so unlike others in Freud. This change in the girl's understanding is almost magically immediate, with no doubt or deferral intervening between the perception, the knowledge, and the desire—seen, knows, wants, in Freud's equally hurried shorthand. But it is the whole life afterwards that then slows down almost to the point of immobility. The boy eventually realizes an

[32] See 'The Infantile Genital Organization', 144–5.

[33] Freud, 'Female Sexuality' (1931), *SE* xxi. 229.

[34] See Toril Moi's stimulating analysis of the different logics ascribed to boys and girls in Freud's account of these developments, 'Representation of Patriarchy: Sexuality and Epistemology in Freud's Dora', in Charles Bernheimer and Claire Kahane (eds.), *In Dora's Case* (London: Virago, 1985), 181–99.

[35] 'Some Psychical Consequences', 252.

unwelcome fact whose acceptance can then set him up along new paths in a new world. The girl, even if she does end up in the normal social role of wife and mother, never finally accepts: 'She acknowledges the fact of her castration, and with it, too, the superiority of the male and her own inferiority; but she rebels against this unwelcome state of affairs. From this divided attitude three lines of development open up.'[36]

It is in fact at the end of the flashy short paragraph that Freud adds the footnote in which he declares he has changed his mind about children's first research question. Here it is:

> This is an opportunity for correcting a statement which I made many years ago. I believed that the sexual interest of children, unlike that of pubescents, was aroused, not by the difference between the sexes, but by the problem of where babies come from. We now see that, at all events with girls, this is certainly not the case. With boys it may no doubt happen sometimes one way and sometimes the other; or with both sexes chance experiences may determine the event.[37]

'I believed . . . we now see': here is another minor conversion experience, from religion to enlightenment. And how ironic that this should be the one place where the girl's case becomes paradigmatic, while boys are indeterminately one way or the other. (As indeed Freud is here himself, for having said that it 'is certainly not' true that the baby-origin problem is first for girls, he ends by saying that perhaps for both sexes it may be either way round. The only certainty seems to be that the baby question has lost its assumed priority, and probably its actual priority in most instances.) There is also the characteristic masculine situation of deferred understanding, here from 'many years ago', in the early years of psychoanalysis, to the present.

The reversal of babies and sexual difference has further implications, which conduct us to the third concern in relation to boys' and girls' enlightenments: the type of knowledge that is acquired. With the question of where babies come from, the false answer—the stork fable—is false in relation to a factual truth. One accurate answer to the question is presumed to be: from inside mothers. Children such as Little Hans are said to have worked this out quite well for themselves: 'The change which takes place in the mother during pregnancy does not escape the child's sharp eyes.'[38] Here a perception of a growing body corresponds to a truth and a kind of solution to the question: babies do come from mothers, and what makes mothers mothers is that they carry babies. The positive distinction, the larger size and the swelling, is in this instance on the female side.

The sharp-eyed researchers of the first first question are thwarted, as we have heard—first, by the grown-ups' counter-story and second, by their ignorance of an invisible origin further back along the line, the existence of the vagina, sometimes associated with boys' assumption that all humans have a penis. The sexual

[36] 'Female Sexuality', 229.
[37] 'Some Psychical Consequences', 252.
[38] 'On the Sexual Theories', 214.

difference question is already implied by the hypothesis that the solution for the baby question is not enough: children are supposed to sense that there is something more to be known that they can't yet imagine, and that is where the unknown vagina enters the frame in its significant absence. Whence the second first question and the second set of solutions, this time sexually differentiated just as the question itself is concerned with sexual difference. And now, what both sexes in their diverging ways perceive, eventually or immediately, is not a fact of nature but a value: not (changing) anatomy but their destiny as men and women in the cultural community. Enlightenment is not, this time, a perception of the truth like Leonardo's precocious hypothesis that 'The sun does not move'; from that point of view sexual difference as castration is no better than the story of the stork.

Yet this realization is not in itself a cause for either protest or resignation. Freud does not ultimately claim castration is a fact of nature; instead, as Laplanche and Pontalis suggested, children's early questions and solutions can be seen as their initiation into the world of human meanings and myths, where empirical truth has no priority. But we may speculate as to whether, or how much, the earth has moved since Freud's day, now that Little Hans's descendants would be unlikely to hear the stork fable other than as a fairy-tale and now that, for better or worse, the imagined destinies of men and women are far less clearly differentiated than they were then. The Freudian child's researches were ultimately marked by either depression and failure or negative illumination, the reluctant observer finally forced to believe. And so, perhaps, we can be curious about possible different outcomes—happier forms of enlightenment—for twenty-first century children, wherever they may come from.

CHAPTER 7

Freud's Theory of Metaphor: *Beyond the Pleasure Principle*, Nineteenth-Century Science and Figurative Language

SUZANNE RAITT

At the beginning of the final lecture in Freud's 1933 publication, *New Introductory Lectures on Psycho-Analysis*, Freud declared summarily and triumphantly that psychoanalysis was a science. 'As a specialist science, a branch of psychology . . . it is quite unfit to construct a *Weltanschauung* of its own: it must accept the scientific one.'[1] This was a view he continued to stress as his career drew to a close. In 1940, seven years after the lecture on the *Weltanschauung*, he noted that psychology was 'a natural science like any other', asking defiantly: 'What else can it be?'[2]

If Freud was defiant, it was partly because he had not always been so sure. One of his earliest publications in 1893, written while he was still seeking a new form for his researches into the unconscious mind, was called *Project for a Scientific Psychology*, and announced his intention to 'furnish a psychology that shall be a natural science: that is, to represent psychical processes as quantitatively determinate states of specifiable material particles'.[3] But twenty years later he acknowledged that the *Project* had failed, remarking that 'every attempt to . . . discover a localization of mental processes, every endeavour to think of ideas as stored up in nerve-cells and of excitations as travelling along nerve-fibres, has miscarried completely'.[4] Taken at face value, this looks like an admission that no one has succeeded in putting depth-psychology on a 'scientific' basis: that is, if 'scientific' means that psychic impulses have yielded to observation, measurement, and categorization. Freud was undaunted, however. If psychoanalysis did not look like a

[1] Sigmund Freud, *New Introductory Lectures on Psychoanalysis* (1933 [1932]), in *The Standard Edition of the Complete Psychological Works of Sigmund Freud*, trans. and ed. James Strachey in collaboration with Anna Freud (London: Hogarth, 1953–74), xxii. 5–182 (158). Hereafter *SE*.

[2] Freud, *An Outline of Psychoanalysis* (1940 [1938]), *SE* xxiii. 139–207 (158, 282).

[3] Freud, *Project for a Scientific Psychology* (1950 [1895]), *SE* i. 281–397 (298).

[4] Freud, *The Unconscious* (1915), *SE* xiv. 159–215 (174).

science, then, he declared, it was our conceptualization of the natural sciences, not psychoanalytic methodology, which was at fault. 'There is a hiatus', he wrote, in our understanding of the biochemical processes underlying psychical states, but it is one that 'at present cannot be filled, nor is it one of the tasks of psychology to fill it. Our psychical topography has *for the present* nothing to do with anatomy . . . our work is untrammelled and may proceed according to its own requirements.'[5] In this passage, it seems not to matter that psychoanalysis failed to represent 'psychical processes' in terms of 'material particles', as the *Project* announced that it would. By 1915, Freud had decided that such a task was simply beyond the scope of psychology—and thus of psychoanalysis—as a 'natural science'. The undertaking that, in the *Project*, was to have established psychoanalysis as a natural science, is now dismissed as irrelevant, and from now on psychoanalysis, still a science but somehow 'untrammelled' by scientific convention, will make its own rules.

Freud was able to continue to argue that psychoanalysis was a 'natural science' almost until the day of his death partly because, for him, 'science' itself was a protean and unreliable concept: he felt justified in changing his mind about exactly what constituted 'natural science'. In part, of course, this attitude simply betrays the slightly reckless opportunism for which psychoanalysis has often been criticized. Just like the hapless patient whose refusal to accept a particular interpretation proves not that the interpretation—and hence the analyst—is wrong, but simply that the idea is subject to particularly intense repression, so Freud's failure to establish the material basis of psychical events is, when it suits him, set aside as irrelevant to the issue of whether or not psychoanalysis can be regarded as a science. It is hard not to be reminded of Freud's own analogy for the over-determination of dreams: a man, accused by his neighbour of damaging a borrowed kettle, 'asserted first, that he had given it back undamaged; secondly, that the kettle had a hole in it when he borrowed it; and thirdly, that he had never borrowed a kettle from his neighbour at all'.[6] But Freud's shifting of the goalposts where science is concerned is, in fact, not simply an opportunist move (although it certainly is that). Rather, critical reflection on the conventions of scientific language and method was one of the principal—though largely unacknowledged—preoccupations of psychoanalysis. If the science of psychoanalysis was forced, in the end, to describe psychical events using analogies (topographically, for example), did that mean that in fact all scientific language operated through analogy and figure? Was there such a thing as a transparently literal language, as positivist science would have us believe? In this chapter I shall read *Beyond the Pleasure Principle*, and in particular the concept of the death drive, as a meditation on scientific language, and, by implication, on the economy of language and of metaphor itself. Freud returns to scientific experiments conducted in the 1890s as a way of exploring the relationship between the literal and the figurative in terms that are oddly reminiscent of literary texts from the same era. Like, for

[5] Ibid. 174–5. [6] Freud, *The Interpretation of Dreams* (1900), *SE* iv. (120).

example, Oscar Wilde's *The Picture of Dorian Gray* (1890), Freud plays with the idea that the opposition between death and life holds only in the realm of representation. Death, which might seem to be the limit of illusion, becomes in these texts the moment at which the distinction between the flesh and the word—or the image—dissolves.

Freud, of course, was not the only thinker to question the scientific status of psychoanalysis. His doubts and reconsiderations were part of a lengthy debate over the identifying features of scientific discourse in general, in which psychoanalysis itself played a leading role. Science had long been dominated by a positivist distrust of figurative language, and a desire to make scientific language as transparent and precise as possible.[7] As Gadamer noted in 1975: 'It is the aim of science so to objectify experience that it no longer contains any historical element.'[8] In the early twentieth century the development of quantum physics encouraged a new self-consciousness on the part of those who worked in science. But it was psychoanalysis that was largely, if indirectly, responsible for one of the most significant scientific paradigm-shifts of the early twentieth century. In 1919, around the time that Freud was working on *Beyond the Pleasure Principle*, a young philosopher, Karl Popper, was also wondering about the particular characteristics of scientific methodologies. His basic puzzlement was, he wrote, about 'demarcation': what made statements of the 'empirical sciences' distinct from all other varieties of statement?[9] Popper's interest in this question stemmed from his engagement with the writings of Freud and Alfred Adler, one of Freud's earliest psychoanalytic colleagues. Not that Freud and Adler's thinking was unconvincing: as Popper puts it, 'the world was full of *verifications* of the theory', verifications that, according to traditional scientific methodologies, gave the theory scientific status. Popper was concerned not because there were not enough verifications but because, in his view, there were far too many.

> I could not think of any human behaviour which could not be interpreted in terms of either theory [Freud's or Adler's]. It was precisely this fact—that they always fitted, that they were always confirmed—which in the eyes of their admirers constituted the strongest argument in favour of these theories. It began to dawn on me that this apparent strength was in fact their weakness.[10]

Popper's revelation prompted him to declare that falsifiability, rather than verification, should be the criterion of scientific status, thereby, in his view, disqualifying psychoanalysis as a science. Psychoanalysis became for Popper the test case of scientific method.

[7] There is an extensive literature on this topic. See, for example, S. Michael Halloran and Annette Norris Bradford, 'Figures of Speech in the Rhetoric of Science and Technology', in Robert J. Connors, Lisa S. Ede, and Andrea A. Lunsford (eds.), *Essays on Classical Rhetoric and Modern Discourse* (Carbondale, Ill: Southern Illinois University Press, 1984), 179–92; and David Locke, *Science and Writing* (New Haven, Conn.: Yale University Press, 1992).

[8] H. G. Gadamer, *Truth and Method* (New York: Seabury, 1975), 311.

[9] Karl R. Popper, *Conjectures and Refutations: The Growth of Scientific Knowledge* (London: Routledge, 1963), 41.

[10] Ibid., 35.

Popper's categorization of psychoanalysis as fundamentally unscientific (an opinion that he was moving towards in the years after the First World War, but that he did not publish until the 1930s) sharpened the debate around the disciplinary status and methodology of psychoanalysis.[11] In 1968 Habermas joined the conversation, arguing that because the aim of analytic treatment was to enlist the participation of the patient in her own transformation, the patient could not be considered the object of a scientific study. Thus for Habermas Freud was mistaken in seeing the analytic situation as 'quasi-experimental in character', and his claims to empiricism—the basis of his claims to be developing a science—were unfounded.[12] Three years after the attack by Habermas, Adolf Grünbaum attempted to rehabilitate psychoanalysis for science, arguing against Habermas that 'when Freud unswervingly claimed natural science status for his theoretical constructions throughout his life, he did so first and foremost for his evolving clinical theory of personality and therapy, rather than for the metapsychology'.[13] Popper's objections, Grünbaum argued, were simply misguided: Freud's 1915 paper, 'A Case of Paranoia Running Counter to the Theory of the Disease', contained a clear example of Freud citing and reviewing evidence that falsified his own theory. Thus Freud's own evasions, and perhaps his opportunism, seem to have guaranteed that the question of the scientific status of psychoanalysis would remain for the moment unanswerable.

One of Freud's texts that provoked the most controversy in this context was *Beyond the Pleasure Principle*, published in 1920 around the same time that Karl Popper was deciding that psychoanalysis could not be a science. Freud's eclectic methods and evidence in this paper are well known. First, he cites in one breath dreams in which traumatized patients continually revisit the scene of their trauma, his little grandson's game with a cotton reel, the analytic transference, and Tasso's *Gerusalemme Liberata* as evidence of the existence of a 'compulsion to repeat'.[14] He then moves into discussion of the structure of living things (in language somewhat reminiscent of the early *Project*), and offers a theory of trauma and of anxiety dreams. Several pages on experiments on unicellular organisms follow, culminating in a citation from Plato's *Symposium*. All this is laced with what Freud calls 'speculation, often far-fetched speculation', 'an

[11] Popper's work on falsifiability was first published in *Logik der Forschung* (*The Logic of Scientific Discovery*) in 1934. The debate over psychoanalysis as a science has a lengthy bibliography. See, for example, Paul Kline, *Fact and Fantasy in Freudian Theory* (1972; 2nd edn. London: Methuen, 1981); Seymour Fisher and Roger P. Greenberg, *The Scientific Credibility of Freud's Theories and Therapy* (New York: Basic Books, 1977); Seymour Fisher and Roger P. Greenberg (eds.), *The Scientific Evaluation of Freud's Theories and Therapy: A Book of Readings* (New York: Basic Books, 1978); and Tullio Maranhão, 'Pyschoanalysis: Science or Rhetoric?', in Herbert W. Simons (ed.), *The Rhetorical Turn: Invention and Persuasion in the Conduct of Inquiry* (Chicago: Chicago University Press, 1990).

[12] Jürgen Habermas, *Knowledge and Human Interests*, trans. Jeremy J. Shapiro (Boston: Beacon, 1971), 253.

[13] Adolf Grünbaum, *The Foundations of Psychoanalysis: A Philosophical Critique* (Berkeley: University of California Press, 1984), 6.

[14] Freud, *Beyond the Pleasure Principle* (1920), *SE* xviii. 1–64 (19). All further references in the text to *BPP* will be to this edition.

attempt to follow out an idea consistently, out of curiosity to see where it will lead' (*BPP*, 24). Even for someone as inventive and inconsistent as Freud could be, *Beyond the Pleasure Principle* was something of a discursive muddle, a series of ingeniously connected references from various different genres and registers thrown together in what could—ungenerously—be seen as a glorious flight of fancy.

Some of Freud's most sympathetic readers were suspicious of his methodologies in *Beyond the Pleasure Principle*. Max Schur, the doctor who treated Freud in his final illness, complained that: 'the inductive reasoning that in Freud's case almost always was solidly grounded on observed phenomena here gives way to speculation'.[15] Otto Fenichel, one of Freud's closest colleagues, noted of *Beyond the Pleasure Principle* that the 'new classification of instincts now proposed by Freud to replace the old one . . . has two roots, one speculative, the other clinical'. Furthermore, he asserted that the sense in which Freud uses (or misuses) the word 'instinct' in *Beyond the Pleasure Principle* threatened the links of psychoanalysis with biology, on which its status as a science depended.[16] Analyst Jean Laplanche objected that: 'The argument progresses through a series of interruptions, obstinately following the details of a scientific debate only in order to abandon it abruptly, like an unlucky gambler who suddenly kicks over the table.'[17] All in all, *Beyond the Pleasure Principle* was agreed to be maddening in its refusal to stay in one register, whether that be anecdotal, scientific, mythological, or clinical. Freud announced his transitions from one to another with little or no embarrassment.

Most puzzling of all, though, were the pages at the beginning of chapter 6 in which Freud details with enormous fascination the experimental work in the 1890s of biologists August Weismann, Lorande Woodruff, Gary Calkins, Alexander Lipschütz, and Émile Maupas, on unicellular organisms, particularly *Paramecium*. These paragraphs are usually passed over in silence in discussions of the text. Peter Brooks, for example, in a meticulous discussion of *Beyond the Pleasure Principle* in *Reading for the Plot*, says simply: 'the complexities of the next-to-last chapter of *Beyond the Pleasure Principle* need not be rehearsed in detail', and never mentions Freud's sources.[18] Laplanche asks exasperatedly: 'what is the function of the recourse to the life sciences, manifest at times as unrestrained speculation, at others, as a series of references to precise experimentation? . . . [Why does Freud] carry back to the biological level two theses that can be justified only in relation to the discovery of psychoanalysis?'[19] Ego-psychologist Heinz Hartmann notes that: 'at least a part of the considerations on

[15] Max Schur, *Freud: Living and Dying* (New York: International Universities Press, 1972), 320.

[16] Otto Fenichel, 'A Critique of the Death Instinct', 1935, repr. in *The Collected Papers of Otto Fenichel*, 1st ser. (New York: Norton, 1953), 363–72 (364, 366).

[17] Jean Laplanche, *Life and Death in Psychoanalysis*, trans. Jeffrey Mehlman (Baltimore: Johns Hopkins University Press, 1976), 110.

[18] Peter Brooks, *Reading for the Plot: Design and Intention in Narrative* (Cambridge: Harvard University Press, 1992), 105.

[19] Laplanche, *Life and Death*, 106, 122.

which Freud bases his speculation in *Beyond the Pleasure Principle* (1920) refers to questions to be discussed and probably to be decided within biology proper, possibly with the help of experimental biologists'.[20] Even Derrida, who comments that this section of the text was the only part that 'was not yet edited at the death of [Freud's] daughter', and on which Freud worked as he mourned the beautiful Sophie Halberstadt, pays attention only to Weismann, and refers disparagingly to these pages as a 'biologistic detour via the genetics of the period'.[21] But exactly because these pages were written with enormous effort during the gloomy weeks following Sophie's sudden and premature death, we cannot regard them simply as a 'detour'. Freud laboured at them as if they somehow contained the key to all his thoughts on the subject. Indeed, the experiments on protista had enough of a hold on Freud's imagination to appear again three years after the publication of *Beyond the Pleasure Principle* in *The Ego and the Id*, where they become a figure for the relation between the ego and the super-ego: 'In suffering under the attacks of the super-ego or perhaps even succumbing to them, the ego is meeting with a fate like that of the protista which are destroyed by the products of decomposition that they themselves have created.'[22] The philosophical ramifications of the theory of the death drive (its links to Schopenhauer, its anticipation of existentialism, and the idea that biological and psychical structures are responsible for their own destruction), unfolded out of work that had been done in scientific laboratories thirty years earlier.

Freud begins chapter 6 of *Beyond the Pleasure Principle* with a summary of German biologist August Weismann's work on the immortality of the germ plasm. By 1920 Weismann's work had been well known for many years, and continued to be cited in biology textbooks well into the 1920s.[23] Weismann asserted in 'The Duration of Life', a lecture delivered to the Salzburg meeting of the Association of German Naturalists in 1881, that unicellular organisms that reproduce by splitting into two new individuals cannot be said to die. 'I am well aware', he declared, 'that the life of the individual is generally believed to come to an end with the division that gives rise to two new individuals, as if death and reproduction were the same thing. But this process cannot be truly called death. Where is the dead body? what is it that dies? Nothing dies; the body of the animal only divides into two similar parts.'[24] Protista were fascinating to Weismann and later to Freud not only because they seemed to have the capacity to live forever, but also—more importantly—because their life cycle called into question the

[20] Heinz Hartmann, 'Notes on the Theory of Aggression', in Heinz Hartmann, Ernst Kris, and Rudoph M. Loewenstein (eds.), *Papers on Psychoanalytic Psychology* (New York: International Universities Press, 1964), 56–85 (58).

[21] Jacques Derrida, *The Post Card: From Socrates to Freud and Beyond*, trans. Alan Bass (Chicago: University of Chicago Press, 1987), 363–4.

[22] Freud, *The Ego and the Id* (1923), *SE* xix. 1–66 (56–7).

[23] See, for example, Lorande Loss Woodruff, *Foundations of Biology* (New York: Macmillan, 1922), 216.

[24] August Weismann, *Essays Upon Heredity and Kindred Biological Problems*, ed. Edward B. Poulton, Selmar Schönland, and Arthur E. Shipley (Oxford: Clarendon, 1889), 25. Hereafter referred to as *Essays Upon Heredity*.

definition of death itself. In their case, death, if it occurred at all, occurred invisibly and left no traces (we remember that Freud described the death instincts as doing 'their work unobtrusively', *BPP* 63). As Weismann noted, in unicellular organisms 'death can only be spoken of in the most figurative sense'. Weismann indulged in an uncharacteristic flight of fancy as he pursued his argument: 'If we can imagine an Amoeba endowed with self-consciousness, it might think before dividing "I will give birth to a daughter", and I have no doubt that each half would regard the other as the daughter, and would consider itself the original parent.'[25] Weismann's recourse to anthropomorphism—to a rhetorical figure—demonstrates how far the discussion has moved from the description of empirical fact. The life cycle of protista exposes death as a philosophical or linguistic, rather than a scientific problem. It was really up to the individual observer to decide whether what she had just witnessed was the death of one individual, or the birth of two more. Indeed among the protista one event seemed to accomplish both tasks; here life and death have the same currency.

But this conundrum is not yet evident in Freud's initial discussion of Weismann in the opening pages of chapter 6 of *Beyond the Pleasure Principle*. Although I shall argue that it was exactly Weismann's emphasis on the figurative dimensions of the life cycle that caught Freud's attention, and ultimately determined many of the arguments of *Beyond the Pleasure Principle*, at this stage Freud, like a good biologist, is apparently intent on citing experimental, empirical observations.[26] He seems in these pages still to be seeking an answer—a *scientific* answer—to the question of whether or not protista are immortal. At first, Freud suggests that his own work and Weismann's come to the same conclusions. Weismann's division of the body into the mortal 'soma' and the immortal 'germ plasm', destined to be handed down indefinitely through the generations, seems analogous to Freud's own identification of two forces at work in living matter, the death instincts and the life instincts. But Freud quickly dismisses the notion that he and Weismann are ultimately saying the same thing. For Weismann stresses the fact that only unicellular organisms are immortal (indeed, the germ plasm could not be immortal if that were not so). Noting that organisms have to return to a unicellular condition in order to begin the process of reproduction, he declared in 1883 that: 'Each individual of any . . . unicellular species living on the earth today is far older than mankind, and is almost as old as life itself.'[27] Death is an adaptive mechanism adopted by multicellular organisms to prevent fatal competition for limited resources. This, of course, is no good for Freud's purposes, since, as he says, 'if death is a *late* acquisition of organisms, then there can be no question of there having been death instincts from the very beginning

[25] Weismann, *Essays Upon Heredity*, 26.

[26] William Coleman notes that as the nineteenth century progressed, 'a self-conscious quest [emerged] to make biology an experimental science' (12) rather than one based on historical explanation and classification. See Coleman, *Biology in the Nineteenth Century: Problems of Form, Function, and Transformation* (Cambridge: Cambridge University Press, 1977), 2 ff.

[27] Weismann, *Essays Upon Heredity*, 72.

of life on this earth' (*BPP* 47). The crucial thing for Freud is to establish whether or not unicellular organisms are immortal, and to do this he turns away from Weismann to laboratory work on protozoa.

Although Freud cites experiments by a number of different scientists, he does not, in fact, seem to have read first-hand accounts of their experiments. His main source was a book by Alexander Lipschütz, published in 1914 and called *Warum wir Sterben* (*Why we Die*). Lipschütz describes a range of experiments on protista undertaken in the previous couple of decades by scientists such as Maupas, a biologist working in Paris in the late 1880s, Gary Calkins, Professor of Protozoology at Columbia, and Lorande Woodruff, Professor of Biology at Yale. Émile Maupas in 1888 undertook a series of experiments designed to disprove Weismann's theory of the immortality of the germ cell (and, by implication, any unicellular organism that reproduces by dividing). Maupas found that after only 300 generations, the microscopic animals started to diminish in size and to suffer from various structural abnormalities. Eventually, all the organisms died.[28] Maupas's findings were replicated by a number of other scientists, with one exception. Woodruff found that: 'Paramecium, under favorable environmental conditions, can continue reproducing indefinitely; at least for fourteen years and some ten thousand generations, without conjugation and without any signs of degeneration.'[29] Freud, reporting on Woodruff's experiments, tells us in *Beyond the Pleasure Principle* that: 'This remote descendant of the first slipper-animalcule [Paramecium] was just as lively as its ancestor and showed no signs of ageing or degeneration.' From this, Freud deduced, 'the immortality of the protista seemed to be experimentally demonstrable' (*BPP* 47). At this stage of the argument, if one paid attention only to Woodruff (Maupas, for his part, was adamant that he had, in fact, 'proved the occurrence of "physiological" death in unicellular organisms') biology seemed to contradict the theory of the existence of death instincts.[30]

Why were Woodruff's findings so at odds with the work of all other scientists who had attempted the same experiment? In a 1922 textbook, written in the fifteenth year of his experiment, Woodruff simply stands by his results, noting only that he has observed a periodical 'internal nuclear reorganization' that 'apparently effects a physiological stimulation' similar to that observed after two cells of Paramecium have been allowed to conjugate (that is, to exchange nuclear material—a kind of fertilization without reproduction).[31] Freud duly mentions this finding in his discussion of Woodruff's experiment in *Beyond the Pleasure Principle*, but he is much more interested in a result that Woodruff simply glosses over in the 1922 textbook. According to Freud, Woodruff discovered that if the nutrient fluid was not changed for each new generation, the organisms did not

[28] See Gary Calkins, *Biology*, 1914 (2nd edn.; New York: Holt, 1917), 70 ff. for discussion of Maupas's work.

[29] Woodruff, *Foundations of Biology*, 246–7.

[30] August Weismann, *Essays Upon Heredity and Kindred Biological Problems*, ed. Edward B. Poulton and Arthur E. Shipley, 2 vols. (Oxford: Clarendon, 1892), ii. 201.

[31] Woodruff, *Foundations of Biology*, 247.

fare so well. They weakened, got smaller, and eventually died, 'injured', Freud says, 'by the products of metabolism which they extruded into the surrounding fluid' (*BPP* 48). In other words, they were poisoned by their own waste (an insight, incidentally, that had fuelled sanitary engineering in fast-growing Victorian cities that were on the point of drowning in the excretions of their inhabitants). Freud was fascinated by the idea that 'it was only the products of its *own* metabolism which had fatal results for the particular kind of animalcule'; they 'flourished in a solution which was over-saturated with the waste products of a distantly related species' (*BPP* 48). Even protista, which could be seen to be immortal, were shown in the end to face inevitable destruction as a result of their own—but only their own—vital processes. In other words, biologically, organisms—as Maupas had predicted—were, Woodruff's results notwithstanding, set up to bring about their own demise. So, Freud concludes triumphantly, 'our expectation that biology would flatly contradict the recognition of death instincts has not been fulfilled' (*BPP* 49). Science seemed to have come to his aid once again.

Freud's preoccupations in *Beyond the Pleasure Principle* were thus entirely continuous with those of late nineteenth-century experimental biology, as well as of such texts as *The Picture of Dorian Gray*, which explored the distinction between the living and the dying. The curious replication of Dorian by the portrait whose existence keeps Dorian's death at bay echoes Weismann's account of the amoeba, whose division 'gives rise to two new individuals, as if death and reproduction were the same thing'.[32] Since he is doubled, Dorian cannot die: *Dorian Gray* thus exemplifies the preoccupations of late-Victorian science in ways that have so far largely been overlooked.[33] Long before Wilde imagined an economy in which one body decays while the other is endlessly rejuvenated, celebrated Victorian biologists such as Herbert Spencer had identified the rhythm of living organisms as a vacillation between the processes of life and the processes of death, what he called 'waste and repair'. 'Repair is everywhere and always making up for waste,' Spencer wrote in 1864.[34] Weismann, commenting on the biochemical underpinnings of death in multicellular organisms, noted that: 'In the mature animal, cell-reproduction still goes on, but it no longer exceeds the waste; for some time it just compensates for loss, and then begins to decline.'[35] Gary Calkins noted in 1914 that waste and repair together made up the metabolism: 'destructive metabolism called *katabolism* being the sum of processes concerned with the breaking down or combustion of protoplasmic substances, and constructive metabolism, called *anabolism*, being the sum of processes having to do with repair

[32] Weismann, *Essays Upon Heredity*, 25.

[33] Critical discussions of Wilde's interest in science include the following: Philip E. Smith II, 'Protoplasmic Hierarchy and Philosophical Harmony: Science and Hegelian Aesthetics in Oscar Wilde's Notebooks', in Regenia Gagnier (ed.), *Critical Essays on Oscar Wilde* (New York: Prentice Hall, 1993), 202–9; and Heather Seagroatt, 'Hard Science, Soft Psychology, and Amorphous Art in *The Picture of Dorian Gray*', *Studies in English Literature 1500–1900* 38 (1998), 741–59.

[34] Herbert Spencer, *The Principles of Biology*, 2 vols. (London: Williams & Norgate, 1864), i. 171.

[35] Weismann, *Essays Upon Heredity*, 32.

and growth'.[36] Spencer's original terms were still current in 1922: Woodruff identifies 'the power of waste and repair' as one of the 'diagnostic' properties of all living things.[37] Freud's fascination with the experiments on unicellular organisms was in part an offshoot of his immersion in late nineteenth-century biological accounts of the physiology of life and, by extension, of death. In a sense, the theory of the death instincts is nothing more than a psychologization of concepts that had long been accepted by biologists and explored by writers such as Oscar Wilde. After all, the experiments that Freud cites with so much enthusiasm had taken place several decades before he was writing. By 1920, even if Woodruff's paramecia were still quietly dividing after fourteen years in their Petri dish, science had moved on to other, more pressing, concerns.

So why did Freud bother to cite the work of Weismann, Maupas, and the others? Partly, in the wake of the carnage of the First World War, he was interested in the history of current scientific explanations of mortality. Returning to the experimental work that determined the content of textbooks such as Woodruff's 1922 *Foundations of Biology*, Freud found himself asking again the questions that had shaped late nineteenth-century artistic and scientific culture.

Partly, too, he wanted to make sure that biology did not disprove the existence of a drive towards death, and Lipschütz's book directed him towards exactly the experiments he needed to establish that his theory was, at the very least, not falsified by experimental science. At the beginning of the section of *Beyond the Pleasure Principle* on unicellular organisms, Freud notes that credence in 'the notion of "natural death" is quite foreign to primitive races', and announces his intention to 'turn to biology in order to test the validity of the belief' (*BPP* 45). At first, he implies that biology will, in fact, show that some organisms are immortal: 'we may be astonished to find how little agreement there is among biologists on the subject of natural death and in fact that the whole concept of death melts away under their hands' (*BPP* 45). However, as we have seen, after discussion of all the various experiments he refutes his earlier statement, and concludes that in fact there is such a thing as natural death: 'our expectation that biology would flatly contradict the recognition of death instincts has not been fulfilled' (*BPP* 49) His flirtation with the possibility that unicellular organisms might have the ability to live for ever turns out to have been something of a rhetorical flourish, a staged debate in which he collects, and coolly dismisses, evidence that contradicts his own conclusions. It seems that even as he was writing those pages, Freud knew that he would end up dismissing almost all the results and conclusions he so painstakingly outlined.

It cannot, then, have been merely the results or the facts of the case that interested him. Rather, his imagination was fired by the audacity of the idea that there might be some animals that have managed somehow to cheat death, and, as Oscar Wilde had done, he allowed himself to linger for some time in the fantasy. Many critics have suggested that *Beyond the Pleasure Principle* was Freud's personal

[36] Calkins, *Biology*, 12. [37] Woodruff, *Foundations of Biology*, 10.

response to a series of bereavements that preceded its writing, most importantly, of course, the death of his daughter Sophie.[38] But Freud insisted that a draft of the text was already written at the time of Sophie's death.[39] If his impulse to explore these experiments was personal, that was only part of the story. It seems more likely that, even before Sophie died, he was attracted to the combination of exuberant hypothesis and meticulous observation that was characteristic of the experiments on protista, which managed to be at once empirical and wildly speculative—exactly Freud's own preferred mode. As he notes with satisfaction in the closing pages of *Beyond the Pleasure Principle*:

> it is impossible to pursue an idea of this kind except by repeatedly combining factual material with what is purely speculative and thus diverging widely from empirical observation. The more frequently this is done in the course of constructing a theory, the more untrustworthy, as we know, must be the final result. But the degree of uncertainty is not assignable. One may have made a lucky hit or one may have gone shamefully astray. (*BPP* 59)

The death drive, exactly because it only ever expressed itself obliquely, in games, compulsive repetitions, and aggressive behaviour, was at best an inference from the observable facts, a hypothesis that could be called upon to explain a wide range of disparate phenomena, but whose actual existence would remain for ever in doubt. Freud noted that we continually feel the force of the life instincts, 'emerging as breakers of the peace and constantly producing tensions whose release is felt as pleasure', but the death instincts 'seem to do their work unobtrusively' (*BPP* 63).

Indeed, when Freud tried to imagine how it was that the human psyche might conceptualize death, he was unable to do so. In *The Ego and the Id*, published, as we have seen, three years after *Beyond the Pleasure Principle*, he notes that distinguishing the fear of death from dread of an object or neurotic anxiety 'presents a difficult problem to psycho-analysis, for death is an abstract concept with a negative content for which no unconscious correlative can be found'.[40] In other words, there is no image or memory by which the unconscious can come to know its own mortality. As Virginia Woolf put it, death is the 'one experience I shall never describe'.[41] The speculative theory of the death instincts was possible exactly because we have no language for our own mortality. We die at the limits of our own observable experience; we can describe death only by allowing empiricism to give way to speculation, exactly in the mode of *Beyond the Pleasure Principle* itself. For an absurd moment Freud even allowed himself to imagine that mortality itself was merely speculative. Perhaps, he wondered, 'this belief in the internal necessity of dying is only another of those illusions which we have created "*um die Schwere*

[38] See, for example, Fritz Wittels, *Sigmund Freud: His Personality, His Teaching and His School*, trans. Eden and Cedar Paul (1924; New York: Dodd, Mead & Co., 1924), 231; and Schur, *Freud*, 332 ff.

[39] See Peter Gay, *Freud: A Life for Our Time* (1988; repr. New York: Norton, 1998), 395, for discussion of Freud's energetic denials of Wittels's claims in his 1924 biography.

[40] Freud, *The Ego and the Id*, 58.

[41] *The Diary of Virginia Woolf*, ed. Anne Olivier Bell and Andrew McNeillie, 5 vols. (New York: Harcourt Brace, 1980), iii. 117 (23 November 1926).

des Daseiens zu ertragen" ["to bear the burden of existence"]' (*BPP* 45). It is possible, he suggested, that we close our eyes to evidence of the random and contingent nature of death, and cling to the belief that 'all living substance is bound to die from internal causes' (*BPP* 44) simply in order to console ourselves for losses that might otherwise be unbearable. 'Perhaps', he mused, 'we have adopted the belief because there is some comfort in it. If we are to die ourselves, and first to lose in death those who are dearest to us, it is easier to submit to a remorseless law of nature . . . than to a chance which might perhaps have been escaped' (*BPP* 45). Our belief in death itself may be founded in fear rather than in fact.

The experiments on unicellular organisms, then, were important not so much for what they proved about the mortality of even the most primitive living things, as for the ways in which, like *The Picture of Dorian Gray*, they seemed to open up the possibility that living and dying were somehow interchangeable processes, distinguishable from one another only conceptually. As Weismann noted in 1889, among unicellular organisms which reproduced by splitting in two, 'death and reproduction were the same thing'.[42] The same insight is expressed in slightly different terms in Freud's description, during his elaboration of the theory of death instincts, of the phenomena of life as merely 'circuitous paths to death' (*BPP* 39). What we understand to be the prolongation of life is in fact merely the efforts of the organism to 'die only in its own fashion' (*BPP* 39). The experiments on protista, then, far from drawing Freud into a world dominated by empirical observation and scientific experiment, actually propelled him away from the world of fact into a world of fiction, where language was the only means of distinguishing life from death. When Weismann imagined the amoeba that divided to produce two 'daughters', each of which believed itself to be the 'original parent', he was also imagining a world, like that of *The Picture of Dorian Gray*, in which life-processes were so ambiguous that life or death were simply figures for one another.[43] In other words, the life cycle of unicellular organisms exposed the extent to which even the most basic of oppositions, that between the living and the dead, masked an underlying unity. In that underlying unity lay the possibility of metaphor, and metaphor thus became the founding trope by which life, and hence meaning, came into being.

Freud understood that what he was writing in *Beyond the Pleasure Principle* was as much an exploration of the conditions of possibility of linguistic meaning, as it was an examination of the forces that conspire to inch us along to our deaths. Towards the end of *Beyond the Pleasure Principle*, he assures his readers that they need not be too concerned about the number of 'bewildering and obscure processes' that are described in his 'speculation upon the life and death instincts'. They are bewildering only, he says, because he has been obliged:

> to operate with the scientific terms, that is to say with the figurative language, peculiar to psychology . . . The deficiencies in our description would probably vanish if we were already in a position to replace the psychological terms by physiological or chemical ones.

[42] Weismann, *Essays Upon Heredity*, 25. [43] Ibid. 26.

It is true that they too are only part of a figurative language; but it is one with which we have long been familiar and which is perhaps a simpler one as well. (*BPP* 60)

Where in 1895, at the beginning of his career, he had identified natural sciences as those disciplines that seek 'to represent psychical processes as quantitatively determinate states of specifiable material particles', by 1920 he has reversed his position almost completely.[44] Now, it is figurative language that is the hallmark of scientific knowledge, here the 'figurative language' that is 'peculiar to psychology'. This also should alert us to the fact that he cited the experiments on protista, not because of their associations with empirical, laboratory research, but because they showed scientists acting like artists, wrestling with the relationship between the figurative and factual. Indeed in those experiments the factual dissolved into the figurative, authorizing the figurative as the idiom of scientific, as well as of artistic knowledge.

To suggest that scientific language is scientific exactly in so far as it is figurative was, for Freud, a radical move. He needed to make it in order to support his insistence that psychoanalysis, that most figurative of disciplines, was in its very idiom a natural science *par excellence*. But to say that was also to say that even the most literal of languages—the language of science—was metaphorical. His defence of psychoanalysis as a science was thus also, obliquely, a critique of the referential properties of language, and a critique of the natural sciences' pretensions to transparency and denotation. It was also a quasi-Nietzschean and -Derridean exploration of the centrality of metaphorical relations to the economy of language itself. If the distinction between life and death could be shown to be merely figurative, then the natural sciences themselves—even, or perhaps especially, biology, the science of life—were themselves no more than the endless elaboration of linguistic figure. Psychoanalysis need no longer apologize for the fact that it was not grounded in controlled experimental research, the late nineteenth-century standard for scientific authority.[45] Instead, its very dependence on language, its foundation in metaphor (a hysterical cough is both memory and symptom, for example), comes to seem its scientific badge of honour. In *Beyond the Pleasure Principle* not only did Freud develop his dualistic theory of the instincts, he also offered final proof that psychoanalysis should be considered a science. The triumph of the text was that he did it by calling late nineteenth-century laboratory science as a witness against its own procedures. The careful experiments of Maupas and of Woodruff in the end counted for nothing against the philosophical problems posed by those tiny paramecia, dividing forever in their little Petri dish.

[44] Freud, *Project for a Scientific Psychology*, 298.

[45] See Coleman, *Biology in the Nineteenth Century*, 1–32, for further discussion of this issue.

CHAPTER 8

On Not Being Able to Sleep

JACQUELINE ROSE

> It is essential to assume that there is such a thing as a state of sleep for the inner life.
>
> We are building out into the dark.
>
> (Freud, *The Interpretation of Dreams* (1900))
>
> A man who falls like a stone into his bed night after night, and ceases to live until the moment when he wakes and rises, will surely never dream of making, not even great discoveries, but the merest observations about sleep.
>
> (Proust, *Sodome et Gommorhe* (1922))[1]

It is not easy to think about sleep. Probably because we assume that when we sleep, we relinquish our thinking selves. 'The dream-work', as Freud put it in a famous comment at the end of chapter 6 of *The Interpretation of Dreams*, 'does not think'. Although the whole task of the book was to restore the dignity of the psyche to the dreamer ('the *dignity—die Würde*—of being a process of the psyche' (65; *Stud.* 100)), and although Freud insisted that dream-thoughts are 'entirely reasonable' and 'formed with all the expense of psychical energy we are capable of', he none the less shared with his predecessors the belief that there is something about the dreaming mind that differs radically from the mind awake: 'It is not that [the dream-work] is more negligent, more unreasonable, more forgetful, more incomplete, say, than waking thought; it is qualitatively different from it, and so at first not comparable to it. It does not think, calculate, judge in any way

This chapter was originally presented as a paper at the centenary for *The Interpretation of Dreams*, Vienna, May 2000. Special thanks to the discussion group at Queen Mary College, London University, with whom I read the new translation of *The Interpretation of Dreams* and whose ideas and insights have contributed much to this chapter: Amanda Dackombe, Matt ffytche, Sandra Lahire, Lisa O'Sullivan, Maeve Pearson, Aimee Shalan, and Drew Shaw.

[1] *The Interpretation of Dreams*, trans. Joyce Crick, with an intro. and notes by Ritchie Robertson (Oxford: Oxford University Press, 1999), ch. 7, p. 387 (*Die Traumdeutung. Studienausgabe* (Frankfurt am Main: Fischer Verlag, 1972), 561; hereafter *Stud.*); ibid. 359 (*Stud.* 524). Marcel Proust, *Sodom and Gomorrah. In Search of Lost Time*, trans. C. K. Scott-Moncrieff and Terence Kilmartin and by Andreas Mayor, rev. by D. J. Enright, 6 vols. (London: Chatto & Windus, 1992), iv. 60 (*A la recherche du temps perdu* (1913–27), Pleiade edn., 3 vols. (Paris: Gallimard, 1954), ii. 651).

at all' (329; *Stud.* 486). That difference is attributable in significant measure to sleep. 'At bottom,' Freud adds in a footnote in 1925, 'dreams are nothing other than a peculiar form of thinking made possible by the conditions of the state of sleep.'[2]

Although Freud will crucially identify all the features of the dream-work, not only in symptom formation, but also in jokes and slips, the dream—through sleep—at least partly escapes the mantle of these forms. It breaks the line which Freud—in a gesture that might be seen as the founding gesture of psychoanalysis—runs from the neurotic to the everyday (the 'approximately normal person' as he famously describes himself in the preamble to the specimen dream). Sleep changes everything. It is a special case; or, to put it another way, it is the state of sleep that makes a special case of the dream: 'a mental disorder occurring during sleep';[3] 'a "hallucinatory wishful psychosis" ' to use the more explicit, stronger, formula of 'A Metapsychological Supplement to the Theory of Dreams' (1915).[4] In sleep, our unconscious thoughts take on a hallucinogenic form. It is sleep therefore that brings dreams closer to psychosis than neurosis, closer to the madness with which the first chapter of *The Interpretation of Dreams* comes to an end. For Kant, Freud observes, 'The lunatic is one who dreams while awake' ('the madman is a waking dreamer'), for Schopenhauer 'the dream a brief madness and madness a long dream'.[5]

The central question of this chapter, then, is: what was sleep for Freud? What kind of problems did it present him with, where did it lead? 'I have had little occasion', Freud asserts on the first pages of his book, 'to occupy myself with the problem of sleep, for this is essentially a problem of physiology.' 'So,' he continues, 'in my account I have disregarded the literature on sleep' (9; *Stud.* 33) (the only literature in the magisterial overview of the first chapter that he pays the compliment of leaving alone). Predictably enough, perhaps, given the boldness of that dismissal, sleep returns and, in the final chapter of the work, it more or less occupies centre stage. 'I am', Freud asserts, 'an excellent sleeper' (177; *Stud.* 236). But in the course of this work it seems as if, against the normal order of things, sleep is the one thing that will not let him rest. 'The dream', he states in chapter 1, 'appears as a reaction to everything that is simultaneously present and currently active in the sleeping psyche' (176; *Stud.* 235). The psyche, one could say, never sleeps. For the interpreter of dreams, there is no sleep for sleep. Then, as if exhausted by his own labours, by what he has himself asked of the psyche to redeem it into meaningfulness, he affirms in the final chapter: 'It is essential to assume that there is such a thing as a state of sleep for the inner life' (less a statement than a plea) (387; *Stud.* 561). Because it is so peculiar and challenging to

[2] *The Standard Edition of the Complete Psychological Works of Sigmund Freud*, trans. and ed. James Strachey in collaboration with Anna Freud, 24 vols. (London: Hogarth Press, 1953–74), v. 506 n; *Stud.* 486.

[3] 'An Outline of Psychoanalysis' (1938), *SE* xxiii. 195 (*Gesammelte Werke*, 18 vols., ed. A. Richards and I. Gubrich-Simitis (Frankfurt am Main: S. Fischer, 1987), xvii. 126; hereafter *GW*).

[4] *SE* xiv. 299; *GW* x. 420.

[5] p. 75; *SE* v. 90; *Stud.* 111.

Freud, so agitating as one might say, sleep—I will be suggesting here—can tell us about far more than the dream.

Of all the chapters in *The Interpretation of Dreams*, chapter 7, the most psychological, at first glance therefore the most un-psychoanalytic, seems to have received the least attention. Ernest Jones describes it as the 'most difficult and abstract of Freud's writings', something which belonged for Freud halfway between an obstruction and a whim: 'Freud gets *held up* by the *impulse* to sketch out the essay on general psychology' (my emphasis), 'Evidently the chapter is giving a great deal of trouble.'[6] In June 1899, Freud wrote to Fliess: 'The chapter is becoming more drawn out', and 'will be neither nice nor fruitful'.[7] But although Freud describes the chapter as drudgery in this letter ('a duty'), almost exactly a year before he describes it to Fliess as something he had composed 'as if in a dream'.[8] According to Jones, Freud's daughter Mathilde confirmed that, while he was writing it, he would arrive at family meals 'as if sleep-walking'.[9] In the last chapter of *The Interpretation of Dreams*, it seems that far from making his exit from the work, Freud cannot get out of his dream.

In fact, chapter 7 was one of two blocks—or gaps—impeding completion of the book: 'the gap in the psychology as well as the gap left by the [removal of the] thoroughly analysed sample [dream]'.[10] By association therefore, the block in the psychology and the omission of the hidden details of Freud's life are linked. From the wording, it sounds as if, at one stage of composition, the whole chapter on the dream of Irma's rejection, not just selective details, had been removed as too revealing from the text. Something more intimate or private than abstraction ('the most difficult and abstract of Freud's writings'), something more like a dream ('composed as if in a dream') seems therefore to have been at stake. In Freud's mind at least, theory and intimacy join hands. Certainly the correspondence suggests that, in the course of writing chapter 7, Freud shifted uneasily between dream and reason, from inside to outside his topic. Perhaps then, it is not the abstract nature of the chapter that is the problem, but the opposite—the effort it took Freud to wrest himself clear from the processes he was trying to describe (the heave into abstraction, abstraction as a 'duty' and impossible goal). Instead of psychology acting as aberration for psychoanalysis, a kind of scientist relic, in this case things appear to have worked the other way around. If chapter 7 of *The Interpretation of Dreams* is difficult, it is because it shows us Freud forced back inside the very realm or space that he was attempting to master for the future of his science—the space that the psychoanalyst, unlike the sleeper, could talk about.

At the start of chapter 7 Freud writes: 'we must be clear in our minds: the stretch of our way that makes for easy going lies behind us'. Remarkably proceeding as if

[6] *Sigmund Freud—Life and Work*, 2 vols. (London: Hogarth Press, 1954–7), ii. 392.

[7] 27 June 1899, in *The Complete Letters of Sigmund Freud to Wilhelm Fliess 1887–1904*, trans. and ed. Jeffrey Moussaieff Masson (Cambridge, Mass.: Harvard University Press, 1985), 357.

[8] 20 June 1898, *Complete Letters*, 318.

[9] Jones, *Sigmund Freud*, 394 n.

[10] 23 October 1898, *Complete Letters*, 332.

everything up to this point has been crystal clear—a world of light, enlightenment and full understanding—he continues: 'Until now, if I am not mistaken, all the paths we have trodden have led us into the light, to enlightenment and to full understanding (*ins Lichte, zur Aufklärung und zum vollen Verständnis*), from the moment we propose to go more deeply into the psyche's inner processes of dreaming, all our ways lead into the dark' (331; *Stud.* 489–90). Nothing will be explained here; the realm he is entering into now is unknown: 'It is impossible for us to get as far as *explaining* the dream as psychical process, for explaining means tracing back to what is already known.' He can only proceed 'speculatively', taking care 'not to spin [the new assumptions] too far beyond their first logical links, for if we do, their worth will vanish into uncertainty' (331; *Stud.* 490, emphasis Freud's).

And yet there is something strange about these disclaimers. It is not as if this is the first (or indeed the last) time in the book that uncertainty appears in Freud's thought. Returning, only a few pages on, to the dream of the burning child with which the chapter opens—the one of which he famously comments that it requires no interpretation—Freud writes: 'Indeed, the dream-thoughts we come upon as we interpret cannot in general but remain without closure, spinning out on all sides into the web-like fabric of our thoughts' (this is the dream's 'navel', the place where it plumbs into the 'Unknown') (341; *Stud.* 503). Not knowing, as he also observes in this part of the chapter, is, of course, the condition of analytic work: 'As long as anyone has not decided to let go of the question of certainty when tracing an element from the dream, the analysis will come to a standstill' (336; *Stud.* 495).

Why then does Freud present the inquiry of this chapter as especially vulnerable to uncertainty (the 'gravest uncertainties' as he will repeat in the 'Metapsychological Supplement to the Theory of Dreams' (*SE* 229; *GW* x. 420)); or as if uncertainty, instead of being an indispensable part of the investigation, were an obstacle? 'The disruption of the action at a point of uncertainty' is, in Gillian Beer's words in her recent introduction to Freud's case histories, the hallmark of analytic work.[11] There seems to be more going on here than the often observed tension between the enlightenment and speculative impulses in Freud's thought—less hesitation than dread. Uncertainty appears here like a night-terror, something a child goes to sleep afraid of confronting in the dark: 'all our ways lead into the dark'. As soon as we take sleep as our focus, then fear of the dark, instead of a metaphor for the limits of knowledge, a salutary caution against psychoanalytic knowingness, turns real, becomes—precisely—fear. It traces itself on the mind—like the 'luminous particles and lines of the retina's own light' which, in one of the early accounts of dreaming to which Freud alludes in chapter 1, sketch out the shape of the dream (29–30; *Stud.* 58). When asked by her mother why she

[11] Introduction to Sigmund Freud, *Case Histories: Little Hans, 'The Rat Man', 'The Wolf Man' together with 'Some Character Types Encountered in Psychoanalytic Work'* (London: Penguin, 2002), p. viii.

had not spoken to the doctor of her fear of the dark, the little girl in Winnicott's famous case, *The Piggle*, replies that she had 'packed all the dark away'.[12] Freud, we could say, is up to something similar. What he seems to be trying to do in chapter 7, though he also knows the task to be impossible, is to build bridges out into the dark: 'We are building out into the dark' ('*ins Dunkle hinaus zu bauen*'—the *Standard Edition* translates this 'we are building our way out into the dark' which renders the phrase somehow safer, as if we could be sure the way was ours and that it had been—or at least would be—found) (359; *Stud.*, 524; *SE* 549).

Could it be then that the greatest fear for the analyst is not the fear of not knowing, one loss of omnipotence, but another, more tangible, more physical, the fear of slipping backwards (regression is of course also central to this chapter) of turning—with awesome, hallucinogenic vividness—into a frightened child? This might have implications both for the way Freud tends to be read in the humanities departments of universities, but also for certain schools of analytic thought. On the one hand, it makes it harder to take the crisis of knowledge, divested of physical, childlike actuality, as the key to Freud's thinking, but then nor would it be possible to raise that same child, its lived experience, to the status of true analytic object to be redeemed by insight, since the analyst herself would have to recognize the very movements of interpretative theory as potentially struck dumb and blind—like the sleeper—with her own fear: 'we are paralysed in dreams'.[13] 'It is', Proust writes, 'the image of sleep which sleep itself had projected'—'My eyes blinded, my lips sealed, my limbs fettered, my body naked'—which then, like Giotto's portrait of envy with a serpent in her mouth, takes on monstrous shapes.[14] '*Unrestrained, exaggerated, monstrous*' are the words Freud also uses to describe the activity—the '*productive*' nature—of the dream (70; *Stud.* 105, emphasis Freud's).

Once you start thinking of it like this, then sleep appears less as a metaphor for, more as a pathway into, something else. When we go to sleep, we close our eyes. 'The task which [sleep] assigns to us,' Proust writes, 'we accomplish with our eyes closed.'[15] This is of course at least partly misleading. As Fritz Perls, psychoanalyst turned founder of gestalt therapy, remarked on the subject of insomnia, closing your eyes is no help, since your eyes close as a consequence of—rather than as a means to—falling asleep. Sleep, as all insomniacs know, cannot be willed. It only comes inadvertently. The best way to fall asleep is to think of anything but sleep. As if sleep were, in Proust's words, 'our other master',[16] and falling asleep one of the ways we pay tribute to the unconscious, to the idea of something vital and uncontrollable in our minds. If sleep cannot be willed, crucially we never know what will happen—or where exactly we are going—when we fall asleep.

[12] D. W. Winnicott, *The Piggle—An Account of the Psychoanalytic Treatment of a Little Girl* (London: Hogarth Press, 1977), 49.

[13] Sigmund Freud, *Project for a Scientific Psychology* (1895), *SE* i. 338.

[14] *Guermantes Way. In Search of Lost Time*, iii. 163 (*Le Côté de Guermantes*, Pleiade ii. 146).

[15] *Time Regained. In Search of Lost Time*, vi. 32 (*Le Temps retrouvé*, Pleiade iii. 717).

[16] Ibid.

At the time of writing *The Interpretation of Dreams*, Freud still instructed his patients to close their eyes: 'In order for [the patient] to give all his attention to self-observation', he writes in chapter 2, 'it is helpful for him to lie down and close his eyes' (81; *Stud.* 121). Not quite sending the patient to sleep, analysis aimed to freeze or suspend the patient, impossibly one might say, at sleep's threshold: 'it is a matter of producing in the psyche a condition sharing a certain similarity in distribution of psychical energy (mobile attention) with the condition present just before falling asleep' (82; *Stud.* 122). Four years later, as a footnote in the *Standard Edition* points out, this requirement has been dropped: 'he does not even ask them to close their eyes'.[17] Along with all the other things that psychoanalysis will relinquish—hypnosis: 'a condition sharing a certain similarity . . . with the condition present just before falling asleep (and certainly with the hypnotic state too)', male hysteria, according to the recent argument of Juliet Mitchell[18]—psychoanalysis, one could say, lets go, or will ask its patients to let go, of sleep.

In the famous letter to Fliess of 2 November 1896 about the death of his father that inaugurates *The Interpretation of Dreams*, Freud writes:

> I must tell you about a nice dream I had the other night after the funeral [. . .] I was in a place where I read a sign:
>
> You are requested
> to close the eyes.
>
> The sentence on the sign has a double meaning: one should do one's duty to the dead (an apology as though I had not done it and were in need of leniency), and the actual duty itself.[19]

What happens, we might ask, when someone closes their eyes? Or to what else might someone be alluding, even if unconsciously, when they are talking about sleep? If we run a line from this letter to the last chapter of Freud's text, then one of the things we are joining, one could say, is father to child. 'In my inner self', he writes in the same letter to Fliess, 'the whole past has been reawakened by this event.'[20] The father dies; the child, inside the son, reawakes. Almost too symmetrically, the last chapter begins with the dream of the burning child—lying dead in reality in the adjacent room and whose shroud and arm have caught fire—who awakes, to reproach his father, from death.

Child or father, something awakes, against the odds, in both cases. What are the conditions from which one does, and does not, awaken? 'There has indeed been death', Proust writes, 'as when the heart has ceased to beat and a rhythmical traction of the tongue revives us.'[21] In a move that blurs the image of psycho-

[17] 'Freud's Psycho-Analytic Procedure' (1904), *SE* vii. 250; *GW* v. 5. Cf. also the note to *Studienausgabe*, 'Auf des Schliessen de Augen (ein Überbleibsel der alten hypnotischen Heilmethode) legte Freud schon wenig spater keinen Wert Mehr' (121 n.).

[18] *Mad Men and Medusas—Reclaiming Hysteria and the Effects of Sibling Relations on the Human Condition* (London: Allen Lane, 2000).

[19] *Complete Letters*, 202.

[20] Ibid.

[21] *Guermantes Way*, 94; Pleiade 88.

analysis as a one-way passage from darkness into light, Freud holds himself and the dreamer, Janus-faced, at the point where they divide; or where light, instead of illuminating, fulgurates (the dream 'blazes up in a moment [. . .] like a firework' (377; *Stud.* 549)). What is an awakening? 'According to Delbeouf', Freud writes in the first chapter, 'there is no valid criterion whether something is a dream or waking reality except . . . the sheer fact of awakening' (*der Tatsache des Erwachens*) (45; *Stud.* 77). Freud is once again being more of a philosopher than he thinks (and more of a philosopher than the subsequent evolution of psychoanalysis has allowed). For 'the literature of philosophy' since Freud, writes Donald Meltzer in *Dream-Life*, one of the crucial questions has remained: 'Can we know that we are dreaming?'[22] There is a world of difference, one could say, between a psychoanalysis that sees its task as waking the soul into reason, and a psychoanalysis that does not know, cannot be sure, whether it is itself awake.

How far are we awake when we sleep? 'I am bound to draw the conclusion', Freud writes on our ability to tell ourselves that we are dreaming, '*that, through the entire state of sleep, we know just as certainly that we are dreaming as we know that we are asleep*' (374; *Stud.* 544, emphasis Freud's). And yet, it is one task of the dream-work, that of secondary revision, 'to lull to sleep a certain agency/instance' (*einer gewissen Instanz*) so as to 'reduce the significance of what has just been experienced and help to make what follows bearable' (to allow the dream to continue so that the dreamer can sleep) (318; *Stud.* 470). The ego, in Freud's later formulation, both sleeps and does not sleep: 'The ego is the mental agency . . . which goes to sleep at night, though even then it exercises its censorship on dreams' (works in its sleep, we could say).[23] Freud's attempt to distribute these features between different mental agencies and functions is only partially successful (the second topography repeats the problem in new guise). Like an insomniac, he never stops counting off the parts of the mind. Sleep remains obdurate and indivisible. And ungraspable. 'Our actual awakenings', Proust writes, 'produce an interruption of memory . . . we describe these states of sleep because we no longer remember them'. When we wake, sleep 'races' against us 'to obliterate' its own traces: 'After all these centuries we still know very little about the matter.'[24]

So does the dreamer sleep? In one of Freud's most celebrated formulas, he asserts: 'The dream is *the guardian of sleep, not its disturber*' (*Der Traum is der Wächter des Schlafes, nicht sein Störer*); '*Every successful dream is a fulfilment of this wish*' (the one 'universal, invariably present, and constant wish' (*allgemeine, regelmäzig worhandene und sich gleichbleibende*)) (180, 181; *Stud.* 240, 261; emphases Freud's). And yet, 'the dream never wastes its time on trifles. We do not allow a mere nothing / trifle to disturb our sleep' (*um Geringes lassen wir uns im Schlaf nicht stören*) (140; *Stud.* 195). Sleep is only ever partial: 'What is repressed in the system *Ucs* does not obey the wish to sleep.'[25] 'None of the evidence that Freud

[22] *Dream-Life: A Re-examination of the Psycho-Analytical Theory and Technique* ([Perthshire]: Clunie, 1984), 24.

[23] *The Ego and the Id* (1923), *SE* xix. 17; *GW* xiii. 243.

[24] *Time Regained*, 32; Pleiade 88.

[25] 'Metapsychological Supplement', 225; *GW* 415.

brought to bear in this respect,' argues Meltzer on the first page of *Dream-Life*, 'argues any more strongly for the thesis that dreams are guardians of sleep than that they are its destroyers' (11). In his correspondence with Fliess, Freud self-disparagingly dismisses his conviction that the dream is the guardian of sleep as a 'platitude' or 'commonplace': '*Tant de bruit*' (as if his own concept were the unwelcome noise getting in the way of his sleep).[26] By the time of *An Outline of Psycho-Analysis*, he has qualified it: 'a dream is invariably an *attempt* to get rid of a disturbance of sleep . . . The attempt may succeed more or less completely; it may also fail.'[27] Although sleep is in the service of the ego ('a narcissistic withdrawal of the positions of the libido onto the subject's own self'[28]), it can in extreme cases renounce its own nature, let go of sleep, 'because of its fear of its dreams': 'They did not dare to sleep because they were afraid of their dreams.'[29]

The dream of the burning child is famous, but there is another moment, less commented on, that seems to me to be as important in this context, and that also turns on a child, on waking or not waking, and on what that transition in itself—the passage between the two states ('the condition present just before falling asleep')—might provoke by way of fear. No dream this time, but a vision that summons up therefore even more boldly the hallucinogenic colour of sleep, which is at the heart of chapter 7: a mother wakes in a room to the presence of her uncle, whom she knows to be in a lunatic asylum, and rushing to her child to protect him from the sight, covers him over with a sheet. Behind the vision is a childhood memory of being told by a nursemaid that her mother had suffered convulsions ever since the time her brother had appeared to her disguised as a ghost. Covering her child to protect him from a vision (which is in fact her hallucination), the mother wards off the dreaded identification with the insane uncle, whom her son uncannily resembles, but of course what she has done is shroud her child (repeat the vision and turn him into a ghost). To what lengths will a parent go to protect a child—to protect herself—from fear?

From the father's dream to the mother's vision, an impossible boundary is crossed. Each time, it is the child who is called on to make the journey. Commentaries, such as that of Lacan, which emphasize death as unspeakable in relation to the father's dream,[30] or which more simply draw attention to the death of Freud's father as the instigator of the work, seem to me therefore to miss a dimension that is more or less staring us in the face. They miss, that is, the passage, the state of transition—the psychic no man's land of waking and not waking—where Freud situates the dream. We might be getting closer to understanding why sleep returns in the final chapter of *The Interpretation of Dreams*.

[26] 9 June 1899, *Complete Letters*, 354.

[27] *SE* xxiii. 171; *GW* xvii. 93.

[28] 'On Narcissism' (1914), *SE* xiv. 83; *GW* x. 149.

[29] 'Wish-Fulfilment', *Introductory Lectures*, *SE* xvi. 218; *GW* xi. 224. 'Metapsychological Supplement', 225; *GW* 416.

[30] Jacques Lacan, *Les quatres concepts fondamentaux de la psychanalyse, Le seminaire XI* (Paris: Seuil, 1973), 53–62; *The Four Fundamental Concepts of Psycho-Analysis*, trans. Alain Sheridan, ed. Jacques-Alain Miller (London: Hogarth Press, 1977), 53–64.

Not because of some scientific-cum-psychological digression (or 'impulse'), but because sleep brings with it some of the most important questions Freud has to ask about the limits, the outermost boundaries, of the mind.

A further question then arises that gathers up all the rest. Does the child sleep? If we read back through *The Interpretation of Dreams*, we could then argue that this is, if not the founding, at the very least one of the central questions of the whole work. Father of six children as well as founding father, Freud could not help but be preoccupied, at the most mundane level, by the problem of sleep (even if we have difficulty imagining him getting up in the middle of the night). On 12 June 1900, *The Interpretation of Dreams* finally completed, he writes to Fliess:

> Do you suppose that someday one will read on a marble tablet on this house:
> Here, on July 24, 1895,
> the secret of the dream
> revealed itself to Dr. Sigm. Freud;

in the same letter, he reports on his 8-year old son: 'Ernst has been ill again with a sore throat and fever for four days. His energy is inexhaustible. Even when he has a temperature of 38.5, he still shouts: "One could not possibly feel better; I want to get up." . . . This manic vivacity and wildness sometimes strike me as uncanny, like that of a consumptive.' That children are inexhaustible is of course the commonest parental complaint (the 'uncanny' is Freud's note, heralding the 1919 paper of the same title in which the traces of the child in the mind of the adult will once again be linked to fear). Freud's letter simply inflates both halves—both perversions—of parenting, pushes them to an extreme. Monumentalizing himself for posterity (the plaque is of course in Vienna today), hovering anxiously over his son's vitality as something to boast of and to fear. Parents may try to control, but what parent, one might ask, can *read* the energy of a child?

Even before chapter 7, the question of whether the child sleeps is present in Freud's text. 'We find', he writes with reference to one set of dreams '*the child, with its impulses, living on in the dream*' (147; *Stud.* 203; Freud's emphasis). No wish will make itself felt in the dream, he argues, unless an unconscious infantile wish has attached itself to it. And yet it is not always clear whether Freud means a childhood wish (a wish *of* childhood), or—as seems far more to be the case—the wish *to be* a child. When Freud speaks of infantile wishes, it seems to be above all the wish *for* infancy of which he speaks. Long before he formulates his account of topographical regression, the theory starts, as it were, to regress: 'Paradise is nothing other but the mass fantasy of the childhood of the individual . . . dreams can take us back to this Paradise every night' (188; *Stud.* 250).

If we dream of nakedness it is because children are not ashamed (in the 'Metapsychological Supplement', Freud describes sleep as an 'undressing of the mind', '*eine ganz analoge Entkleidung*'—222; *GW* 412). If we dream of the death of loved ones, it is because children are uninhibited in their murderousness towards rivals. We are not ashamed of our repudiated wishes; more than anything we wish to be uninhibited and unashamed: 'these processes which I have called irrational are not falsifications of normal processes, not intellectual errors, but modes of activity of the psychical apparatus that are free of inhibition' (400; *Stud.* 574). There are, one could almost say, no contents to these wishes (another reason why trying to pin them down to a content feels like something of a wild goose chase), nothing to be achieved, other than the desire to fulfil—as in propel—oneself; to be gloriously unfettered, ruthless, unabashed: 'Children aim quite ruthlessly at . . . satisfaction.' In this the dreamer is like the hysteric who wants nothing so much as to be a naughty child: 'the similarity of what we call the hysterical character with that of a naughty child is really quite striking' (193; *Stud.* 256).

Pitch the child adrift on a sea of plenty, of multitudinous wishes jostling and clamouring for satisfaction, you are already on the wrong track (sentimental or punitive, or both). The image is a decoy, even if the wishes are real, and for the child overwhelming, enough. They all need to be swept up under the adult wish that, when we think of the child in this way, they are being used to conceal. The overwhelming wish of the dreamer is—no more or less than—to wish: 'the first psy. system is utterly incapable of taking anything unpleasant into the context of its thoughts. All the system can do is wish' (396; *Stud.* 570). By the time Freud gets to his account of regression in chapter 7, the ground has therefore been prepared well in advance. Perceptual identity represents the wish as fulfilled. But it is also a way that the system has of remembering itself. Repetition is a wish: 'The repeating of what had been experience in that period is in itself a fulfilment of a wish.'[31] Above all the wish seeks representation, to be actualized, to come to life in the mind. 'Dreaming is a part of the—surmounted—childhood life of the psyche' (370). The *Standard Edition* translates: 'a piece of infantile mental life that has been superseded' (567); in the German 'child-soul/psyche-life' is one word, '*des überwundenen Kinderseelenlebens*' (540). It is the living psyche, the soul, of the child that awakens in the dream.

The image of the child at work in *The Interpretation of Dreams* will of course be subject to dramatic modifications; the whole edifice of infantile sexuality is yet to come. But this might be one of those moments in Freud's work where, precisely because something has not yet been uncovered, because the child is not yet abrim with a sexuality that draws all the attention, that something else is able to slip through the net. Freud is working with a literal idea of a remembered scene: 'a renewal of a visual stimulus that once actually happened, brought alert by the memory of it' (358; *Stud.* 522); (in the letter of 10 March 1898 he distributes dreams, fantasies and psychoneuroses according to what was, in the 'prehistoric' period of 1 to 3, *seen*, *heard*, *sexually experienced* (302)). But those who use this moment to discredit Freud's later views, most famously Jeffrey Masson, can only do so by stressing the scene itself at the expense of the metapsychological questions that accompanied it at this stage of his work. To make it a real scene, or only a real scene, they have to divest it of its hallucinatory quality; lift it clear out of the

[31] 10 March 1898, *Complete Letters*, 302.

visualizing mind. In fact, they do not give the scene *too much* reality, but *not enough* (Masson, not that this perhaps needs to be stated, has missed the point).

The problem, then, is not the arrival of fantasy in Freud's thought, not the loss of a reality no less banal for being cruel (the famous seduction theory of neurosis), but the opposite. It is the loss of the journey that, in chapter 7 of *The Interpretation of Dreams*, Freud sets us on into the darkest, but also most vivid, hallucinatory recesses of the mind. Subsequent psychoanalytic theory will distinguish between hallucinogenic wish-fulfilment and hallucination, will insist on the distinction between inside and outside the mind (Meltzer: 'the differentiation between internal world phenomena and external ones is a *sine qua non* of sanity, just as the differentiation between good and bad is central to mental health'[32]), clearer on these matters than Freud, tidying up the dark even more than Freud, one could say. Every night, Freud writes, the adult relives a 'prehistoric' way of being with compelling sensory vividness. This comment could almost (apart from the style which is of course everything) have been lifted out of Proust: 'even in people who do not normally have a visual memory, their earliest childhood recollections are characterized by a sensory vividness that is retained and persists into later years' (357–8; *Stud.* 521).

Remember too that one of the other things to which the body of the sleeper is most fully awake, more finely attuned than the waker, is the innermost vitality—the sights and sounds—of the body itself: 'those agile vegetative powers whose activity is doubled while we sleep';[33] in dreams, 'all the current bodily sensations assume gigantic proportions'.[34] When we dream we are all hypochondriacs, a condition to which Proust was also more than attuned. 'World of sleep—in which our inner consciousness, subordinated to the disturbances of our organs, accelerates the rhythm of the heart or the respiration . . . as soon as we have embarked on the dark current of our own blood.'[35]

'I find it distressing to think', Freud writes on the subject of absurd dreams, 'that many of the premises at the basis of my psychological solution will produce incredulity and laughter once I have published them.' When he tells his patients that impressions from the first or second year of life leave 'lasting traces in their emotional life', they are, he says, in the habit of 'parodying this newly acquired insight' by declaring that they are 'ready to look for memories from the time when they were *not yet in the land of the living*' (291–2; *Stud.* 436; emphasis Freud's). As if going along with the absurdity, Freud himself at this stage of his writing referred to the period of infancy re-evoked by the dreamer as 'prehistoric': 'Biologically, dream life seems to me to derive entirely from the residues of the prehistoric period of life (between ages one and three)'.[36] But, in his own words from the same section of his work: 'dreams are often at their most profound when they appear at their craziest' ('*so ist der Traum oft am tiefsinnigsten, wo er am tollsten*

[32] *Dream-Life*, 115.
[33] Proust, *Guermantes Way*, 93; Pleiade 88.
[34] 'Metapsychological Supplement', 223; *GW* 413.
[35] *Sodom and Gomorrah*, 185; Pleiade 760.
[36] 10 March 1898, *Complete Letters*, 302.

erscheint') (285; *Stud.* 429). In their ridicule, Freud's patients may be up to more than they think. For if the part of the mind that travels back ('regressively back transformed', '*regredienter Gedanken ver wandlung*' (357; *Stud.* 521)) is unconscious to us, how can we possibly be sure, when we sleep, where it might take us, just how far back in fact we go?

A passage from Proust, which is I think worth quoting in full, makes the point:

> We possess all our memories but not the faculty of recalling them . . . What then is a memory which we do not recall? Or indeed let us go further. We do not recall our memories of the last thirty years; but we are wholly steeped in them; why then stop short at thirty years, why not extend this previous life back to before our birth? If I do not know a whole section of the memories that are behind me, if they are invisible to me, if I do not have the faculty of calling them to me, how do I know whether in that mass which is unknown to me there may not be some that extend back much further than my human existence? If I can have in me and round me so many memories which I do not remember, this oblivion (a *de facto* oblivion, at least, since I have not the faculty of seeing anything) may extend over a life which I have lived in the body of another man, even on another planet.[37]

Not for the first time outstripping Freud on the same ground, Proust could be said here to be taking him beyond the point he is willing to travel. While giving the mind infinite extension, he also shatters its continuity even more radically than Freud, concluding the passage: 'The being that I shall be after death has no more reason to remember the man I have been since my birth than the latter to remember what I was before it.'[38] Compare this sentence, which Freud added to the last page of *The Interpretation of Dreams* in 1909 on unconscious wishes in their 'most fundamental and truest shape': 'we shall have to remember, no doubt, that psychical reality too has more than one form of existence' (*SE* 620; *Stud.* 587). This curious, ghostly addition—which allows us to envisage a psyche, more like that of Proust, extending, multiplying, dividing itself—is then replaced in 1919 with the following: 'we shall have to conclude, no doubt, that *psychical* reality is a particular form of existence not to be confused with *material* reality' (ibid.). The beauty of the first formula was that the question of the nature of the psyche's reality—concrete, material, infinitely plastic—was left completely in suspense; as indeed was the question of with what it should, or could, be confused. 'Perhaps,' Proust suggests, 'there are other worlds more real than the waking world.'[39]

In *On Not Being Able to Paint*, Marion Milner's autobiographical account of teaching herself to paint, she describes the anxieties aroused by space—space, that is, lifted to the surface of consciousness, actualized and brought to life: 'I remembered a kind of half-waking spatial nightmare of being surrounded by an infinitude of space rushing away in every direction for ever and ever. In the same way, the term "vanishing point" aroused vistas of desolation.' As soon as she tried to think about space, 'the whole sensory foundation of the common sense world

[37] *Sodom and Gomorrah*, 444; Pleiade 985.

[38] Ibid.

[39] *The Captive. In Search of Lost Time*, v. 132; *La Prisonnière*, Pleiade iii. 123.

seemed to be threatened'.[40] Milner is creating an analogy between painting and a waking dream (or nightmare since it involves the fear of being mad). In fact it is less analogy than a recovery, a way of bringing 'dream imaginings up above (inside)' (21). Not a retreat from the adult world, more 'like a search . . . a going back to look for something', a kind of 'uncommon sense', a 'primary madness' which all of us 'have lived through and to which at times we can return' (11, 28). 'The dream', Milner writes, 'comes first', because 'we are not born knowing the difference between thoughts and things' (27).

Milner's account of painting, or rather of what has to be confronted in order for anyone to paint, bears at moments uncanny similarities to everything I have been describing here in relation to *The Interpretation of Dreams*. Her book reads like an artistic or poetic, above all a waking, transfiguration of the same space, something which she has decided to summon, or grasp in her hands. Painting, like the dream, draws on the part of the mind that feels like a body. It threatens the same loss of the waking self. A picture, Cézanne wrote, 'is an abyss in which the eye is lost'; 'All these tones', he asks, 'circulate in the blood, don't they?' (25) (compare Proust: 'as soon as we have embarked on the dark current of our own blood'). By closing her eyes, Milner generates a 'meeting which destroyed neither the dark possibilities of colour nor dimmed the light of consciousness' (the possibility of colour on the inner eye is 'dark', only consciousness lays claim, without hesitation or scruple, to the light). In her Postscript, Milner describes the experience as ecstatic, a 'blissful surrender', a harking back to the 'all-out body giving of infancy'. But in the main text what she describes is more like a journey through her own fears: 'of embracing, becoming one with, something infinitely suffering, fears of plunging into a sea of pain' (25). Compare, again, Proust on sleep:

> As soon as I succeeded in falling asleep, at that more truthful hour when my eyes closed on the things of the outer world, the world of sleep . . . reflected, refracted the painful synthesis of survival and annihilation, in the organic and now translucent depths of the mysteriously lighted viscera.[41]

'Sleep', writes Proust, is 'the only source of invention'.[42] Freud was famously reticent about artistic invention, but on the earliest pages of his work, citing Egger, he allows that we all become artists in the very process of retrieving, retranscribing, and transforming our remembered dreams: '*on devient artiste à son insu*' (Freud citing Egger (41; *Stud.* 71)). On the very last pages of *The Interpretation of Dreams*, he asks: 'Do not the unconscious impulses revealed in dreams possess the value of real forces in our inner life? Is the ethical significance of our suppressed wishes to be treated as an unconsidered trifle (*gering anzuschlagen*), for as they create dreams, they may one day create other things?' (411; *Stud.* 587). (Remember: 'We do not allow a mere trifle to disturb our sleep', '*um Geringes lassen wir uns im Schlaf nicht stören*'). Freud is attempting to absolve the dreamer

40 (London: Heinemann, 1950), 12,
41 *Sodom and Gomorrah* 760; Pleiade 184–5.
42 *The Captive*, 133; Pleiade 124.

of her moral anxieties; attempting too, no doubt, to dissuade his readers from retreating appalled from the unconscious desires he claims to have uncovered in the dream. But he is also, it seems to me, saying something else. That although dreams are not prophetic, they are generative, forward-looking, not in the predictable but in the unpredictable sense. Precisely because they lead us back into the deepest recesses of the psyche—to the point where for Proust the psyche goes beyond the psyche—they lead us forward into something else.

The only quote on the back of the new English translation of Freud's work (not that Freud needs a publicity blurb) is these lines from a letter of 1899 which wonderfully convey how little is elucidated or settled—all paths are open—even if you think you have left the dark behind.

> The whole thing is planned on the model of an imaginary walk. First comes the dark wood of the authorities (who cannot see the trees). Then there is a cavernous defile through which I lead my readers . . . and then, all at once, the high ground and the open prospect and the question: 'Which way do you want to go?'

Things can gain from being unclear. In his last introductory lecture on dreams Freud cites the Chinese proverb: 'Little what see much what wonderful' to make the point.[43] Much of Milner's book is a critique of present-day educational methods, which she believes are stultifying the creative energies of the child. Likewise Freud seems to be saying, in a rare comment on creativity in that last plea of *The Interpretation of Dreams*, that ethical piety in recoil from the night is killing. It pulls us away from the world of the dream, keeps us awake, and stops us from being able to paint.

[43] 'Uncertainties and Criticisms', Lecture XV, *Introductory Lectures on Psycho-analysis*, Part 2, *Dreams*, *SE* xv. 231; *GW* 238.

CHAPTER 9

'Brownie' Sharpe and the Stuff of Dreams

MARY JACOBUS

> The dream is the matrix from which art is developed. 'We are such stuff as dreams are made of.'
>
> Analytic technique is an applied art and as in all art its principles are conditioned by the limitations of its medium.
>
> (Ella Freeman Sharpe, *Dream Analysis* (1937))[1]

What is the medium of dream analysis? And what is the 'stuff' of dreams? One answer, at least since *The Interpretation of Dreams*, would be metaphor. But there are others. For Winnicott, it might be a holding medium, such as oil, which helps the wheels of the dream to go round; for Bion, alpha-function, making raw sensations and perceptions available for dreaming; or for André Green, negative states trying to accede to symbolization.[2] Formulations such as these—whether visceral or highly abstract—offer ways to imagine the always incomplete transformation

[1] Ella Freeman Sharpe, *Dream Analysis: A Practical Handbook for Psycho-Analysts* (1937; rpt. London: Karnac, 1988), 59, 124; subsequent references in the text (cited as *DA*) are to this edition. Sharpe's book was given as a series of lectures to students in training at the London Institute of Psycho-Analysis during the 1930s. The Shakespearian reference is to *The Tempest*, IV. i. 156–7.

[2] See, for instance, D. W. Winnicott, 'Withdrawal and Regression' (1954), *Through Pediatrics to Psycho-Analysis* (1958; rpt. New York: Bruner Mazel, 1992), 256–7: Winnicott tells his patient, 'you imply the existence of a medium' to which his patient replies: 'Like the oil in which wheels move'. For the friendly spaces of intrauterine existence, see also Michael Balint, 'Flying Dreams and the Dream Screen', *Thrills and Regressions* (New York: International Universities Press, 1959), 75–6. For Bion's views on dreaming, alpha-function, and the contact barrier, see, for instance, W. R. Bion, *Learning from Experience* (1962; rpt. London: Karnac, 1988), 6–7: 'As alpha-function makes the sense impressions of the emotional experience available for conscious and dream-thought, the patient who cannot dream cannot go to sleep and cannot wake up' (ibid. 7); see also ibid. 24–7 for the contact barrier, and Gérard Bléandonu, *Wilfred Bion: His Life and Works 1897–1979* (London: Free Association Books, 1994), esp. 151–5. See also André Green, 'Negation and Contradiction', *On Private Madness* (Madison, Conn.: International Universities Press, 1986), 254–76, and 'The Intuition of the Negative in Playing and Reality', *The Dead Mother: The Work of André Green*, ed. Gregorio Kohon (London: Routledge, 1999), 205–21. The central preoccupations of André Green's work are well expressed by Adam Phillips: 'Dreams and affect, and states of emptiness or absence have been the essential perplexities of Green's work because they are the areas of experience (or anti-experience) in which the nature of representation itself is put at risk'; see 'André Green and the Pragmatics of Passion', ibid., 165.

of bodily experience and affect into unconscious phantasy and dreaming. I want to explore the part played by the body and its metaphors in Ella Freeman Sharpe's psychoanalytic writing, focusing on her 1937 sequel to Freud's dream-book, *Dream Analysis: A Practical Handbook for Psycho-Analysts.* Donald Meltzer credits Sharpe's *Dream Analysis* with taking such 'a quietly divergent view from Freud's . . . that hardly any notice has been taken of it'.[3] Perhaps this is because of its practicality. But an intriguing cluster of associations suggests that 'Brownie' Sharpe (as she was affectionately known to her friends and colleagues) was more than the first practical critic of psychoanalysis. A former teacher of English literature, she practises an art that is 'applied' in the sense of bringing the techniques of literary-critical analysis to bear on the poetics of dreaming.[4] However, she thinks of her medium in ways that go well beyond the formal or tropological analysis of poetic diction for which she is often cited (simile, metaphor, metonomy, and so on).[5] For her, the language of poetry and dreams is always concrete, and corporeal. Metaphor may be what we get in exchange for ceasing to express emotions through the body, but the origin of metaphor in forgotten bodily experience leaves its visceral imprint on dream-language.

In his introduction to the 1977 reprint of *Dream Analysis,* Masud Khan praises Sharpe both for taking on the grammatical legacy of Freud's dream book ('the grammar of how man dreams') and for anticipating Lacan's '*l'inconscient est structuré comme un langage*' (*DA* 10, 9).[6] But André Green, among others, has taken issue with Lacan, objecting that the Freudian unconscious is structured not by (or 'like') a language but by thing-presentations; affect—'the signifier of the flesh'—both seeks and resists linguistic representation.[7] Khan himself writes

[3] See Donald Meltzer, *Dream-Life: A Re-examination of the Psycho-Analytical Theory and Technique* ([Perthshire]: Clunies Press, 1983), 27. Meltzer fruitfully suggests that Ella Sharpe's work 'builds a bridge to the field of aesthetics in general' (ibid. 113), as well as to an investigation of the dream as an aesthetic object.

[4] Sharpe taught at the Hucknell Pupil Teachers Training Centre for boys and girls aged 15 to 18 years old from 1904 until 1916. In 1917, she sought psychoanalytic help for anxiety and depression and began analysis with Jesse Murray and James Glover. She went to Berlin in 1920 for a period of analysis with Hans Sachs, continuing with him during summers for several years. She became a full member of the British Psychoanalytic Society in 1923. She is said by Sylvia Payne to have undertaken more training analyses than any other analyst in England. At the time of her death, she was writing a novel. For an overview of Sharpe's work, see Carol Netzer, 'Annals of Psychoanalysis: Ella Freeman Sharpe', *Psychoanalytic Review* 69 (1982), 207–19; and see also Payne's obituary, *International Journal of Psychoanalysis*, 28 (1923), 54–6.

[5] See, for instance, Meltzer, *Dream-Life*, 27: 'Her central creative contribution to the theory of dreams was to point out the mountains of evidence that dreams utilize what she chose to call the "poetic diction" of lyric poetry. By this she meant that dreams imply the many devices of simile, metaphor, alliteration, onomatopoeia etc. by which the language of poetry achieves its evocative capacity.'

[6] Khan was Sharpe's last analysand; see *DA* 9.

[7] 'Lacan is saying the unconscious is structured like a language . . . [But] when you read Freud, it is obvious that this proposition doesn't work for a minute. Freud very clearly opposes the unconscious (which he says is constituted by thing-presentations and nothing else) to the pre-conscious. What is related to language can only belong to the pre-conscious; Lacan tried to defend himself by emphasizing that he said "like a language", but that's not true. In other instances he says the unconscious is language'; see 'The Greening of Psychoanalysis: André Green in dialogues with Gregorio Kohon', Kohon (ed.), *The Dead Mother*, 24. On Green and affect, see also Phillips, ibid. 163–72.

revealingly of Sharpe's recognition that metaphor 'is a *collage* of mind and body' and insists on her 'uncanny' knowledge that 'all language is born of the body' (*DA* 9, 10). His own metaphors suggest that Sharpe troubles a strictly Lacanian account of the language of the unconscious. While *collage* implies the layering of material surfaces, her knowledge is 'uncanny' in revealing what ought to remain hidden (for instance, a female birth-body). Sharpe's writing, in fact, frequently represents dream-thoughts and phantasies concretely in terms of their manufacture, and even as a woven texture—literally, as textile—rather than as a linguistic text. Nor is the dream-text, for her, necessarily cut from the same fabric as the literary text. Analytic technique, she writes, has to take into account the individual's relation to the environment if it is to be 'a subtler instrument than the yardstick which measures every type of cloth' (*DA* 124). Her technique in *Dream Analysis* is attuned to the ideolect of the dreamer and its manifestation 'in individual settings'.[8] This is the expressive medium in which she works, a medium as material as 'Oils and water colours, clay and stone, violin and piano, lyric and novel' (*DA* 124).

After Sharpe's death, a colleague recalled that she treated her patients 'as if she were handling some fragile piece of pottery'—hinting at the concreteness of Sharpe's psychoanalytic world as well as her careful handling.[9] Even the analytic setting, as Sharpe evokes it, has a unique texture to which her analysands bring their own tastes, styles, and emotions. Every analysand is aware that the room in which the analyst works is charged with as much phantasy and feeling as exchanges with the analyst herself. Sharpe's view of analysis as an applied art emphasizes its situatedness, along with its materiality. But at the same time—cannily—she asks questions about the corporeality of metaphor. If metaphor is born of bodily experience, as she proposes, is this because all language bears the traces of affect and the drives (our own or others')? Or is it merely by virtue of the pervasiveness and persuasiveness of the corporeal analogy itself? Sharpe's body-based model makes bodily continence the price paid for language, and language itself a form of materiality. Her 1940 essay on 'Psycho-Physical Problems Revealed in Language' argues (however improbably) that metaphor arises when the infant achieves control over its messy products and bodily orifices. Instead of faecal matter, the child produces language ('the immaterial expresses itself in terms of the material'). Motions become emotions when they take the form of verbal representations. Sharpe posits a mysterious 'subterranean passage between mind

[8] The same attunement to the nuances of her analysands' communicative styles distinguished her as a practitioner; see, for instance, the former analysand cited by Netzer: 'The most striking impression to me was Ella Sharpe's great sensitiveness to the nuances of verbal and non-verbal expression, and particularly to the implications hidden in the use of words and phrases which characterize most clichés' ('Annals of Psychoanalysis', 207). For a less sympathetic view, however, see Margaret Little's account of her disastrous analysis with Sharpe in 'Psychotherapy with Ella Freeman Sharpe, 1940–1947', *Psychotic Anxieties and Containment: A Personal Record of an Analysis with Winnicott* (London: Jason Aronson, 1990), 31–8.

[9] Quoted by Wahl, in Franz Alexander (ed.), *Psychoanalytic Pioneers* (New York: Basic Books, 1966), 267.

and body [that] underlies all analogy', and defines speech as an 'avenue of "outerance" present from birth'. Her coinage, 'outer-ance', equates the achievement of sphincter control with linguistic discharge: 'The activity of speaking is substituted for physical activity now restricted at other openings of the body, while words themselves become the very substitutes for the bodily substances.'[10] Words, then, have the qualities of substitute substances; they can function as projectiles, weapons, gifts, or magical performances. Speech is metaphorical 'outer-ance'.[11]

Sharpe's account of utterance as the unmediated substitution of words for bodily substances short-circuits what Lacan, in his seminar on 'The Object and the Thing' in *The Ethics of Psychoanalysis*, calls 'the problem of sublimation'. Lacan is briskly dismissive of a Kleinian aesthetics of reparation involving the maternal body.[12] He is equally contemptuous of the 'puerile results' yielded by Sharpe's papers on sublimation.[13] Although Sharpe calls sublimation 'the very woof and weft of civilization', she (by contrast) tends to see it as magical, primitive, and delusory.[14] Still, Lacan is intrigued that her paper, 'Certain Aspects of Sublimation and Delusion' (1930), should site the first artistic production in an underground cave. For Lacan, cave-painting defines the elusive 'place of the Thing' as a 'construction around emptiness'—a way of mastering what lies beyond signification via 'the figuration of emptiness' and painting's mastery of the illusion of space.[15] Sharpe herself vividly compares cave-painting to the darkened theatre of the 1930s 'moving picture', and (like other modernists such as

[10] 'Psycho-Physical Problems Revealed in Language: An Examination of Metaphor' (1940), in *Ella Freeman Sharpe, Collected Papers on Psycho-Analysis*, ed. Marjorie Brierley (London: Hogarth Press, 1950), 156–7. For a perceptive recent discussion of Sharpe's writing on aesthetics, figuration, and incorporation, see also Lyndsey Stonebridge, *The Destructive Element: British Psychoanalysis and Modernism* (London: Macmillan, 1998), 85–93.

[11] 'So we may say speech in itself is a metaphor, that metaphor is as ultimate as speech' (ibid. 157); Sharpe is paraphrasing John Middleton Murry here: 'Metaphor is as ultimate as speech itself, and speech as ultimate as thought' (ibid. 155 and n.).

[12] 'I can tell you right away that the reduction of the notion of sublimation to a restitutive effort of the subject relative to the injured body of the mother is certainly not the best solution to the problem of sublimation'; Jacques Lacan, 'The Object and the Thing', *The Ethics of Psychoanalysis 1959–1960: The Seminar of Jacques Lacan* (Book VII), ed. Jacques-Alain Miller, trans. Dennis Porter (London: Routledge, 1992), 106. Lacan returns to the problem of Klein, sublimation, reparation, and aesthetics in the next seminar, 'On Creation ex nihilo', referring to Klein's account of the artist Ruth Kjar; see ibid. 116–17.

[13] Ibid. 107.

[14] 'If for us the idea of the dead is freed from the cruder superstitions and fears of past ages, it is because we are phalanxed right and left, behind and before, by a magical nullification of fear in sublimation that is the very woof and weft of civilization'; see 'Certain Aspects of Sublimation and Delusion' (1930), *Collected Papers on Psycho-Analysis*, 136. Sharpe claims, among other things, that art 'springs from the same root as the delusion of persecution' (ibid. 131); the 'magical performance' of dancing is her main example of an omnipotent phantasy of incorporation of, and primal identification with, parental power.

[15] *The Ethics of Psychoanalysis*, 139–40. For a succinct account of 'the Thing' in Lacan's seminar of 1959–60, see Dylan Evans, *Dictionary of Lacanian Psychoanalysis* (London: Routledge, 1996), 204–5; 'the Thing' is both the beyond of the signified and the forbidden object of sexual desire. The same entry takes issue with those (such as André Green) who question the linguistic aspect of the unconscious, emphasizing Lacan's distinction between *das Ding* and *die Sache* in Freud's account of thing-presentation (*Sachvorstellungen*).

H. D. and Bryher) celebrates the cinema, calling it 'the most satisfying illusion the world has ever known'.[16] Here Lacan may after all point the way to a rereading of Sharpe's body-based theory of signification. Sharpe connects the illusion of the movies—'The great figures will move and live . . . as they did even in life'—with the cave-artist painting his bison; for her, both film and painting aim to 'reconstruct . . . life that has passed away'.[17] The moving image stands between us and the absent object. For all its focus on poetic language and metaphor, *Dream Analysis* is underpinned by Sharpe's recurrent appeal to technologies of material production and visual projection. In this respect, her writing anticipates Lewin's dream screen, since the late 1940s one of the organizing metaphors of contemporary dream-theory. However problematic, Sharpe's emphasis on the concreteness of psychic representation—what she calls 'concrete image thinking' (*DA* 58)—gives a distinctively modernist inflection to her aesthetics. This is the aspect of her work that I want to explore. The tendency of Sharpe's writing is to modify the prevailing assumption that concreteness (if not concrete thinking) is antipathetic to symbolization. Indeed, her work implies that some degree of concreteness is foundational to all symbol-formation, even if in the last resort symbolization also involves the mourning and linguistic letting-go of the object that Kleinian aesthetics associates with the depressive position.[18] This bodily understanding of metaphor informs her reading of the dream text.

'*REALLY*, MISS SHARPE'

Sharpe's 1940 essay on metaphor ends by demonstrating how an apparently meaningless word—'really'—points to the core of a serious illness. Her patient, a

[16] *Collected Papers on Psycho-Analysis*, 136. H. D. and Bryher, like Sharpe before them, also became analysands of Sachs later in the 1920s. For their involvement in the cinema, and their contributions to the journal *Close Up*, see *Close Up 1927–1933: Cinema and Modernism*, ed. James Donald, Anne Friedberg, and Laura Marcus (Princeton, NJ: Princeton University Press, 1998); for the links between cinema and psychoanalysis, as well as Sachs's collaboration with Pabst, see especially Laura Marcus, ibid. 240–6 (and, for Sachs's contributions to *Close Up*, see ibid. 254–6, 262–7). Virginia Woolf also makes the link between cinema and primitive experience in her 1926 essay 'The Cinema'; see Rachel Bowlby (ed.), *The Crowded Dance of Modern Life* (Harmondsworth: Penguin, 1993), 54–8; Barbara Low, another British analyst who was Sachs's analysand, regards cinema as similarly connected with primitive thinking and magical wish-fulfilment; see 'Mind-Growth or Mind-Mechanization? The Cinema in Education' (1927), reprinted in Donald, Friedberg, and Marcus (eds.), *Close Up 1927–1933*, 247–50.

[17] Ibid. 136, 125.

[18] For a brief statement of Kleinian aesthetics in relation to symbolization and the depressive position, and the relation between 'concrete thinking' and the so-called 'symbolic equation', see Hanna Segal, 'The Function of Dreams', in Sara Flanders (ed.), *The Dream Discourse Today* (London: Routledge, 1993), 101: 'Briefly stated, the theory is that when projective identification is in ascendance and the ego is identified and confused with the object, then the symbol, a creation of the ego, becomes identified and confused with the thing symbolized. Only when separation and separateness are accepted and worked through does the symbol become a representation of the object, rather than being equated with the object.' Cf. also Cecily de Monchaux, 'Dreaming and the Organizing Function of the Ego', ibid. 207: 'De-differentiation between the thing symbolized and the symbol is the basis for the concrete thinking of the schizophrenic; for the cathexis of words . . . as if they were objects.'

poet and translator, 'was the last person *really* to use meaningless words, since they were the stuff of imagination for him.' Studying his use of 'really', Sharpe observes: 'Whenever he was surprised into saying something critical about me, my belongings, or the analysis, he put up his hands in a beseeching way and said in an apologetic deprecating voice: "*Really,* Miss Sharpe . . .".' Sharpe comes to understand that this deprecating 'really' represents an underlying infantile situation, repeated in the transference, involving criticism, anger, and fear. These emotions reproduce others originally aroused, for instance, by 'the appearance of a new baby, the awareness of parental intercourse . . . the sight of the female genital, the sight of menstrual blood', and even his own emotions; as in 'Another baby, really?—Really!' or 'Made by father and mother, really?—Really!' or 'A person without a penis, really?—Really!', or even 'I feel like killing, really?—Really!' In this patient's idiolect, 'Really!' signifies disavowal of his perceptions, thoughts, and feelings: 'I see these things, know these things, but they are not real; I feel like this, but I mustn't feel like this, not really.'[19] Both external and psychical reality are negated by this unconsciously freighted expression of disbelief.

I want to notice another apparently innocuous word—so neutral as to seem equally meaningless—that recurs in Sharpe's own idiolect. This word is 'material', as in: what the analysand gives to the analyst; the raw material of art, drama, psychoanalysis; everything that is kept in the storehouse of memory and experience; language itself. How material is 'the material' for Sharpe, *really*? What does the stuff of reality (and imagination) consist of for her? Among the early meanings of 'stuff', in the sense of 'equipment, stores, stock', is both an auxiliary force or reinforcement and a defensive padding used under or in place of armour (other meanings include an army's stores, baggage, provisions, moveable property, and furniture; *OED*). Sharpe's 'stuff' could be equated with 'supplement' or even 'defence', as well as the everyday baggage and furnishings of our lives. In another and more familiar sense, the word 'stuff' includes the idea of a substance to be wrought or matter of composition (building materials; what persons—or dreams—are made of; and even 'material for literary elaboration'). Finally, 'stuff' in a more specific sense has the meaning of woven material, a textile fabric; woollen cloth. A pervasive feature of Sharpe's own usage is the way in which the 'stuff' of dreams gets tugged towards a specific association with woven material. Paradoxically, this form of materiality also lends itself to literary elaboration, including the fiction that we could really finger or even munch on the texture of metaphor. If dreams, by a submerged pun, are the 'matter' as well as *mater* (the matrix) of art, Sharpe imagines the stuff of which they are made as a homespun forerunner of what we know today, in the wake of Lewin's dream screen, as the dream envelope—the name Didier Anzieu gives to the visual dream film, the fine, ephemeral membrane that replaces the tactile envelope of the ego's vulnerable skin under the stress of daytime and instinctual excitation.[20]

[19] *Collected Papers on Psycho-Analysis*, 167–8.

[20] See Didier Anzieu, 'The Film of the Dream', rpt. in Flanders (ed.), *The Dream Discourse Today*, 137–50.

Anzieu's metaphor for the repair-work of dreams is weaving; whereas Penelope unravels at night the 'shroud' she weaves by day, 'The nocturnal dream . . . re-weaves by night those parts of the skin ego that have become unravelled by day.'[21] Early on in *Dream Analysis*, in a chapter called 'The Mechanisms of Dream Formation', Sharpe establishes weaving as a symbolic representation of the primal scene. A woman patient has an infantile memory of visiting the country; she and her parents slept in an old-fashioned four-poster bed: 'The big four-square loom was equated with a bed, the flying shuttle with the penis, the thread with semen, the making of the material from the thread with a child' (*DA* 55–6).[22] Subsequently, the silk-weaving looms of this rural district gave Sharpe's patient a chance to see 'the swift flinging of the shuttle', and the weaver at his work took over in phantasy from the scenes witnessed by the small child in the four-poster bed. Local silk-weaving becomes a screen memory for the scene of parental intercourse, whether witnessed or phantasized. For this patient, thread, silk, cotton, and string—milk, water, and semen—are at once the material (the matrix) of dreams and unconscious phantasy, and what goes into making a baby, symbolically represented as 'woven material'.[23] Indeed, Sharpe notes: 'There was not an operation relevant to the work of weaving that did not appear as symbolic of unconscious phantasies'—and also, one might add, of the production of unconscious phantasy itself. For Sharpe, the work of weaving stages the primal scene of the dream's manufacture. Her metaphor for the dream-space is the bed-loom of a home-based technology (entirely in keeping with a modernist emphasis on the homemade crafting of objects).[24] Pontalis—who uses the phrase 'dream machine' for this homely mode of production—suggests that the dream also fosters an illusion. This illusion consists 'of being able to reach that mythical place where nothing is disjointed: where the real is imaginary and the imaginary real, where the word is a thing, the body a soul, simultaneously body-matrix and body phallus'.[25] For Sharpe's patient, and perhaps for Sharpe herself, the place of the mythical rejoining of body-matrix and body-phallus is an old-fashioned four-poster bed.

The loom's association with the arts-and-crafts movement, as well as modernist modes of production, makes it an apt figure for Sharpe's own applied

[21] Ibid. 141.

[22] The word 'loom', as Sharpe would have known, once meant a tool or implement, and only later took on the meaning of an instrument for weaving; see George Willis, *The Philosophy of Speech* (New York: Macmillan, 1920), 56. Sharpe cites Willis ('those bridges of thought which are crossed and recrossed by names in their manifold mutations', ibid. 55) on the importance of naming; see *DA* 29, 39.

[23] See n. 30 below for a series of dreams linked by this 'all-dominating theme of thread' (*DA* 182).

[24] See the account of modernism offered by Douglas Mao in *Solid Objects: Modernism and the Test of Production* (Princeton, NJ: Princeton University Press, 1998). Mao argues for 'a variety of ways in which modernists were affected by certain images of production—above all, the image of the individual maker crafting the individual object' (12). See also Hugh Kenner, *A Homemade World: The American Modernist Writers* (New York: Knopf, 1975).

[25] J.-B. Pontalis, 'Dream as an Object', in Flanders (ed.), *The Dream Discourse Today*, 120–1; Pontalis also compares the dream screen to Freud's 'protective shield' (ibid. 119).

art.[26] Sharpe herself is most vividly communicative when it comes to the good-enough-to-eat, mouth-watering woven material that appears in her patients' dream associations. Here is how she describes a woman (presumably the same one) for whom different types of material produce intense bodily sensations: 'an oatmeal coloured material had a "crunchy" feeling', making her teeth tingle, while

> A cherry coloured silk will make her mouth water and she longs to put her cheeks gently on its surface. The range of colours for this patient are in terms of cream, butter, lemon, orange, cherry, peach, damson, wine, plum, nut brown, chestnut brown. Materials can be crunchy like biscuits, soft like beaten white of eggs, thick like cake. Threads can be coarse like the grain of wholemeal bread, shine like the skin of satin. I do not let any reference to colour or material or to dress escape me in the dreams this patient brings. (*DA* 92–3)

(One thinks of the derivation of Lewin's 'dream screen' from the sight and feel of the mother's breast as the well-fed baby falls asleep.)[27] In this rich cornucopia of taste-sensations, Sharpe conveys her own pleasure in colour, surface, and texture. But I want to notice how she signs her writing. When she questions her patient, ' "How do you *feel* a *name* lying about?" ' (*DA* 100), we begin to notice how the feel of a name literally colours the dream-associations of this exceptionally clothes- and colour-conscious patient (Payne's obituary, incidentally, recalls Sharpe herself giving her paper on sublimation in 'a soft brick-red dress' that threw her dark colouring into relief).[28] In an engaging approximation to her patient's mindless stream of consciousness—'Free association means for her a recounting of all that has happened in reality' (*DA* 98)—Sharpe highlights the colour 'brown' until we register its transferential associations with the dark-complexioned 'Brownie' Sharpe herself:

> The flowers in that vase of yours are lovely. The flowers in the other are not as good, the colour of the vase isn't right; it's the wrong brown, but the others are lovely. . . . I'm longing to get on with my jumper. I'd like it finished at once. I've a new stitch to do. I want to see how the jumper looks when it's made. The trouble is about wool. Strange how shops don't stock the right brown, the brown I want. You see the wool in made-up things—in jumpers already knitted—but one can't get the actual colour in wool and make it oneself. *I want wool like the dark brown of your cushions.* (*DA* 100; my italics)

And so on. In case we missed it, Sharpe interprets: 'the mother transference within the analysis is shown by reference to the brown colour of the cushion

[26] Visitors to the Freud Museum can see the loom on which Anna wove, sometimes incorporating the hair of the Freuds' chow into her weaving.

[27] See Bernard Lewin, 'Sleep, the Mouth, and the Dream Screen', *Psychoanalytic Quarterly* 15 (1946), 419–34, and 'Reconsideration of the Dream Screen', *Psychoanalytic Quarterly* 22 (1953), 174–99.

[28] This was the dress in which Sharpe gave her paper on 'Sublimation and Delusion' in 1928: 'She was capable of being a great actress . . . On this particular occasion she wore a soft brick-red dress; her dark hair and dark eyes, and rather dark complexion were thrown into relief by the warm colour.' Sylvia Payne goes on to note that 'the paper was not an intellectual communication, but a living thing to which she was giving birth'; see *International Journal of Psychoanalysis* 28 (1923), 55. Elsewhere, Sharpe reads her own name in her patients' submerged play on 'sharp' and 'flats'; for 'Sharpe' she reads 'flat' or 'block of flats' (*DA* 38).

which she wants for herself and to the flowers in the vases . . .' (*DA* 105). For her, this patient's feminine aesthetics—her minute attention to colours, clothes, knitting, and furnishings—are the sign of a phantasy-life strongly under repression. Bion might see it as an approximation to the beta-element screen, a dream-like state that is not a dream, but a confused jumble of sensations and perceptions.[29] The patient's preoccupation with the world of flower-vases, knitting wool, cushions, and, of course, shopping (the locus of modernist aesthetic consumption and consumer desire) presents a particular difficulty for the analyst: 'she is immensely occupied by reality' (*DA* 98).[30] Preoccupation with reality functions here, like '*Really!*', as a defence against unconsciously disturbing thoughts and feelings (for instance, the idea of a sexual link between her parents). As Sharpe writes elsewhere, clearly alluding to the same patient: 'just because she is intensely occupied with reality I must be aware of her denial of reality' (*DA* 152).[31] Sharpe's patient is like Mrs Dalloway without a mind of her own—let alone a room.[32]

The medium of Sharpe's applied art, she tell us, is the displaced and variable manifestation of human emotions and idiolects 'in individual settings'. The brown cushions and flower-filled vases of her consulting room—the analytic 'setting' in a different sense—are brimming with unconscious phantasy. This metaphorical filling provides the material for Sharpe's object-lesson mode of interpretation:

> I said, 'You see the wool you want for your work in jumpers that are already made. It is to be had, people do have it and yet it is inaccessible to you. You want to make something, you want to have what others can have. *I have cushions of the colour you want.*
>
> . . . Then I gathered together the references that indicated an unconscious phantasy concerning the making of babies, the brown wool for jumpers, the putting of things together neatly . . . (*DA* 103; my italics).

One plain, one purl. Sharpe herself neatly knits together feminine desire, envy of the analytic breast, the small child's sudden evacuation of her bowels on a long-forgotten holiday, and her patient's unconscious phantasy about the making of babies: '*I want wool like the dark brown of your cushions.*' It is her patient's 'want' that allows Sharpe to furnish her mind. What does a woman *really* want when she

[29] See *Learning from Experience*, 22, where Bion defines the screen of beta-elements 'as indistinguishable from a confused state and in particular from any one of those confused states which resemble dreaming'; for Bion, the beta-element screen may be designed, among other things, to destroy the analyst's potency, to withhold information, or to prevent common-sense interpretations occurring to the patient, and is apt to elicit a desired counter-transferential response from the analyst (ibid. 23).

[30] See Mao, *Solid Objects*, 40: 'Many readers of *Mrs. Dalloway* . . . have noted that Clarissa's gift extends to that still purer form of consumption known as shopping.'

[31] This patient 'will not believe there was any connection between her parents so long as she blots out real situations of her own life' (*DA* 154).

[32] See Robert Caper, *A Mind of One's Own* (London: Routledge, 1999), esp. 111–26. Ella Sharpe had been Adrian Stephen's analyst from 1926 to 1927; she also refers to and quotes from *A Room of One's Own* (the Manx cat) in her lectures on 'The Technique of Psycho-Analysis' (1930), *Collected Papers on Psycho-Analysis*, 95; see also Elizabeth Abel, *Virginia Woolf and the Fictions of Psychoanalysis* (Chicago: University of Chicago Press, 1989), 16, 19.

knits a new woollen jumper, or shops for wool of a colour she associates with her analyst?—'It is to be had, people do have it, and yet it is inaccessible to you.' Should this be coded as the everyday form of Kleinian envy—or, alternatively, as Lacanian desire? As Lacan puts it mischievously, 'the signifier of the phallus plays a central role beneath a transparent veil'.[33] He makes this well-known remark in the context of his back-handed tribute to none other than Sharpe herself. Sharpe is praised for having given pride of place to literary culture in analytic training, while Lacan deplores her failure to recognize the veiled role of the phallic signifier (he has salmon in mind, not knitting). But it is Lacan himself, in his reading of Freud's dream of the witty butcher's wife, with her unsatisfied desire for smoked salmon (when she really prefers caviar sandwiches), who insists that 'Desire must be taken literally.' The butcher's wife expresses a gratuitous desire—for something she doesn't really want or need. Was there ever a wittier argument for a new sweater?[34]

'OUR PRIVATE INNER CINEMA'

Sharpe's aesthetics are not reparative in the strictly Kleinian sense, although they do turn out to involve the maternal body. What interests Sharpe, however, is the Kleinian dream-theatre. Invoking *The Psycho-Analysis of Children*, she writes: 'as Mrs. Klein has shown, the child plays its dream, develops apprehension, enacts roles' (*DA* 59).[35] Sharpe defines drama much as she defines dreaming, as the overcoming of painful external reality and the playing out of instinctual fears and internal dangers in a manageable form; drama and dreams both have their origin in trauma and unconscious phantasy. Sharpe's reference to the child's theatrical-

[33] Jacques Lacan, 'The Direction of the Treatment and the Principles of its Power', *Écrits: A Selection*, trans. Alan Sheridan (New York: Norton, 1977), 251.

[34] Sharpe later records a series of dreams and nightmares linked by 'this all-dominating theme of thread', each representing a stage towards the unravelling of her patient's ongoing psychic struggle: ' "I found a piece of cotton in my mouth and began to pull it out. After pulling a long time I dared pull no longer for I felt it was attached to some inner organ which might come out with it. I awoke in terror." ' ' "I said to you I only understand the process of introjection by thinking of the muscles which run into and form the eyeball itself." ' ' " . . . dream in which the patient was again taking a hair out of her mouth. It came out quite easily, it was not attached to anything, and no anxiety was felt in the dream" ' (*DA* 182). Sharpe comments that 'The element of "cotton," "thread," "hair" . . . bridged the unconscious phantasy with real experiences from early infancy to late childhood' (*DA* 182). Her term, 'bridged', recalls her quotation from George Willis's *The Philosophy of Speech* (1919), 55: 'those bridges of thought which are crossed and recrossed by names in their manifold mutations'; cf. Sharpe: 'the bridges of thought are crossed and re-crossed by names and names have manifold mutations' (*DA* 29; cf. *DA* 39). Willis also writes that 'language develops along the framework of the real'—a framework described in the concrete terms of Sharpe's dream-loom.

[35] Sharpe is referring to 'The Significance of Early Anxiety Situations in the Development of the Ego', in *The Psycho-Analysis of Children* (New York: Free Press, 1975), 176 as well as to 'The Psychological Foundations of Child Analysis', ibid. 8. In 'A Note on "The Magic of Names" ' (1946), Sharpe is critical of the tyranny of the terms 'good object' and 'bad object' on the grounds of their unconsious appeal to the super-ego and to an unconcious belief in white and black magic—apparently a reaction against Kleinian terminology.

ity and use of personification as a means to master anxiety coincides with her professional interest in, and extensive writings on, Shakespeare, and even her own well-attested gift for dramatic expression.[36] But in *Dream Analysis*, drama slides imperceptibly towards a specifically modern medium—film. Alongside the phantasy of the bed-loom, 'The Mechanism of Dreaming' includes a section on 'dramatization' (the same chapter includes sections on condensation, displacement, and symbolism). Sharpe defines 'dramatization' in cinematic terms as the dreamer's projection of visual images onto an interior screen: 'A film of moving pictures is projected on the screen of our private inner cinema. This dramatization is done predominantly by visual images' (*DA* 58).[37] 'Our private inner cinema' involves what Sharpe calls 'the reversion to concrete image thinking' (*DA* 58). For her, drama and poetic diction—along with psychosis, hallucination, and dreaming—share their reversion to the concrete image. Already a vehicle for modernist fascination with the image, film technology provides a handy projective analogy for dreaming.[38] But Sharpe's cinematic analogy also draws on the material properties of film itself, considered as a medium that combines surface with transparency. The image is the product of a light-diffusing, light-sensitive membrane—Anzieu's 'pellicule'.[39] The cinematic dreams of Sharpe's patients adhere to this permeable, flexible, clinging film-surface.

Sharpe's theory of drama envisages the distribution of different parts of the dramatist throughout the dream-scenario: 'Could we analyse a play in terms of

[36] See, for instance, her essays on *Hamlet* (1929, and unfinished) and 'From *King Lear* to *The Tempest*' (1946) in *Collected Papers on Psycho-Analysis*. Payne writes in her obituary that Sharpe 'knew Shakespeare's plays in the same way as a devoted priest knows his Bible', comparing her to her analyst Hans Sachs. Payne also pays tribute to Sharpe's dramatic skills in delivering papers ('She was capable of being a great actress if other things had not interfered') and in her seminars: 'she used unconsciously her acting gifts and could reproduce a session with a patient in an unique way' (55).

[37] By 'dramatization', Sharpe seems to mean more than Freud's 'dramatizing' of an idea, or 'the transformation of thoughts into situations' (*Standard Edition of the Complete Psychological Works of Sigmund Freud*, trans. and ed. James Strachey in collaboration with Anna Freud, 24 vols. (London: Hogarth Press, 1955–74), iv. 50, v. 653); rather, she accords cinema the status of hallucination ('dreams hallucinate . . . they replace thoughts by hallucinations' (*SE* iv. 50)), emphasizing the pictorial and literal aspects of both dreaming and cinematic visualization. See also L. Saalschutz, 'The Film in its Relation to the Unconscious' (1929), in Donald, Friedberg, and Marcus (eds.), *Close Up 1927–1933*, 256–60, for an emphasis on the regressive aspect of the cinema: 'we may first liken the film generally to the dream process called Regression. The dreamer is usually looking on at the dream enactments as a spectator surveys the stage . . . and this is called Regression by Freud; we call it cinema' (ibid. 258).

[38] See, for instance, Laura Marcus's account of the relation between dream theory and film theory in her introduction to *Sigmund Freud's The Interpretation of Dreams*, ed. Laura Marcus (Manchester: Manchester University Press, 1999), esp. 33–43. Marcus examines the history of psychoanalysis's interest in the cinematic apparatus as well as cinema's interest in the mental apparatus by which the mind represents its unconscious workings to itself. Touching on Lewin's and Anzieu's writing, as well as Sharpe's, Marcus also notices the shift to an emphasis on 'the play of sign on surface' in more recent theoretical writings. For the sustained engagement of psychoanalysis with cinema, see also Laura Marcus, introduction to Donald, Friedberg, and Marcus (eds.), *Close Up 1927–33*, part 6, 'Cinema and Psychoanalysis', 240–6.

[39] See 'The Film of the Dream', ibid. 137: 'the French term "pellicule" designates a fine membrane which protects and envelops certain parts of plant or animal organisms . . . "pellicule" also means the film used in photography; that is the thin layer serving as a base for the sensitive coating that is to receive the impression. A dream is a "pellicule" in both these senses.'

the inner life of the dramatist, we should find the plot and all the characters taking part in it to be . . . projections of himself into imaginary characters' (*DA* 59).[40] Prospero—a magician with the power to create and dissolve airy spectacles and to dispense sleep—becomes the type of the film director: 'We are such stuff | As dreams are made of . . .' Sharpe views dramatic aesthetics in Romantic terms, as the magical resolution of discord ('a unity of creation within which discordances resolve into harmony', *DA* 60). Dreams, she writes, can be 'a kind of abortive drama' and dream-mechanisms are an attempt to make unity out of 'the raw material of conflicting forces'. But when it comes to her use of the cinematic analogy, Sharpe focuses instead on the mechanisms of image production and the materiality (and iterability) of film. In a telling illustration, 'The Mechanisms of Dreaming' enlists cartoon-technology to illustrate how a stereotyped dream symbol becomes 'real and fresh again'. The dream of a train from which the patient and other passengers get out, but '*I never saw them inside, I never saw them get out or get in*' is initially associated by the dreamer with a feeling of boredom. But later in the same session, Sharpe's patient gives an enthusiastic description of Mickey Mouse jumping into a giraffe's mouth: 'The long neck had a series of windows down it, and one could see Mickey all the time, you didn't lose sight of him, you saw him go in, and come out' (*DA* 57). We know how Mickey gets in (through the giraffe's mouth), although how he gets out is left to the imagination. This 'new and exciting world without and within' includes both the discovery of the inside of the human body (the train) and the mysterious process by which babies and symbols are made. But the windows in the giraffe's neck also allude, cinematically, to the serial frames of a film-strip. The stereotypical symbol of the body as a train becomes new and exciting when it animates cartoon culture; moving films revitalize tired old Freudian symbolism.

Dream Analysis posits the primal scene as the privileged site not only for the production of phantasy, but for the infant director's first film. Talkies in the age of mechanical reproduction introduce a new dimension to dreaming. Sharpe's next chapter narrates an anxiety dream: '*A man is acting for the screen. He is to recite certain lines of the play. The photographers and voice recorders are there. At the critical moment the actor forgets his lines. Time and again he makes the attempt with no result. Rolls of film must have been spoilt*' (*DA* 75–6). Sharpe explains: 'The photographers and voice recorders cannot get the actor to perform although they are all assembled for that purpose. He forgets his lines' (*DA* 76). The infantile situation reconstructed on the basis of the dreamer's associations involves a reversal: 'the dreamer was once the onlooker when his parents were "operating" together. The baby was the original photographer and recorder and he stopped his parents in the "act" by noise. *The baby did not forget his lines!*' (*DA* 76; my italics). Sharpe tells us that the one thing the infant onlooker could do by way of interruption was

[40] In this respect, Sharpe's view of dramatization in dreams is similar to the account of fantasy given by Jean Laplanche and J.-B. Pontalis in 'Fantasy and the Origins of Sexuality'; see Victor Burgin, James Donald, and Cora Kaplan (eds.), *Formations of Fantasy* (London: Routledge, 1986), 5–34. See also Sharpe's reading of *The Tempest* in 'From *King Lear* to *The Tempest*'.

'make a mess and a noise that brought the operations to a standstill' (*DA* 77). Readers of 'The History of an Infantile Neurosis' (1918)—Freud's *tour de force* of dream-analysis—will recall that in his reconstruction of the primal phantasy behind the Wolfman's enigmatic dream, the baby similarly interrupts his parents.[41] Here too ' "rolls of film [i.e. "a huge amount of faecal matter"] must have been wasted" ' (*DA* 76). Sharpe's own account of the dream's mechanisms underlines the importance of the cinematic dream screen: 'The modern invention of the screen of the cinema is pressed into service as the appropriate symbol, the screen being the modern external device corresponding to the internal dream picture mechanism' (*DA* 77). But her interpretation of the dream itself recycles Freud's text, reminding us of the iterable quality of film. In Sharpe's cinema, nothing is wasted, nothing spoiled; the analyst does not forget her lines, even if the dreamer prefers visualization to sound.

Film in *Dream Analysis* also clings to and denotes the body, especially the female body. Another of Sharpe's patients, a man whose main problem is his terror of the maternal body, has the following dream of a semi-naked woman: '*I saw a lady who had black stuff round the top of her body covering the breasts and black stuff round her hips hiding her genitals, only the middle part of her body was naked*' (*DA* 107). (Sharpe notes that the dream itself was an anal gift after the weekend.) But the patient quickly turns to what really interests him—a long story whose scientific technicalities take half an hour to tell—which involves an account of the production of physiological illustrations 'drawn by a medical colleague for the purpose of making slides for a lantern' (*DA* 108). The illustrations are reproduced first on large sheets of paper, coloured red, then reduced in scale. But 'when the slides were eventually shown the colour was wrong. Then the patient laughed and said "only if it had been possible to use a black light would those diagrams have shown red on the sheet".' Interpreting this close-up as a cover-up, Sharpe underlines the change of colour: ' "From black we can infer red"—to which her patient replies impatiently: "It can be what bloody colour you like." ' (*DA* 108). His angrily repeated 'bloody' functions for Sharpe as the return of the repressed, the red, raw, and inflamed appearance of the mother's genitals, hugely magnified to a child's eye: 'The expletive is symbolical of the thing ejected at the being of whom he was and is afraid' (*DA* 109)—including, presumably, his female analyst. Language is hurled like a missile against a bit of embodied reality. When the matrix of dreams is no longer a metaphor, but a feared and needed mother ('genitals and breasts as nursing homes'), language takes on fiercely primitive energy. As Bion reminds us, 'bloody' ('By Our Lady') belongs both to the archaic turmoil of the angry infant and to the realm of the sacred.[42] Uncovering a piece of concrete thinking, Sharpe shows how the maternal body gets under the

[41] See also Marcus (ed.), *Sigmund Freud's The Interpretation of Dreams*, 36–7, who similarly notices the resemblance to the primal scene of looking that underlies the Wolf Man's childhood dream.

[42] 'On a Quotation from Freud' (1976), *Clinical Seminars*, 307.

skin-ego of the baby and becomes Lacan's 'Thing'—the prehistoric, unattainable object of incestuous desire.[43]

Sharpe brings *Dream Analysis* to a close with an eloquent and moving 'last dream' (so called by her) that contains an emblem of her recomposed poetics. The penultimate chapter has illustrated the way in which analysed persons are able to 'knit together' (Sharpe's phrase) body-ego and psychical ego (*DA* 195). The psyche finally receives its due; analysis is the technician's art, but 'the new synthesis is brought about by the forces within the psyche itself' (*DA* 199). Now Sharpe links Psyche to Eros and Thanatos. This 'last dream', dreamed by an 83-year-old woman three days before her own death, movingly condenses the vicissitudes and illnesses of a long life into a single image: '*I saw all my sicknesses gathered together and as I looked they were no longer sicknesses but roses and I knew the roses would be planted and they would grow*' (*DA* 200). Sharpe lets us see how symbolization and mourning go together. In her dreamer's bouquet, a sickness is a rose is a rose-bed—imagery at once horticultural, pastoral, and commemorative. Sharpe's 'The Magic of Names', disagrees with the saying that 'a rose by any other name would smell as sweet' on the grounds that 'Poetic words have always an individual significance.'[44] The dream pays tribute not just to the psyche, not just to psychoanalysis, but to the significance of these particular roses—and to the dreamer's individuality too: 'She shared in her old age the interests of youth and any movement that promised fairer and better conditions for mankind in the future appealed alike to her mind and heart, and among these was psycho-analysis.' The dream reveals the knowledge that sustains and consoles her in the face of death. In the words that end *Dream Analysis*, 'It is Eros alone who *knows* that the roses will be planted and grow' (*DA* 201). With this elegantly understated gesture, Sharpe lays her own psychoanalytic knowledge at the feet of Eros: 'Eros . . . *knows*.' Eros, as others have pointed out, is an anagram for 'rose'.[45] Eros is also the nightly visitor who woos Psyche under cover of sleep and darkness and prohibits looking. Recomposing the letters of the rose as an allegorical figure—a verbal representation of thought—requires a momentary fading of the concrete image and a letting go of the object, so that its multilayered associations can bloom (beauty, poetry, life, mortality . . . and love). For Ella Sharpe '*imagiste*', the rose takes on a spectral afterlife when it flourishes as a linguistic symbol.

*

In Sharpe's culturally gendered aesthetics, psychoanalysis is a legacy to be passed on via the maternal line. This 'last dream' represents Sharpe's own dream of psyche as matrix—memorialized as a progressive and psychoanalytically-minded old woman, who still finds comfort in her dreams. *Dream Analysis* ends by

[43] See 'Das Ding', *The Ethics of Psychoanalysis*, 53, 67.

[44] *Collected Papers on Psycho-Analysis*, 107.

[45] See Lyndsey Stonebridge, 'Bombs and Roses: The Writing of Anxiety in Henry Green's *Caught*', *Diacritics* special issue: Trauma and Psychoanalysis, 28: 4 (winter 1998), 25–43.

redefining the matrix of art, not just as a womb or a place of origin, but also a 'medium in which something is "bred", produced or developed' (the *OED* definition). For Sharpe, the matrix *is* the medium. We might recall Pontalis's account of dreaming as 'above all an effort to maintain the impossible union with the mother'.[46] With this impossible union in mind, I want to revisit Sharpe's statement that dream analysis is an art 'conditioned by the limitations of its medium' (*DA* 124). Here are the words with which Sharpe begins *Dream Analysis*: 'Dreaming is . . . a psychical activity inseparable from life itself, for the only dreamless state is death' (*DA* 13). Sharpe's 'last dream' marks the limit before dreamlessness intervenes. Pontalis defines the unplumbed navel of the dream as the vanishing-point of the seeable, and the unseeable as the face of death: 'The dream is the navel of the "seeing-visible" (*voyant-visible*: Merleau Ponty). . . . Death, as we all know, is not something to be looked at in the face.'[47] Sharpe's dreamer allows us to see life as if through the eyes of the still-dreaming dead—as a posthumous after-image on the dreaming eye. The dream makes roses bloom, if only for a moment, between us and the unimaginable emptiness beyond. The visual economy of *Dream Analysis* quietly revises Freud's view that the function of dreams is to prevent the dreamer from waking. Instead, Sharpe implies that one function of dreams may be to plant images between us and the faceless, dreamless state of death. But the allegory enacted by her 'last dream' also reminds us that (as Eros knows) the object can only be recovered as a symbol when it has been let go as a Thing.

[46] 'Dream as an Object', 119. [47] Ibid.

CHAPTER 10

On Not Knowing Why: Memorializing the Light Brigade

TRUDI TATE

Near the beginning of *To the Lighthouse* (1927), Mrs Ramsay is sitting by the window with her little boy James. Her husband is outside, walking up and down the terrace, reciting poetry to himself. Mrs Ramsay listens to the rhythmical sound, 'half said, half chanted . . . something between a croak and a song'.[1] Occasionally a line bursts out: 'Stormed at with shot and shell'; 'Someone had blundered'. She finds the sound calming, soothing, assuring her that all is well. Lily Briscoe, quietly working on her painting, is alarmed to feel Mr Ramsay bear down upon her, 'arms waving, shouting out, "Boldly we rode and well", but mercifully, he turned sharp, and rode off to die gloriously she supposed upon the heights of Balaclava' (25).

Tennyson's 'The Charge of the Light Brigade' (1854) was too familiar to need naming, and Woolf could, I think, rely on many of her readers noticing the errors she introduces. Mr Ramsay misquotes one line; it is not 'we' but 'they' who ride boldly in the charge:

> Stormed at with shot and shell,
> Boldly they rode and well,
> Into the jaws of Death,
> Into the mouth of Hell
> Rode the six hundred. (lines 22–6)[2]

The other mistake is Lily's. The Light Brigade perished not on the heights, but on the plain—'All in the valley of Death'—at Balaklava.

A lifetime after the event, Tennyson's poem was still familiar to many of Woolf's readers in 1927, especially those who remembered the First World War, when the poem enjoyed a resurgence of popularity and was recited on some public occasions (from the top of a tank, for example, at Southampton[3]).

[1] Virginia Woolf, *To the Lighthouse* (1927; Oxford University Press, 1992), 24.

[2] Alfred Tennyson, 'The Charge of the Light Brigade' (1854), in Christopher Ricks (ed.), *Tennyson: A Selected Edition* (Harlow, Longman, 1989), 508–11.

[3] Speech given from the top of a tank bank at Southampton, reported in the *Southern Daily Echo*, 4 January 1918.

For Woolf, the poem is comforting yet unsettling, frightening yet rather absurd. 'Never was anybody at once so ridiculous and so alarming,' thinks Lily Briscoe as Mr Ramsay gallops off on his imaginary horse. But what does the Charge of the Light Brigade mean to Mr Ramsay, a middle-aged philosopher? He imagines himself in the cavalry of fifty years earlier, riding with a group of aristocrats in short jackets and tight red trousers ('cherry bums'); men who are praised by Tennyson for their ability to act rather than to think or speak:

> Theirs not to make reply
> Theirs not to reason why
> Theirs but to do and die:

This is a curious locus of identification for a middle-class intellectual in the early twentieth century, and we perhaps need to recover a sense of its strangeness. As always, Woolf is alert to the complexity of Victorian fantasy, even as she mocks it. Nor is it obvious what the poem and its misremembered history mean to Lily, struggling with her own work of representation. Much of Virginia Woolf's writing engages with the problem of warfare in the twentieth century, and it is striking that she often looks back beyond her own memories of the First World War to the Crimean War of 1854–6, a conflict that took place nearly thirty years before she was born. There is nothing unusual about writers turning to wars of the past, but the Crimea has a particular resonance in both Victorian and modernist literature. It troubled many thinkers and writers of the 1850s and continued to irritate well into the twentieth century.

The sheer familiarity of 'The Charge of the Light Brigade' is deceptive, leading some readers to assume that the poem is an obvious device to make Mr Ramsay look foolish, casting him as a naïve Victorian deluded by visions of patriotic grandeur and heroic virtue.[4] But Woolf's novel is altogether more subtle than this. As Gillian Beer has shown, Woolf, like many other modernists, does not reject her Victorian predecessors so much as argue with them. Woolf absorbs their culture—which was her culture, too, for the early part of her life—into her own writing, where it can take on some surprising shapes and meanings.[5]

In May Sinclair's autobiographical novel, *Mary Olivier: A Life* (1919), a young

[4] See, for example, Susan Stanford Friedman, *Mappings: Feminism and the Cultural Geographies of Encounter* (Princeton, NJ: Princeton University Press, 1998). Despite its title, Friedman's book is not really interested in 'encounter' with Victorian writers; in my view, she does not give due weight to Woolf's knowledge of her literary predecessors.

[5] Gillian Beer, *Arguing with the Past: Essays in Narrative from Woolf to Sidney* (London: Routledge, 1989), 138–40. Beer shows how Woolf was, on the one hand, interestedly engaged with her Victorian predecessors, and, on the other, always alert to events of her own time. Woolf was an avid reader of the newspaper for most of her writing life, and more work remains to be done on the references to newspaper material scattered throughout Woolf's fiction. Beer's work helps us to avoid falling into the trap of reading Woolf through the lens of 'presentism'—of imposing our own preoccupations upon earlier texts—and shows us the importance of paying due attention to Woolf's own concerns. Exemplary essays in this spirit include 'The Island and the Aeroplane: The Case of Virginia Woolf' and '*Between the Acts*: Resisting the End', both in Gillian Beer, *Virginia Woolf: The Common Ground* (Edinburgh: Edinburgh University Press, 1996).

man repeatedly chants lines from 'The Charge of the Light Brigade' to avoid talking about a painful subject. Here the words block conversation, suggesting a kind of hopelessness and protesting against a lack of love.[6] By 1919, the poem is familiar yet almost meaningless, comforting like a nursery rhyme yet unsettling precisely because it is incomprehensible. What do the words mean? What did the action mean? Why do people still remember the poem, Woolf and Sinclair seem to wonder, seventy years after the event? How can we get the noise out of our heads—and do we want to?

The Crimean War was the first war to be reported first hand in the newspapers, the first to be photographed, the first to be painted by official war artists, and the first to make use of the new technology of the telegraph. It was the first modern war in the sense that it took place partly at the level of representation. More than this: representations of the war had a tangible effect upon the conduct of the conflict and the politics which surrounded it.[7]

*

These are large claims, and I hope to support them through a discussion of 'The Charge of the Light Brigade'. This chapter has two aims. First, I want to explore what poetry in the 1850s thought it had to offer in representations of the war. Secondly, I want to offer some thoughts on the troubled relationship between literature and the newspapers during this period, and to suggest that the effects of this dynamic persisted well into the twentieth century. Underlying these aims are questions about knowledge during wartime. How are citizens informed about warfare; how do they make judgements about the conflict? And how is the war memorialized in writing, both at the time and afterwards? The Crimea marks a turning point in modern warfare and especially in modern *representations* of war. It is an important context for reading both Victorian and modernist literature, and can help us to think critically about representation and knowledge of warfare in our own time.

'The Charge of the Light Brigade' raises questions of knowledge and interpretation that were immediately contentious and continue to exercise historians right up to the present day. Tennyson's poem might seem to have little to offer to our understanding; even the most subtle of critics tend to treat the poem as self-evidently patriotic: an unambiguous celebration of military courage. Christopher Ricks, for example, demonstrates that Tennyson's techniques in the poem are

[6] May Sinclair, *Mary Olivier: A Life* (1919; London, Virago, 1980), 205–7.

[7] On the history of the Crimean War, see Olive Anderson, *A Liberal State at War: English Politics and Economics during the Crimean War* (London, Macmillan, 1967); Winfried Baumgart, *The Crimean War 1853–1856* (London: Arnold, 1999); Kellow Chesney (ed.), *Crimean War Reader* (1960), rev. edn (London: Severn House, 1975); J. B. Conacher, *Britain and the Crimea, 1855–56: Problems of War and Peace* (Basingstoke: Macmillan, 1987); David Goldfrank, *The Origins of the Crimean War* (London: Longman, 1994); Andrew Lambert, *The Crimean War: British Grand Strategy, 1853–56* (Manchester: Manchester University Press, 1990); Andrew Lambert and Stephen Badsey, *The War Correspondents: The Crimean War* (Stroud: Alan Sutton, 1994).

sophisticated, but he treats its ideas as fairly straightforward.[8] Writing from a different perspective, Joseph Bristow finds complexities in gender and sexuality in Tennyson's war poetry, but he treats the attitudes towards nation and warfare as transparently simple—even simple-minded.[9] In what follows, I want to suggest that Tennyson has a more nuanced attitude towards difficult political questions than is usually recognized. Tennyson is rarely treated as a profound thinker on matters political, even by his greatest supporters. Ricks remarks that Tennyson liked battles because he thought they were simple, not difficult.[10] Perhaps so, yet 'The Charge of the Light Brigade' unsettles precisely this assumption.

*

On the morning of 25 October 1854, the British Light Cavalry Brigade advanced across the plain at Balaklava, right into the mouths of the Russian artillery. Guns fired upon them from both sides as well as from the front. More than 600 men set off. By the end of the day, fewer than 200 were left standing. They had charged in the wrong direction. In military terms, the charge of the Light Brigade was a minor incident on the way to the siege of Sebastopol. Its greatest significance was cultural, and it also had curious political implications. Precisely what occurred was much debated at the time and has never really been resolved. The charge was, and remains, a problem of interpretation.

It took nearly three weeks for the first reports to appear in the British press.[11] From mid-November, the newspapers were full of eyewitness accounts, official dispatches, letters, analyses, and debates. The charge was celebrated and lamented in poetry, paintings, lithographs, and cartoons. By early 1855, the story of the Light Brigade had been absorbed into middle-class entertainment: you could visit a 'diorama' of the charge at the Royal Gallery of Illustration, sing 'On to the Charge', and dance the Cardigan Galop.[12] Some of the representations of Balaklava were by witnesses or participants in the action. Others were produced far away, in England, by people who saw nothing of the event. We might expect the most powerful and lasting works to be the first-hand accounts. Certainly William Howard Russell's articles in *The Times* were gripping and had far-reaching effects. But the work which was to persist longest in cultural memory was not eyewitness journalism, but poetry; written not at Balaklava, but on the Isle of Wight, by a poet who had never seen a battle.

*

[8] Christopher Ricks, *Tennyson*, 2nd edn. (1972; Basingstoke: Macmillan, 1989).

[9] Joseph Bristow, 'Nation, Class, and Gender: Tennyson's *Maud* and War', *Genders* 9 (1990), 93–111.

[10] Ricks, *Tennyson*, 226.

[11] The first reports appeared in *The Times*, 13 and 14 November 1854.

[12] Advertised in *The Times*, 13 January, 27 January, 26 February 1855.

Tennyson was fascinated by the Crimean War. From his home at Farringford on the Isle of Wight, he could hear the sound of cannon practising across the water at Portsmouth and see the troop ships departing for the front.[13] Tennyson followed the progress of the war in *The Times* and was particularly moved by the story of the Light Brigade.[14] A few weeks later, he published one of his most memorable poems.[15]

> Half a league, half a league,
> Half a league onward,
> All in the valley of Death
> Rode the six hundred.
> [...]
> Cannon to right of them,
> Cannon to left of them,
> Cannon in from of them
> Volley'd and thunder'd;
> Stormed at with shot and shell,
> Boldly they rode and well,
> Into the jaws of Death,
> Into the mouth of Hell
> Rode the six hundred. (stanzas 1 and 4, 1854 version)

Hallam Tennyson later remembered his father sitting down to write about the charge immediately after reading Russell's electrifying prose, but in fact there was a gap of about two and a half weeks. The poem appeared in the *Examiner*, a weekly newspaper, under the initials 'A.T.'. Astute readers recognized the author; the poet Sydney Dobell wrote to Tennyson that 'no man living but yourself could have written the first verse and the "cannon" verse'.[16] For John Forster, editor of the *Examiner*, Tennyson was the only poet who could do justice to the war, for only he could match the 'pitch' of the soldiers' actions in writing.[17]

'The Charge of the Light Brigade' was an immediate success in Britain, although some readers were critical of specific aspects of the poem. Tennyson, always sensitive to criticism, revised the work several times for subsequent publications.[18] In August 1855, an army chaplain requested that the poem be printed on

[13] F. B. Pinion, *A Tennyson Chronology* (Basingstoke: Macmillan, 1990), 67.

[14] For a detailed discussion of the composition of the poem, see Ricks, Appendix B, *Tennyson*, 2nd edn. Ricks notes that the earliest draft did not include the opening lines quoted above, but they do appear in the first published version. 'The Charge of the Light Brigade' was substantially revised for *Maud and Other Poems* (published July 1855), then emended again for the sheets sent to the soldiers at the front in August 1855. See also Robert Martin, *Tennyson: The Unquiet Heart* (Oxford: Clarendon Press, 1980).

[15] *The Times* reports appeared 13 and 14 November; Tennyson wrote the poem early December. Ricks, *Tennyson*, 2nd edn., 325–7.

[16] Sydney Dobell to Alfred Tennyson, 7 February 1855; quoted in Cecil Y. Lang and Edgar F. Shannon, Jr. (eds.), *The Letters of Alfred Lord Tennyson*, 3 vols. (Oxford: Clarendon Press, 1987), ii. 104 n. 1.

[17] Forster, letter to Tennyson, 9 December 1854, *Letters of Tennyson*, ii. 102.

[18] Among other changes, Tennyson removed the two references to 'blunder'd' for the 1855 publication in *Maud and Other Poems*; then reinstated the famous phrase, 'Some one had blunder'd', in later versions. See Ricks, *Tennyson*, Appendix B.

single sheets and distributed to the troops at the Crimea. 'It is the greatest favourite of the soldiers, half are singing it and all want to have it in black and white.'[19] Tennyson was delighted and quickly arranged for a thousand copies to be sent to the British troops outside Sebastopol; a further thousand followed later.[20] Soldiers who had witnessed the charge, and were now suffering the privations of trench warfare, were it seems moved by a poem written far from the action—a work based not on experience, but on articles in the newspaper. Why is this important? The popularity of Tennyson's 'Charge' tells us something about the power of representation in wartime. At its best, poetry can speak of, and appeal to, the *fantasy* investment in war.[21]

War's fantasmatic appeal during the Crimean War was a constant theme in *Punch* during these years. Weirdly conservative, trivializing, and banal in many respects, the journal occasionally produced some shrewd commentary upon the events of the day. It devoted several items to the battle of Balaklava, including a full-page cartoon entitled 'ENTHUSIASM OF PATERFAMILIAS, On Reading the Report of the Grand Charge of British Cavalry on the 25th' (Fig. 10.1). The scene is a middle-class drawing-room. The stout and balding father stands by the fireplace waving a fire-iron above his head as he reads from the newspaper. Drooping at the table is the mother with a large handkerchief at her face. She might be weeping at the sad news, or, more likely, holding a cold cloth to her forehead as Pater's enthusiasm makes her head hurt. Crowding the room are their children: three adult daughters, a girl and boy of about 14, and two small boys. Two of the boys wave objects—a table napkin, a knife, a broken plate. One adult daughter, arms folded, looks to heaven in exasperation, or possibly sorrow. The youngest daughter and a small boy read the newspaper over the father's shoulder. A tiny child at the mother's knee watches the father solemnly while eating a piece of bread-and-butter. On the wall is what might be a map of the Crimea (maps were popular objects in exhibitions and the illustrated press). The hearthrug is in disarray and objects have fallen off the table. No wonder Mater is getting a headache.

The cartoon suggests, characteristically, that men and women occupy completely different imaginative realms. What delights father is painful to mother. *Punch* is full of pieces about the supposed peculiarities of women, but here men are its main target. The cartoon mocks the father for his childish enthusiasm, yet he is also applauded, in some respects, for thinking beyond the claustrophobic atmosphere of the drawing-room, and for inspiring his sons to do the same. Of the girls, only the youngest is responsive; she perhaps represents a new

[19] B. L. Chapman, letter to Tennyson, 3 August 1855, *Letters of Tennyson*, ii. 117.

[20] F. B. Pinion, *A Tennyson Chronology* (Basingstoke: Macmillan, 1990), 71.

[21] The argument that follows differs from Jerome McGann's well-known reading of the poem in 'Tennyson and the Histories of Criticism', *The Beauty of Inflections: Literary Investigations in Historical Method and Theory* (Oxford: Clarendon Press, 1985). For McGann, the poem is engaged with the heroic tradition in French painting; in my view, we do not need to look so far afield, for Tennyson engages rather more closely with the history of his own time than McGann recognizes. Disappointingly, McGann's own historicist argument contains several errors of fact about the Crimean War and the newspaper reports, and he misquotes one line from the poem.

ENTHUSIASM OF PATERFAMILIAS,
On Reading the Report of the Grand Charge of British Cavalry on the 25th.

FIGURE 10.1 'Enthusiasm of Paterfamilias, On Reading the Report of the Grand Charge of British Cavalry on the 25th', *Punch*, November 1854.

generation of spirited women (like those who worked with Florence Nightingale, of whom *Punch* thoroughly approved). Or perhaps she is simply too young to realize that she is a woman. *Punch*'s main point, however, is to satirize middle-class enthusiasm for the war—and for the newspapers—while acknowledging the pleasure that war fantasies can bring.

It is never easy to make sense of a war—imaginatively, emotionally, politically, or intellectually—and people often turn to representations for help. This is both the strength and the weakness of war literature—it can simplify, dissemble, reassure, and normalize war, making it bearable and even pleasurable, at least for those at a safe distance. But powerful representations of war can also do the opposite; they can discomfit readers and provoke us to think critically about the conflicting emotions and issues at stake in warfare.

Tennyson's 'The Charge of the Light Brigade' is particularly striking in this context. Often regarded as a simple-minded piece of patriotism, it is in fact a subtle and even anguished reflection upon the Crimean War. Far from merely celebrating the blind courage of the soldiers, Tennyson expresses the ambivalent cultural significance of the event, which generated pleasure and excitement (as the *Punch* cartoon suggests) as well as fear and sorrow. Yet the poem concludes by urging readers not to reflect further, not to think, but simply to pay homage to the participants, both dead and alive:

When can their glory fade?
O the wild charge they made!
  All the world wondered.
Honour the charge they made!
Honour the Light Brigade,
  Noble six hundred! (stanza 6, final version)

The charge was not 'wild'—quite the opposite. The cavalry had to advance slowly at first, so as not to tire the horses; it was only at the end of the charge that they moved from a disciplined trot to an equally disciplined, if desperate, gallop.

Tennyson's poem suggests that war, and representations of war, have complex and contradictory meanings within a society. Poets since Homer have been conscious of this, of course, but the Crimea brought questions of representation into focus in new ways. Nowhere is this more vividly played out than in the charge of the Light Brigade and the writings that surrounded it.

HOW THE LIGHT BRIGADE WAS LOST

What caused the elite British cavalry to hasten to its own demise? The charge of the Light Brigade was one of four cavalry charges at Balaklava, in a battle that lasted one day.[22] Balaklava is on the south coast of the Crimea, on the Black Sea, to the south of Sebastopol. The Balaklava landscape consists of hills, two large valleys, and a high escarpment. The valleys are separated by a low ridge, along which Turkish troops were stationed in rudimentary fortifications known as redoubts. The Commander-in-Chief of the British forces, Lord Raglan, directed activities from the Sapoune Heights, from where he had a panoramic view. The cavalry was 600 feet below, under the command of Lord Lucan. The ridge between the valleys blocked the view; soldiers in the north and south valleys could not see one another. As is often the case in warfare, few people had an overall view of any part of the battle; this is partly why so many different histories circulated afterwards.

Early in the morning of 25 October 1854, the Russians attacked the Turkish redoubts on the ridge dividing the valleys. The Turks defended themselves with 12-pounder naval guns belonging to the British. They stood their ground bravely, but were overcome, and eventually took flight. The Russians now had possession of the British cannon. No army liked losing their weapons to the enemy; apart from the fact that they did not want to arm their opponents, the big guns represented the pride and honour of the nation.

Far in the distance, from his position on the heights, Lord Raglan saw the Russians preparing to remove the British guns from the redoubts. By this time

[22] The following account is drawn from Mark Adkin, *The Charge: Why the Light Brigade Was Lost* (London: Leo Cooper, 1996) and from Alan Palmer, *The Banner of Battle: The Story of the Crimean War* (London: Weidenfeld & Nicolson, 1987), ch. 10.

(shortly before 11 a.m.), Raglan, normally an even-tempered man, was beside himself with rage and frustration. He had sent an order some 45 minutes earlier, instructing Lucan to advance (or so he thought). Unfortunately, Raglan found it difficult to make his orders clear. He disliked the bluntness of orders, preferring to make courteous suggestions. These were then written down by General Airey, who was not good at translating Raglan's politeness into military instructions. The order was so unclear that Lucan interpreted it, reasonably enough, as meaning precisely the opposite of what Raglan intended. Raglan wanted the cavalry to advance; Lucan kept them still.

Annoyed at Lucan's inaction, Raglan sent a further order, which would launch the fateful charge. He regarded the matter as urgent, and chose as his messenger Captain Nolan, a cavalry expert and excellent horseman who despised Lucan. Nolan rode at high speed down the sheer escarpment and delivered the written order to Lucan: 'Lord Raglan wishes the cavalry to advance rapidly to the front—follow the enemy and try to prevent the enemy carrying away the guns—Troops Horse Artillery may accompany—French cavalry is on yr. left—Immediate.'[23] But from his position below, Lucan could not see the guns in question. 'When Lord Lucan received the order from Captain Nolan', wrote William Howard Russell in *The Times*, 'he asked . . . "Where are we to advance to?" Captain Nolan pointed with his finger to the line of Russians, and said, "There are the enemy, and there are the guns, sir, before them; it is your duty to take them." ' Or words to that effect, says Russell. Here, too, there are conflicting accounts of what took place, but the evidence suggests that Nolan had pointed in the wrong direction.[24]

Lucan reluctantly passed the order to Lord Cardigan. 'The noble earl', writes Russell, 'did not shrink', and he led the cavalry in what he believed to be the direction of the order, towards a Russian battery of 6-pounders, facing down the valley. The historian Alan Palmer writes that 'Cardigan, resplendent in his blue and cherry-coloured uniform with gold trimmings across the chest and shoulder, took up his proper place ten yards ahead of the brigade, and ordered 673 men and horses to advance.'[25] They had to cover a distance of about a mile and a quarter.[26]

Half a league, half a league,
   Half a league onward,
All in the valley of Death
   Rode the six hundred.
'Forward, the Light Brigade!
Charge for the guns!' he said:
Into the valley of Death
   Rode the six hundred.

[23] Adkin, *The Charge*, 125–7; Palmer, *Banner of Battle*, 128.
[24] Adkin, *The Charge*, 132–4.
[25] Palmer, *Banner of Battle*, 129–30.
[26] Colin Robins, 'Lucan, Cardigan and Raglan's Order', *Journal of the Society for Army Historical Research* 75 (1997), 89.

'Forward, the Light Brigade!'
Was there a man dismayed?
Not though the soldier knew
Some one had blundered:
Theirs not to make reply,
Theirs not to reason why,
Theirs but to do and die:
Into the valley of Death
Rode the six hundred. (stanzas 1 and 2; final version)

More than 600 set off; fewer than 200 could be mustered at the end of the day.[27] 'You have lost the Light Brigade!' said Raglan to Lucan that afternoon. In fact, the final death toll was not as high as was feared. The figures are still disputed. Historian Mark Adkin calculates a total of nearly 300 soldiers (45 per cent) killed, wounded, or taken prisoner. Of these, around 110 died in the charge or later—a surprisingly low figure under the circumstances, though this was not known at the time. Early reports suggested that 300 to 400 were dead. The wounded men and horses had a difficult night; it was bitterly cold and there was a shortage of food, shelter, and medical care.[28] Some of the men died later from wounds or disease. Many of the horses were shot or died from cold, starvation, or exhaustion.[29]

What went wrong? Was Raglan's order ambiguous; or was it misunderstood? Was the error caused by topography? Precisely who said what and what it meant has never been fully established. And what the charge itself meant was a matter of dispute, both at the time and since. Were the cavalry heroes, or idiots? Did the charge advance the British cause, or hinder it? Who, if anyone, was to blame? The charge raised questions of knowledge and interpretation that were to trouble both the literature and the politics of the 1850s.

In military terms, the Light Brigade was destroyed for no good reason. Commentators were struck by the disparity between the deed and the outcome: 'Never did cavalry show more daring to less purpose'; 'Causeless as the sacrifice was, it was most glorious'; 'Never was exploit more useless, and never more brilliant'.[30] It seemed to be completely pointless, and for many people, the event was emblematic of the incompetence of the army leadership. Yet, at the same time, the charge was regarded as a unique act of martyrdom—what *The Times* called 'a splendid self-sacrifice'—played out in front of a large audience.

Two great armies [said *The Times* editorial], composed of four nations, saw from the slopes of a vast amphitheatre seven hundred British cavalry proceed at a rapid pace, and in perfect

[27] Adkin says 664 charged; Palmer says 673.

[28] Palmer, *Banner of Battle*, 132–3. Adkin, *The Charge*, 217.

[29] Adkin notes that the 11th Hussars' cavalry division was reduced from 2,000 to 200 horses by December 1854: *The Charge*, 216.

[30] 'The Story of the Campaign; Written in a Tent in the Crimea', part II, *Blackwood's Edinburgh Magazine*, January 1855, 118; *The Times*, editorial, 13 November 1854, 6; *Reynolds's Newspaper*, 19 November 1854, 1.

order, to certain destruction. Such a spectacle was never seen before, and we trust will never be repeated. . . .

It is difficult not to regard such a disaster in a light of its own, and to separate it from the general sequence of affairs. Causeless and fruitless, it stands by itself, as a grand heroic deed . . .[31]

Like many commentators of the day, *The Times* was fascinated by the charge as a spectacle—an event which, it remarks extravagantly, took place 'under the eyes of the whole world'. Similarly, the *Morning Chronicle* asks breathlessly 'whether the history of the world affords a similar instance of daring bravery'.[32] 'All the world wondered', writes Tennyson.

> Flash'd all their sabres bare,
> Flash'd as they turn'd in air
> Sabring the gunners there,
> Charging an army, while
> All the world wondered. (stanza 4, final version)

Why was the event so astonishing to contemporaries? Partly because of the physical courage displayed by the men who rode, in strict formation and with perfect discipline, into the mouths of the Russian guns. And partly because the cavalry itself was a remarkable spectacle; not only did the men have splendid, colourful uniforms, but the cavalry was an elite aristocratic institution. The commander of the cavalry, Lord Lucan, Earl of Bingham, had bought his command for £25,000. Lord Cardigan, commander of the Light Brigade, was also an earl; his private yacht was moored in the bay of Balaklava. (The men were brothers-in-law who thoroughly disliked one another.[33]) The Commander-in-Chief, Lord Raglan, and many other high-ranking officers, were of noble birth or had close links with the aristocracy. Both the newspaper reports and Tennyson's poem are acutely conscious of the significance of class in the charge of the Light Brigade. And both are shaped by middle-class hostility towards the aristocracy, though neither takes a simple view of the matter.

THE ARISTOCRACY AND THE MIDDLE CLASSES

What underlies these representations of the aristocracy in 1854? According to David Cannadine, after a flourishing period between 1780 and the 1820s, the aristocracy's power was much diminished and under further attack in the mid-nineteenth century, following the Reform Bills of the 1830s. None the less, in the 1850s, they were still a force to be reckoned with in politics, and they faced a lot of pres-

[31] *The Times*, editorial, 13 November 1854, 6.

[32] 'What is the meaning of a spectacle so strange, so terrific, so disastrous, and yet so grand?', editorial, *The Times*, 14 November 1854, 6. *Morning Chronicle*, 14 November 1854, 4.

[33] Cecil Woodham Smith, *The Reason Why* (London: Constable, 1953).

sure from an increasingly powerful middle class.[34] War was the aristocracy's oldest profession,[35] and by 1854, the army was the last bastion of aristocratic power.

Two opposing views emerged during this period. On the one hand, the war was regarded as an opportunity for the aristocracy to recover some of its lost prestige. Many people felt that middle-class commercialism had begun to dominate all aspects of British life. Success in the war might reintroduce values associated with the aristocracy, such as heroism, selflessness, sacrifice, tradition, devotion to an ideal. On the other hand, those who favoured reform were frustrated that the army was still run according to traditional principles. These were not efficient in a modern war. The army leadership was notorious for its bungling. At the very least, the men needed adequate food, shelter, and medical care if they were to be fit for battle, but the aristocratic leaders proved hopelessly incompetent at these practical aspects of the war. As *Punch* remarked sharply, war involved 'a good deal of sheer *business*'. In order to fight, the army needed to be as well organized as a successful manufacturing industry or railway company.[36] To this way of thinking, the army should be modernized along industrial lines. This in turn would have the happy effect of diminishing the aristocracy's control over the army, to the benefit of the middle classes.

The Crimean War gave new impetus to the struggle between the middle and upper classes, and this was played out with great intensity in the newspapers. *The Times* in particular regarded itself as 'the organ and representative of bourgeois power' during this period; the voice of the middle classes, struggling to establish themselves as the dominant force at all levels of British life. 'The [Crimean] war came to be regarded as a test of the aristocracy's power to rule in a modern world,' says *The Times*' official history, and it clearly took the view that they had failed.[37] This opinion was widely shared. In mid-1855, *Punch* published 'The Voice of the Omnibus', an open letter from Mr Punch to the Prime Minister, Lord Palmerston. Mr Punch advises his Lordship to start travelling on the omnibus and listening to the complaints of the English middle classes. He would learn that they were 'very much disgusted with affairs at home,—very much humiliated by affairs abroad', and seriously dissatisfied with his leadership.

> The Voice of the Omnibusses would also inform your Lordship that the incredible imbecility, incompetence, and mismanagement which have attended every branch of operations carried on by the Government in connection with this War, as well as the diplomacy which preceded and has accompanied it, have led to grave doubts of the exclusive right to governing authority of that order to which your Lordship belongs.[38]

[34] David Cannadine, *Aspects of Aristocracy: Grandeur and Decline in Modern Britain* (New Haven, Conn.: Yale University Press, 1994), 10; see also David Cannadine, *The Decline and Fall of the British Aristocracy* (New Haven: Yale University Press, 1990).

[35] Dominic Lieven, *The Aristocracy in Europe, 1815–1914* (Basingstoke: Macmillan, 1992), 181.

[36] G. R. Searle, *Entrepreneurial Politics in Mid-Victorian Britain* (Oxford: Oxford University Press, 1993), 90–4; Anderson, *A Liberal State at War.*

[37] Stanley Morison, *The History of the Times*, ii. *The Tradition Established, 1841–1884* (London: The Times, 1939).

[38] 'The Voice of the Omnibus', *Punch*, May 1855, 179.

The aristocracy have failed in both diplomacy and war; they have proved themselves unfit for power. For *Punch*, the alternative can be found upon the omnibus, among people who are not 'revolutionary, or democratic, or subversive, or socialist. They are none of these things, but they are business-like.' And this is precisely what is needed during these difficult times.

> [For] war involves a great deal of sheer *business*,—such as contracting for, and forwarding stores and supplies; taking up, stowing, and dispatching ships, and so forth. In fact, when the omnibusses take the war to pieces—apart from the fighting, the one thing which has been well done—they find it to be an aggregate of such acts as most of the passengers are daily doing in the carrying on of their own daily concerns.

For *Punch*, fighting is the least of the war. The matter at stake is not battle, but business; not courage, but commerce.

In another article, *Punch* rudely criticizes 'Aristocratic vermin' and 'Incompetent Nobility', calling for 'every Lord incapable, and every booby Duke' to resign from public office.[39] More moderately, *Fraser's Magazine* commented: 'The frightful misconduct of the war has revived the dangerous agitation, which had slept since the Reform Bill, of the aristocratic question.'[40] The constitution of the army, it argues, is 'the least defensible part of the aristocratic system', and it complains that the Cabinet in early 1855 is dominated by aristocrats and their associates. Unlike some middle-class publications (notably *The Times*), *Fraser's* declares itself no enemy of the aristocracy, but it argues that people of noble birth 'are merely on a level with the other educated classes', and ought to compete on equal terms.

It is certainly true that the British army was badly organized, especially in the early months of the war. But we also need to be aware that bourgeois critics had their own class interests, and stood to benefit from the aristocracy's incompetent leadership of the army. The worse the army looked, the greater the impetus for reform, and the better for middle-class aspirations to control the machinery of the state.

At the same time, the opposing view carried considerable weight. By the 1850s, commerce and industrialization had brought enormous changes to British society. The railways, large-scale mining and manufacturing, and other modern innovations had moved millions of people into the cities, and had made Britain the wealthiest and most powerful nation in the world. But industrial modernity had also produced foul slums, bad health, social injustice, adulterated food, and many other hardships, especially for poorer people. The march of commerce brought harm as well as good, and some hoped that the war would be able to check some of its excesses, and would bring other values to the fore.

Should war become a branch of commerce? people wondered. Clearly, middle-class business skills could help the army to modernize. But if the British military

[39] 'Prince Albert's Example', ibid. 199.

[40] 'The Government, the Aristocracy, and the Country', *Fraser's Magazine* 51 (March 1855), 358.

services were reformed along commercial or industrial lines, what would inspire the soldiers to heroic and daring deeds? Would they be willing to suffer intense hardship and discipline and risk their lives in the name of commercial efficiency? Could commerce represent 'the nation' as the aristocracy had done in the past? The old system produced martinets such as Cardigan—flamboyant, bullying, a man who loved his regiment as much as (perhaps better than) his family, and spent £10,000 a year of his own money in clothing them. His men might not have liked him much, but they respected his military skills and would follow him into the mouth of hell, as the charge of the Light Brigade had shown. Could the middle class produce this kind of obsessive, charismatic leader—or would meritocracy lead to mediocrity, and lose wars?

Somehow, modernizers needed to retain some traces of aristocratic values in the army even as they challenged those values and removed the aristocrats themselves from positions of power. The aristocracy still had something powerful to offer the nation—it provided an idea; a glamour; a devotion to duty. Aristocratic sang-froid (or stupidity) supposedly made them fearless, like Cardigan who, according to the *Morning Chronicle*, was 'magnificent in his cool contempt of danger'. The tradition of aristocratic military leadership went very deep, and had a long and successful history, most recently against the French in the Napoleonic Wars. It could not be discarded lightly. Ultimately, however, the Crimean War failed to revive aristocratic prestige, partly due to the incompetence of the army leadership, and partly—perhaps largely—due to middle-class pressure, especially in the newspapers. The war increased the power of the commercial middle classes, and brought some of their values into the armed services. As Olive Anderson points out, this was a decisive period in the consolidation of middle-class consciousness.[41]

In this political context, *The Times*' reports and Tennyson's poem about the charge of the Light Brigade are quite uncomfortable, and uncertain, about the event. What did it mean? Why did it produce directly opposing emotions—admiration and mockery; inspiration and despair? The ambivalence at work here has its roots in middle-class attitudes towards the aristocracy, the glamorous, heroic fools who led the army. *The Times*' editorials mock and criticize the cavalry for its aristocratic attitudes. The attendance on royalty, exemption from colonial service, and of course the splendid uniforms, says *The Times*, all make the cavalry a 'favourite resort' of the aristocracy, and have 'infected it with the weakness of caste'. The paper hints darkly that the cavalry had not been pulling its weight at the Crimea—a question that had been much discussed in the previous weeks of the war.[42]

Russell, too, comments that the cavalry had been heavily criticized since the war began. In fact, it was not their fault they were inactive; Raglan had deliberately chosen to hold them back. Russell goes on to suggest that perhaps the heroic

[41] Anderson, *A Liberal State at War*, 101–4.
[42] *The Times*, editorial, 14 November 1854, 6.

and foolhardy charge was a response to unjust criticism; perhaps the cavalry seized this opportunity to demonstrate their courage and 'shame their detractors for ever'.[43] Russell calls them 'rash and reckless' but 'gallant fellows who prepared without thought to rush on [to] almost certain death'. The aristocracy are represented as men of action—spontaneous, brave, reckless, stupid—men with bodies rather than minds.

Yet *The Times*' reports also present a spectacular martyrdom which only the aristocracy could provide. Russell describes how they sweep 'proudly past, glittering in the morning sun in all the pride and splendour of war'. 'A . . . fearful spectacle', he adds, of men who are about to be 'annihilated by their own rashness'.[44] Russell recuperates, even as he mocks, the vision of Britain's ancient leadership sending itself to its own death. Like *The Times*' editorials, the eyewitness reports are deeply ambivalent towards the spectacle and suffering of the aristocracy, and this ambivalence is also expressed in Tennyson's poem. Both kinds of writing are drawing upon—and entering into—political debates about aristocratic power in parliament as well as in the army.

*

Britain in 1854 was faced with a paradox. The first major war for forty years showed that army reform was urgently needed. Yet those reforms might destroy precisely the strengths that were regarded as essential for military success. Tennyson's 'Charge of the Light Brigade' enacts this paradox. It celebrates the cavalry, and also mourns its demise; it cherishes the aristocratic elements even as in some sense it shares *The Times*' desire to destroy them. And the emotion is further complicated by the fact that the cavalry had destroyed itself by the very values that the poem seeks to preserve. These complexities would have been familiar to readers at the time, as the nation struggled with questions of reform. How could the good aspects of the past be preserved, and the bad expelled? What happens if the good goes, too; what will remain? These questions echo far into the twentieth century and underlie Woolf's representation of Mr Ramsay in *To the Lighthouse*, reciting fragments of Tennyson's poem.

MEMORIALIZING THE LIGHT BRIGADE

Alongside these concerns were questions about representation, for how was the fearful spectacle of the charge of the Light Brigade to be preserved in cultural memory? In September 1855, an anonymous reviewer in *Fraser's Magazine* urged poets to follow Tennyson's example. The story of the charge might seem 'short and simple', he commented, but it is in fact 'transcendental and inexhaustible'

[43] [Russell], 'The Cavalry Action at Balaklava', *The Times*, 14 November 1854, 7.
[44] Ibid.

and the participants deserve 'a monument worthy of their deed'.[45] Poetry could be such a monument. Indeed, for this reviewer, *only* poetry could offer reparation to the soldiers for their suffering. Like the charge itself, poetry is anti-utilitarian. It serves no practical purpose, and that is precisely its strength. On the other hand, he argues, poetry does have a useful effect, keeping alive the memory of the charge and inspiring the British army to emulate the 'most striking moral deed of our day'. The cavalry's 'absolute and mere self-sacrifice' rouses 'a dread in the hearts of our enemies, a respect in the hearts of our friends'—an action that is best remembered—monumentalized—through poetry. *Fraser's* reviewer, like other commentators, understands the charge and its poetry in similarly contradictory terms. They are at once valuable and useless; above military functionalism yet necessary for military success. But *Fraser's* is alert to this paradox. Like the charge itself—and indeed like the aristocracy, in many views—poetry can inspire by its beautiful and disciplined purposelessness.

A number of poets did write about Balaklava, although few produced the really memorable works *Fraser's* reviewer had hoped for. The charge was also a topic for popular and dogmatic verse, most of it based closely on the press reports. Helen MacGregor's 'Balaklava' retells the newspaper version across fifteen stanzas, emphasizing the smallness of the cavalry group:

While, by that fatal order stirred,
The Light Brigade advance.
Tell ye the tale who saw it,
Six hundred strove that day,
From the grasp of twenty thousand,
To bear those guns away!

The poem concludes with the claim that everyone, from peasants to the nobility, is united in mourning the Light Brigade, and is respectfully conscious of the class affiliations of the cavalry:

The noblest blood of England
Around those guns had flowed,
Ere, dealing death like victors round,
The remnant backwards rode.[46]

MacGregor's tribute to the aristocracy is somewhat compromised, perhaps, by the unfortunate image of horses galloping in reverse. An anonymous 'Retired Liverpool Merchant' published his tribute to the cavalry at his own expense. This aspiring poet shares *Fraser's* hope that the Light Brigade will remain precious in British memory:

O souls of the living—O shades of the dead—
When the word, 'Balaklava', shall sound on the ear

[45] 'Tennyson's Maud', *Fraser's Magazine*, 52 (September 1855), 273.

[46] Helen MacGregor, 'Balaklava', *Lays of the Crimea* (London, Longman, Brown, Green & Longmans, 1855).

> What a halo of glory around it will shed,
> To light up the hearth and illumine the bier![47]

Shortly after the first articles appeared in *The Times*, *Punch* cobbled together 'The Battle of Balaklava', drawing closely upon Russell's reports.

> But who is there, with patient tongue the sorry tale to tell,
> How our Light Brigade, true martyrs to the point of honour, fell!
> ''Twas sublime, but 'twas not warfare', that charge of woe and wrack,
> That led six hundred to the guns, and brought two hundred back!
> Enough! the order came to charge, and charge they did—like men;
> While shot and shell and rifle-ball played on them down the glen.
> Though thirty guns were ranged in front, not one drew bated breath.
> Unfaltering, unquestioning, they rode upon their death!
> ...
> But still like wounded lions, their faces to the foe,
> More conquerors than conquered, they fell back stern and slow;
> With dinted arms and weary steeds—all bruised and soiled and worn—
> Is this the wreck of all that rode so bravely out this morn?
> ...
> Whose was the blame? Name not his name, but rather seek to hide
> If he live, leave him to conscience—to God, if he have died:
> But for you, true band of heroes, you have done your duty well:
> Your country asks not, to what end; it knows but how you fell![48]

Like Tennyson, *Punch* concludes its 'sorry tale' with a defiant refusal to *think* about the matter. And like *Fraser's*, it tries to celebrate the pointlessness of the charge: 'Your country asks not, to what end; it knows but how you fell!' The value of the event lies in its sheer spectacle, which, for most people, is not seen, but imagined.

In what terms is that spectacle imagined? For one anonymous poet, the Light Brigade is like a mighty wave while the Russians are transformed, absurdly, into marine vegetation:

> We dashed among the foe
> Like billows hurled upon the shore
> All white with foam and spray,
> Like seaweed strewn along the beach
> The gunners trampled lay.[49]

Later, the metaphor changes; the cavalry becomes a lion that fights most fiercely 'when he turns to bay'. 'Spasmodic' poets Alexander Smith and Sydney Dobell bring the two ideas together in one of their war sonnets of 1855:

[47] A Retired Liverpool Merchant, 'Balaklava', *The Battle of Inkermann: A Ballad, with Balaklava, Alma, Sinope, &c.* (London: Published for the author by Arthur Hall, Virtue & Co., 1855).

[48] 'The Battle of Balaklava', *Punch*, November 1854.

[49] Anon., 'Balaklava', *Alma, Balaklava* [etc.] (London: Helen Nash, 1855).

> Aye, ye heavens! I saw them part
> The Death-Sea as an English dog leaps o'er
> The rocks into an ocean. He goes in
> Thick as a lion, and he comes out thin
> As a starved wolf [. . .]

Smith's and Dobell's poetry in this collection is uneven at best,[50] but it produces some arresting images. In another piece about the Light Brigade, they invoke a speaker trying to imagine the sheer pleasure of force and speed:

> 'Who would not pay that priceless price to feel
> The trampling thunder and the blaze of steel—
> The terror and splendour of the charge?'
> . . .
> 'At the word they sprung
> In one wild light of sword and gleaming corse,
> Death stood dismayed. Jove! how the cowards shook
> When on them burst that hurricane of horse!'[51]

One of the few respected poems about the charge was *The Death-Ride: A Tale of the Light Brigade* by playwright and critic Westland Marston, published as a pamphlet in 1855.[52] Marston tried to imagine the event from the inside, from the point of view of those who took part. Where Tennyson says 'Theirs not to reason why', Marston explores how 'we' might feel upon receiving Raglan's order:

> Pursue them!—What, charge with our hundreds the foe
> Whose massed thousands await us in order below!
>   Yes, such were his words. To debate
> The command was not ours; we had but to know
>   And, knowing, encounter our fate. (stanza 5)

Where Tennyson finds incompetence—'Some one had blunder'd'—Marston sees insanity—'mad was the summons'. But 'DUTY' beckons the men, who follow her to the end. Duty offers the men neither glory nor triumph; 'She only said "Die!"—and they died' (stanzas 15–17).

Here, as in many of the poems, the men emerge as martyrs who perish for an idea. Marston imagines them as curiously passive, dying on command, while Tennyson's poem conveys a greater sense of struggle. Both poets praise the Light Brigade for their discipline and willingness to follow an order that is clearly wrong (or mad); to act without reasoning why. And both explore how difficult it must have been for the cavalry to ride and fight with great energy while submitting to the idea of duty—to be, paradoxically, at once active and passive; heroic and

[50] Coventry Patmore finds *England in Time of War* 'upon the whole' to be bad poetry. *Edinburgh Review*, 104 (October 1856), 349.

[51] Both poems are entitled 'The Cavalry Charge'. Alexander Smith and the author of 'Balder' [Sydney Dobell], *Sonnets on the War* (London: David Bogue, 1855), 21, 22.

[52] Westland Marston, *The Death-Ride: A Tale of the Light Brigade* (London: C. Mitchell, 1855). The *Dictionary of National Biography* mentions *The Death-Ride* as one of Marston's 'happy inspirations'.

meek; leaders and followers—and to accept their fate as victims not of the enemy but of their own side.

Following Marston's poem is a note in which he pays tribute to the newspapers' coverage of the war: 'The masterly Records of the War which now appear in our crowded journals—records which are at once histories and poems—leave to formal poetry only this task—to adopt their descriptions and to develop their suggestions; to comment, as it were, upon their glorious texts'(8). If the press reports are both history and poetry, what is left for literature to do? For Marston, 'formal poetry' is relegated to a commentary or a gloss upon the newspapers. Without the journalists' reports, the war poet would have no material. The press makes poetry possible, even as it makes it redundant.

Almost all the poems are faced with the problem of representing an event the poet has not seen, and all rely heavily on Russell's eyewitness reports in *The Times*. Reading the poetry and the newspapers together, we can see a struggle between the two forms of writing to be the authentic voice of the event. We can also see the extent to which the poetry draws upon the press for its information; knowledge is increasingly to be found in the commercial realm of the newspapers. But is it reliable? What about the press's own interests; how do they affect the reporting of events? What is the newspaper not telling us? At its best, the newspaper writing is so powerful that some writers wonder if it has supplanted poetry altogether.[53]

Tennyson's 'Charge of the Light Brigade' was taken up at all levels of society, by soldiers as well as civilians, and it circulated among people of all classes in later wars. Particularly striking is the way in which the aristocratic cavalry were perceived as both victims and martyrs, as Richard Trench puts it, 'With a battle-field for altar, and with you for | sacrifice'.[54] As victims, they provide an imaginary point of focus for other groups who feel oppressed by a powerful but incompetent authority.[55] 'Someone had blundered'.

> 'Forward, the Light Brigade!'
> Was there a man dismay'd?
> Not tho' the soldier knew
> Some one had blundered.

The clumsiness or stupidity of the authority that blundered—in its thinking, its words, the order it issued—is set against the perfect discipline of the cavalry, which is the body of action, not thought or speech:

[53] 'The poet, catching the enthusiasm [from the press], burns to sing of the war. Fancy and invention he need not call on for aid, as those elements of poetry have already done their utmost in the columns of the newspaper he subscribes to.' 'Poetry of the War: Reviewed before Sebastopol', *Blackwood's Edinburgh Magazine*, 77 (May 1855), 531.

[54] Trench, 'Balaklava', in *Alma: and Other Poems*, 2nd edn. (London: John W. Parker & Son, 1855), 19. Trench insists here that the sacrifice is not in vain; the rest of the nation is learning about duty from the heroes and martyrs of the Light Brigade (19–20).

[55] This structure reappeared 150 years later, in 1997, around the death of Diana, 'the people's princess'—the very term uncannily echoing the language used to describe the Crimea in the 1850s—the 'people's war' (Anderson, *A Liberal State at War*, 85). The aristocrat is imagined as a victim with whom all other classes are invited to identify.

Theirs not to make reply,
Theirs not to reason why,
Theirs but to do and die:
Into the valley of Death
  Rode the six hundred.

The poem, like the news reports, half-imagines—or wishes—that action might be unambiguous. The physical act of the charge arrests, or prevents, interpretation. At the same time, Tennyson recognizes that action, too, produces contested meanings. Two contradictory impulses drive Tennyson's poem and other writings about the charge: to interpret, endlessly, and to bring an end to interpretation. It must stop; it cannot stop.

According to Mark Adkin, a cavalry charge has a remarkable degree of momentum. 'Once launched, nothing short of the death of horse or rider could halt the frenzied gallop. . . . Men and animals were swept inexorably forward. As somebody once said, "It is difficult to be a coward in a cavalry charge." '[56] The writings about the Light Brigade are fascinated by the unstoppable momentum of living bodies. But it also horrifies them. The war employs the technology, and the social organization, of the Napoleonic past;[57] but flesh and blood and sabres are faced with increasingly mechanized cannon and guns—exploding 'shrapnel shells filled with powder and bullets' and roundshot which 'could slice men in half or disembowel horses'.[58]

The poetry finds it difficult to register this mechanical violence; 'Then they rode back, but not | Not the six hundred', says Tennyson, but he is unable to describe precisely what happened to those who fell, even though some of the newspaper reports are surprisingly explicit about the physical effects of the war—Russell's dispatches from the fall of Sebastopol in September 1855, for example, are extremely graphic. In the 1850s, the British newspapers, which are uncensored, give detailed accounts of the bodily suffering of war, while the literature says very little about it. By the First World War, the situation is reversed: the press is rigorously censored and it is *only* in literature that the violence of war can be described.

Above all, what makes Tennyson's poem memorable is its ambivalent sense of mourning, and its sense of helplessness in the face of the enigma of interpretation. What did the action mean? In the end, Tennyson is not sure; the Light Brigade can be honoured and remembered, but not understood. Their action can be mourned, but not interpreted. The poem submits to its own injunction not to reason why, and it registers both pleasure and muted terror at the spectacle of its own passivity. Finally, Tennyson expresses a melancholic yearning—for a history that is lost even as it occurs, for something prior to and beyond the contemporary world of commercial efficiency—even as its object is pointlessly destroyed.

[56] Adkin, *The Charge*, 4–5.

[57] 'The British went to war in the Crimea with cannon that had been used at Waterloo': Adkin, *The Charge*, 86; see also Palmer, *Banner of Battle*.

[58] Adkin, *The Charge*, 87.

Literature, like the society as a whole, is faced with new sources of knowledge in the Crimean War. Something that emerges very strongly in the writings of the 1850s is the fantasy investment in the daily press during the Crimean War—especially *The Times*. This is played out in relation to fantasies about spectacle, heroism, and suffering that are sometimes directed in unexpected ways. Literature can offer a critique of these fantasies, as Isobel Armstrong argues in relation to Tennyson's *Maud*;[59] but it also keeps them in circulation. The literature is constantly in dialogue with the daily press and is conscious of a new temporal relationship between the events of war and the act of representation. Above all, the Crimean War generates new anxieties that it is the newspapers, and not literature, that have possession of the heart, the soul, and the intelligence of the nation.

These anxieties persist, and intensify, in the early twentieth century, which is perhaps why Tennyson's poem turns up in modernism as well as in war propaganda in the First World War. For Woolf and Sinclair, 'The Charge of the Light Brigade' still seems to have something to say, even as it is invoked, in fragments, as a kind of nonsense. The poem is also used to mourn for a history—the history of modern warfare—which even by 1919 is partly lost, though its legacy continued to be felt throughout the twentieth century.

[59] Isobel Armstrong, *Victorian Poetry: Poetry, Poetics and Politics* (London: Routledge, 1993).

CHAPTER 11

Sounds of the City: Virginia Woolf and Modern Noise

KATE FLINT

Between 1880 and 1937—the calendar span of *The Years*—the sounds of the city changed. The human and animal cacophony of the streets gave way to a mechanical roar and hum, the product, above all, of the internal combustion engine on the ground, and—more intermittently—the drone and throb of the aeroplane in the sky above. The shift was between the acoustic ambience of the streets where 'musicians doled out their frail and for the most part melancholy pipe of sound',[1] and the 'deafening' noise of London, the persistent hooting of car horns, 'the dull background of traffic noises, of wheels turning and brakes squeaking', that greets North on his return to London from Africa in the present day.[2] But whilst Virginia Woolf was far from impervious to the intrusive and disturbing potential of modern noise,[3] she was also exhilarated by it. Moreover, her sensitivity to the acoustic environment, and to the place of the listening, perceptive body within it, is an important component in her representation of the process of consciousness: one which makes her stand out from contemporary fictional observers of London life.

The nineteenth century increasingly characterized noise as nuisance. In 1821, a contributor to the *New Monthly Magazine* could write cheerfully on the topic, seeing the making of noise as a natural human attribute. The writer sympathetically presents the point of view that our love of noise proceeds from an instinctive aversion to our own thoughts. 'There may be reason in this; melancholy is the natural ally of meditation—joy, on the contrary, is made up of noise.'[4] We are

[1] Virginia Woolf, *The Years* (New York: Harcourt, Brace, 1937), 3.

[2] Ibid. 308. The impress of sound on the unaccustomed ear is even more pronounced in the case of Orlando's encounter with nineteenth-century urban life: 'to her ears, attuned to a pen scratching, the uproar of the street sounded violently and hideously cacophonous'. Virginia Woolf, *Orlando* (New York: Harcourt, Brace, 1928), 27.

[3] On 9 February 1924, Woolf contemplated the property at 52 Tavistock Square into which she and Leonard were planning to move: 'We have been measuring the flat. Now the question arises Is it noisy? No need to go into my broodings over that point. Fitzroy Sqre rubbed a nerve bare which will never sleep again while an omnibus is in the neighbourhood.' *The Diary of Virginia Woolf*, ed. Anne Olivier Bell, asst. Andrew McNeillie (New York: Harcourt Brace Jovanovich, 1978), ii. 291.

[4] 'Noise', *New Monthly Magazine* II (1821), 260.

naturally affected by noise—the argument goes—whether music, thunder, cataracts, shouting, or bells, and 'so naturally agreeable is the sound of noise to the ear, that even its most terrific notes have a proportion of pleasing in them'.[5] This was not, however, to prove a majority point of view as the century progressed. John Picker, in a useful article on Victorian street noises,[6] produces some of the most noteworthy examples of those who possessed an apparent pathological sensitivity to the clamour of their environment, such as Thomas Carlyle, attempting to muffle noise by taking refuge in his soundproofed study (though 'the rattle of a barrel organ and the raucous shouts of street hawkers came through walls whose double thickness distorted but by no means excluded the sound'[7]), and Charles Babbage, who took particular exception to these same sounds—not to mention bagpipes, brass bands, and penny whistles;[8] and who was convinced that his neighbours bribed German bands to play outside his house. Within fiction, mechanical sounds could provoke revulsion against the modernity they seemed to stand for. ' "Those railways!" ', exclaimed Lady Dunstane, in *Diana of the Crossways* (1885). ' "When would there be peace in the land? Where one single nook of shelter and escape from them! And the English, blunt as their senses are to noise and hubbub, would be revelling in hisses, shrieks, puffings and screeches, so that travelling would become an intolerable affliction." '[9]

To be sensitive to noise was, for the later Victorians—as for Lady Dunstane—a means of demonstrating one's superiority in relation to one's fellow creatures. The psychologist James Sully, in a sustained discussion of the topic in 'Civilisation and Noise' (1878), writes in highly similar terms:

> The sufferings which afflict the sensitive ear in our noisy cities are largely due to the general dulness of people with respect to disagreeable sounds. That most persons have not as yet reached a high degree of this sensibility is shown plainly enough in the fact that the rents of houses in the suburbs of London tend to be higher in the neighbourhood of railway stations. This proves that to most people the advantages of rising ten minutes later in the morning are of more account than the discomfort arising from all the shriekings and crashings which are wont to make night hideous in the vicinity of our suburban railways.[10]

For Sully, it was a mark of human evolution—one of the 'glorious gains of civilisation', as he put it—that the 'cultivated European' could enjoy the varied delights of Western music, by contrast to the noises in which a 'savage' took pleasure, produced by rattling snake vertebrae or beating on drums made of untanned hide. He acknowledged, however, that this Spencerian view of the melodic means

[5] Ibid. 263.

[6] John M. Picker, 'The Soundproof Study: Victorian Professionals, Work Space, and Urban Noise', *Victorian Studies* 42 (2000), 427–54. See also Peter Bailey's highly suggestive chapter, 'Breaking the Sound Barrier', *Popular Culture and Performance in the Victorian City* (Cambridge: Cambridge University Press, 1998), 194–211.

[7] Virginia Woolf, 'The London Scene. III. Great Men's Houses' (1932), in Rachel Bowlby (ed.), *The Crowded Dance of Modern Life. Selected Essays: Volume Two* (London: Penguin, 1993), 118.

[8] See Charles Babbage, *A Chapter on Street Nuisances* (London: John Murray, 1864).

[9] George Meredith, *Diana of the Crossways* (1885; London: Constable, 1910), 50.

[10] James Sully, 'Civilisation and Noise', *Fortnightly Review* OS 30 (1878), 715.

of gauging human development had another side to it, since although the 'savage' may lack the capacity of responding empathically to the sounds which 'our more highly developed ear' enjoys, 'he is on the other hand secure from the many torments to which our delicate organs are exposed'.[11] Sully himself, as he tells us in his autobiography, was acutely aware of the interference potential of sound in the modern world, and this article is full of testimony to his edgy irritation. Following Helmholtz, his guide in issues of aural perception as he was in terms of visual response, Sully quotes his mentor on the way in which noise differs from sound. Noise involves the rapid alternation of different kinds of sensations of sound, which 'are irregularly mixed up, and, as it were, jumbled about in confusion'.[12] They thus agitate the ear, which is constituted to prefer smooth tones: irregularity, whether manifested through suddenness, or through jerky beats (especially when accompanied by high-pitched sounds), disturbs—according to Helmholtz—the fibres of this organ.

The effects of unpleasant sound were, for Sully, quite literally painful, and he was far from alone in his susceptibility. The art critic Philip Hamerton, for example, came close to envying his wife her deafness: 'I suffer from the opposite inconvenience of hearing too well. When I am unwell'—and he experienced prolonged bouts of nervous debilitation—'my hearing is preternaturally acute'.[13] This hyper-awareness spilled over into his imagination, so that he could surmise what it would be like to hear as if one's ear came equipped with a microphone, magnifying the sound of his pencil scraping on the page, the footsteps of a fly sounding like the tread of a dray horse.[14] This is strongly reminiscent of George Eliot's famous metaphoric formulation, in *Middlemarch*, of the perils of oversensitivity, which would include hearing the grass grow. In addition to the physiological phenomena that Sully termed 'sensuous pains',[15] he was acutely aware of the power of noise to disturb and distract. Jonathan Crary has recently elaborated on fragmentation and interruption of vision as a key element in the visual aspects of modernity:[16] the same claim can be made in relation to sound, that an awareness of its dislocating and disruptive effects was crucial to perceptual relationships in the modern environment. For Sully, discontinuous sound had the ability to destroy not just concentration, but mental health; he is agonizingly aware that not only the more 'civilized' a human being, but the more given to scholarly or meditative thought she or he might be, the more damaged she or he will sustain by noisy interruptions. Vulnerability goes hand in hand with evolutionary advance.[17]

[11] Ibid. 704.

[12] Ibid. 705.

[13] Philip Gilbert Hamerton, *An Autobiography, 1834–1858, And a Memoir by His Wife, 1858–1894* (London: Seeley & Co., 1897), 586.

[14] Philip Gilbert Hamerton, 'On the Perfection of the Senses', *The Quest of Happiness* (Boston: Roberts Brothers; London: Seeley & Co., 1898), 159.

[15] Sully, 'Civilisation and Noise', 707.

[16] See Jonathan Crary, *Suspensions of Perception. Attention, Spectacle, and Modern Culture* (Boston: MIT Press, 2000).

[17] Sully does acknowledge, however, that this state of things may be temporary: 'If all organic development means increased adaptation to external circumstances, and if noises are a permanent—

Part of the problem with sound, for Sully and others, lay in the very fact of its inescapability. Whilst one can, he notes, 'at will shut off completely, or nearly so, the avenues of the eye . . . nature has, in the case of man, left the ear without any power of self-protective movement'.[18] Edmund Gurney explained in 1880, 'We carry about an habitual instinct of having around us a certain amount of space in which we are alone, and any sudden violence to this instinct is very unnerving.'[19] Particularly disorienting is sudden noise, or that which comes from a direction that one cannot determine (and which hence severs the interpretative relationship between ear and eye). These are variables that underline the perceiver's vulnerability. The assault on personal space is the more pronounced when one considers that noise quite literally invades the body. As Bruce Smith eloquently puts it: 'About hearing you have no choice: you can shut off vision by closing your eyes, but from birth to death, in waking and sleep, the coils of flesh, the tiny bones, the hair cells, the nerve fibers are always at the ready.'[20] Whilst vision, as he goes on to explain, fixes objects out there, away from the perceiving self, the sounds that one hears reverberate inside one. John Tyndall alluded to our apprehension of noise as a physiological presence within our own bodies when, in his influential 1867 book *Sound*, he wrote of how 'Noise affects us as an irregular succession of shocks. We are conscious while listening to it of a jolting and jarring of the auditory nerve.'[21]

*

Tyndall's description of noise as sound resulting from stimuli that cannot be resolved into periodic vibrations was to be frequently repeated into the early decades of the twentieth century, but it was increasingly felt to be inadequate. Physiological definitions remained, but social and psychological ones acquired a new prominence. F. C. Bartlett, professor of experimental psychology at Cambridge, and author of the symptomatically entitled *The Problem of Noise* (1934), explained bluntly that 'Noise is any sound which is treated as a nuisance.'[22] In this book, he addresses the fact that the world is increasingly full of noise, as well as of studies of the perceived problem: studies which variously address

not to say continually increasing—factor in our environment, we must suppose that in time man's organism will become modified so as no longer to suffer from these sources': 'Civilisation and Noise', 711–12.

[18] Ibid. 710. Ruskin was a notable dissenter from this viewpoint, asking his reader to remember that 'the eye is at your mercy more than the ear. "The eye, it cannot choose but see." Its nerve is not so easily numbed as that of the ear, and it is often busied in tracing and watching forms when the ear is at rest': John Ruskin, *Seven Lamps of Architecture* (1849), *The Complete Works of John Ruskin*, ed. E. T. Cook and Alexander Wedderburn, 39 vols. (London: George Allen, 1903–12), viii. 156.

[19] Edmund Gurney, *The Power of Sound* (London: Smith, Elder & Co., 1880), 37.

[20] Bruce R. Smith, *The Acoustic World of Early Modern England. Attending to the O-Factor* (Chicago: University of Chicago Press, 1999), 6. I am greatly indebted to the opening chapter of this book for illuminating my thinking concerning the phenomenology of sound.

[21] John Tyndall, *Sound* (1867; 3rd edn., London: Longmans, Green & Co., 1875), 48.

[22] F. C. Bartlett, *The Problem of Noise* (Cambridge: Cambridge University Press, 1934), 2.

attempts that might be made towards noise abatement, the physiological effects of exposure to too much noise (the deafness of airmen, of tube-train drivers, of boiler-makers), and the psychological problems resulting from noise's effects. This last category, Bartlett says, constitutes for the ordinary person the most important and interesting one. To some extent, his conclusions are very similar to those reached by the commentators of the later nineteenth century: that the producer of a noise is far less likely to be affected and irritated by it than its auditor, and that continuous or regular noise is less provocative than intermittent noise. But he also maintains that its disturbing effects are at their maximum for those who are not only occupied at mental rather than manual work, but also for some reason already wearied with their task, or in some way

> tired, run down, bored, maladjusted, uninterested . . . It is not too much to say that whenever, in any community, a sweeping and passionate condemnation of noise is popular, there, within that community, are almost certainly a lot of people who are ill-adjusted, worried, attempting too much, or too little. The complaint against noise is a sign, sometimes, of a deeper social distress.[23]

To be sensitive to noise, in other words, is to demonstrate a form of neurosis produced by one's very contemporaneity. Rather than exhibiting a finely honed evolutionary condition, within the space of fifty years oversensitivity to noise was characterized as revealing an imperfect capacity of adjustment to one's surroundings and circumstances.

Just as in the later decades of the nineteenth century, in the 1930s noise was seen as synonymous with urban modernity. It was incessant:

> With modern developments in mechanical transport, street noise continues intermittently, even in residential districts, until the early hours of the morning, and starts again shortly afterwards with, say, the earliest of the many milkmen who serve the locality. Slammed car doors, noisy adieux, and cars accelerating violently frequently disturb the hours about midnight, and general night traffic by train and motor vehicle contributes a concatenation of jarring sounds on through routes. One locality may be disturbed by the squeal of trams rounding a corner, and another may experience the lumbering process of heavy lorries on long night journeys between distant industrial centres of the country.
>
> During the day in a great city noise arises from the general roar of traffic, and the city dweller adjusts himself in conversation, by raising his voice without knowing it. He takes cognizance, perhaps, of only the particular contribution of some excessively irritating motor horn or motor-cycle, of the shattering clatter of a pneumatic drill, or of the obliterating crescendo of the tube train in which he is seated. In offices the modern mechanical calculators drive some operators to distraction, and force managements to take steps to reduce the din; in some factories the noise may be so great that conversation is almost impossible, and the clamour in boiler factories reaches the limit at which hearing merges into pain.
>
> In some residential quarters, in such quiet as the evening possesses, aircraft pass noisily overhead, and some owners of gramophones and radio-sets seem oblivious of the fact

[23] Ibid. 52–3.

that their instruments are so grotesquely over-loud that speech items are audible and intelligible for several yards up and down the street.[24]

The potential sources of unpleasant noise were almost endless: turbines, electrical motors, steam hammers, pneumatic drills, pile drivers, cement mixers, miniature rifle ranges, electrical substations, and—a recurrent source of complaint—the riveting machines used to make scaffolding secure. Hand in hand with this cacophony went an increasing amount of legislation aimed against the producers of noise (particularly in London, 'the city of din'[25]), from long-standing by-laws prohibiting the excessive ringing of church bells, the use of noisy instruments to announce entertainments or obtain alms; limiting the practising of bagpipes to certain hours of the week, or forbidding the use of whistles to call cabs (a regulation originally intended to benefit the wounded of the First World War), to laws designed to deal with newer developments. Thus injunctions were issued prohibiting the night-time demolition of buildings; to curtail the outdoor use of gramophones; to ban the use of the motor-car horn in built-up areas between 11.30 p.m. and 7 a.m.; to tax solid-tyred motor-vehicles more heavily; and—in 1929 and 1930—legislation was introduced that banned the use of excessively noisy cars on public roads. The formation of the Anti-Noise League in 1934 brought together various medical, legal, scientific, engineering, and architectural authorities, and employed both exhibitions and deputations to the government to publicize their views on noise suppression. It did not go without ironic comment that the advances in mechanical and electrical science that enabled the measurement of noise—the apparatus used to measure the decibels produced by a car engine, for example—were part of the same technological phenomenon that produced this escalation in unpleasant sounds in the first place.

Unsurprisingly, this urban hubbub does not go unremarked in the fiction of the period. In *Backwater* (1916), for example, Dorothy Richardson's Miriam has problems conversing as she travels into London from the suburbs: 'Through the jingling of the trams, the dop-dop of the hoofs of the tram-horses and the noise of a screaming train thundering over the bridge, Miriam made her voice heard'—but only just.[26] Here noise has achieved the volume one associates with interference, but far more frequent is a sense that it is just something that is there: a dull rumble, an urban constancy. Even though characters in Evelyn Waugh's, Aldous Huxley's, and Elizabeth Bowen's London may rush around in motor cars, displaying a self-conscious modernity in their fretful social restlessness, one rarely hears the sound of the engines, of brakes, of horns (although, outside fiction, *The Waste Land* offers a notable exception). Rather, there is simply an occasional acknowledgement of the fact that—to quote E. M. Forster—'the thoroughfare roared gently—a tide that could never be quiet'.[27] One is far more aware of sound, in the

[24] A. H. Davis, *Noise* (London: Watts & Co., 1937), 1–2.

[25] The phrase is that of Dan McKenzie, *The City of Din, a Tirade against Noise* (London: Adlard & Son, 1916).

[26] Dorothy Richardson, *Backwater* (1916), *Pilgrimage*, 4 vols. (London: J. M. Dent, 1967), i. 192.

[27] E. M. Forster, *Howards End* (1910: London: Edward Arnold, 1973), 42.

fiction of the early twentieth century, when it momentarily ceases, than when it is being made. This is true, too, in early Woolf, as when Katherine and Ralph, towards the close of *Night and Day*, come to an intimate understanding in the silent streets of night-time London. But after her first couple of novels—as her fiction becomes more experimental both in form and in representation of consciousness—so Woolf's attitude towards noise also seems to change. By contrast with her contemporaries' work, Woolf's fictional treatment of the sounds of the city is frequently a celebratory one. More than that, she continually makes her reader aware of the acoustic environment.

Nowhere is this celebratory stance more apparent than in *Mrs Dalloway*. For Elizabeth Dalloway, taking a bus into the City of London and walking around its streets, temporarily free of Miss Kilman, and feeling simultaneously both a sense of adult independence when it comes to decision-making and a tie to a paternal heritage of public service, the urban roar is empowering: 'She liked the geniality, sisterhood, motherhood, brotherhood of this uproar. It seemed to her good. The noise was tremendous; and suddenly there were trumpets (the unemployed) blaring, rattling about in the uproar; military music; as if people were marching . . .'.[28] The narratological endorsement of this scene is not completely wholehearted. Elizabeth's exhilaration prevents her from responding to the fact that the post-war unemployed, searching for work, might have some serious grievances against society, for example: the fact that the circumstantial detail concerning the trumpet-blowers is provided for the reader gently underscores Elizabeth's position of privilege. Moreover, the narrative voice suddenly slips out of Elizabeth's consciousness to hypothesize that if someone were to open the window of a room in which a woman had just died, the music would come up to him as 'consolatory, indifferent'. This seems, on the surface, a strange dislocation of point of view, jarring against the girl's awareness of youth and possibility (although it is a dislocation that implicitly links her with her mother, into whose quivering alertness to life the fact of death recurrently intrudes). Yet it serves a wider purpose. Woolf emphasizes that 'this voice, pouring endlessly, year in year out'—and it's not quite clear whether she refers just to the militaristic music, or to the whole roar of the city—'would take whatever it might be; this vow; this van; this life; this procession, would wrap them all about and carry them on, as in the rough stream of a glacier the ice holds a splinter of bone, a blue petal, some oak trees, and rolls them on'.[29]

City noise, for Woolf, implies continuity: even the interruptive sounds that so annoyed her contemporaries and forebears can be assimilated, like fragments of urban archaeology, into a broader continuum, whether diachronic or synchronic. This process is made particularly clear with the episode near the beginning of *Mrs Dalloway* where the pistol-shot-like sound of a car backfiring causes Clarissa to jump and the deferential flower-shop saleswoman Miss Pym to apologize, and,

[28] Virginia Woolf, *Mrs Dalloway* (New York: Harcourt, Brace & Co., 1925), 209.

[29] Ibid. 208–9.

above all, serves to focus the attention of a random cross-section of London's population on the vehicle. Startled into auditory consciousness by the sudden report, they become as if made one by the invasive properties of traffic noise: 'The throb of the motor engines sounded like a pulse irregularly drumming through an entire body.'[30]

The capacity of one individual to merge and blend a whole range of sounds is commented on by Louis in *The Waves*:

'The roar of London,' said Louis, 'is round us. Motor-cars, vans, omnibuses pass and repass continuously. All are merged into one turning wheel of single sound. All separate sounds—wheels, bells, the cries of drunkards, of merry-makers—all churned into one sound, steel-blue, circular. Then a siren hoots. At that shores slip away, chimneys flatten themselves, the ship makes for the open sea.'[31]

In the novel's final section, Bernard's response to the sounds of the city emphasizes less the connective possibilities of sensory alertness to the present (although, as ever with his desire to hold and merge, these are apparent too) than the interlayering of past and present. He claims to be

drawn irresistibly to the sound of the chorus chanting its old, chanting its almost wordless, almost senseless song that comes across courts at night; which we hear now booming round us as cars and omnibuses take people to theatres. (Listen; the cars rush past this restaurant; now and then, down the river, a siren hoots, as a steamer makes for the sea.)[32]

The echoing of Louis's earlier phrase performs the connection in a somewhat artificial way, but the underlying desire for the linkage between ages that may be formed by noise is a much broader one. Woolf employs the concept of wordless, senseless song to suggest continuity in human emotion when, in *Mrs Dalloway*, she introduces the 'battered old woman' frailly warbling her ages-old lovesong outside Regent's Park tube station.[33] As these examples indicate, Woolf's welcoming of noise of various kinds is repeatedly bound up with the desire to acknowledge human connections. Awareness of sound is unwilled; similarly, our links with others may not be welcomed, but they are as inescapable as is the cacophony of the city. One's response to noise may, therefore, in Woolf's fiction, be read as an index to a character's degree of comfort with that condition. In *Jacob's Room*, a novel concerned from the first page with epistolary correspondence, the anxiety to communicate that lies behind the penned message is linked with other forms of exchanging words that technological development has made possible:

And the notes accumulate. And the telephones ring. And everywhere we go wires and tubes surround us to carry the voices that try to penetrate before the last card is dealt and the days are over. 'Try to penetrate,' for as we lift the cup, shake the hand, express the hope,

[30] Ibid. 20. Compare the effect (visual and aural) of the aeroplane in this novel, and elsewhere in Woolf's writing: see Gillian Beer, 'The Island and the Aeroplane: The Case of Virginia Woolf', in Homi K. Bhabha (ed.), *Nation and Narration* (London: Routledge, 1990), 265–90.

[31] Virginia Woolf, *The Waves* (New York: Harcourt, Brace & Co., 1931), 135.

[32] Ibid. 246.

[33] Woolf, *Mrs Dalloway*, 123.

something whispers, Is this all? Can I never know, share, be certain? Am I doomed all my days to write letters, send voices . . .?[34]

Virginia Woolf's concern with sound and technology has started to receive a certain amount of attention, although much of this has clustered around the conspicuous political threat signalled by the ominous 'halls and reverberating megaphones' that so trouble North in the Present Day section of *The Years*, and the implications of the ticking of the gramophone (and the anxieties caused by worn grooves or unexpected blaring) in *Between the Acts*—a novel that is composed as an aural collage of interpenetrating rural, human, and mechanical sounds.[35] In two illuminating essays, Gillian Beer has written of the links between the composition and ideas informing Woolf's later fiction and the ways in which wave-particle theory caught the imagination in the late 1920s and early 1930s.[36] Melba Cuddy-Keane has also written interestingly on the author's relationship to sound technologies. She suggests that what she calls a 'new aural sensitivity' was coincident with the emergence of the gramophone and wireless.[37] Although I have been arguing that the context of this sensitivity was in fact far wider, Cuddy-Keane is highly suggestive about the ways in which an expanded listening audience, actualized with the advent of broadcasting, may have enabled a conceptualization of the city as consisting of many different potential listening points. She usefully reminds us that, whilst 'mainstream developments in broadcast technologies did indeed inspire justifiable fears about state control, hegemonic dominance, and a passive public',[38] Woolf herself was by no means uniformly pessimistic about these developments. Indeed, in her diary she records challenging Harold Nicholson's views on Empire by suggesting that new technologies might well collapse narrow concepts of nationhood. ' "But why not grow, change?" I said. Also, I said, recalling the aeroplanes that had flown over us, while the portable wireless played dance music on the terrace, "can't you see that

[34] Virginia Woolf, *Jacob's Room* (New York: Harcourt, Brace & Co., 1922), 150.

[35] See Michele Pridmore-Brown, '1939–40: Of Virginia Woolf, Gramophones, and Fascism', *PMLA* 113 (1998), 408–21; Bonnie Kime Scott, 'The Subversive Mechanics of Woolf's Gramophone in *Between the Acts*', in Pamela Caughie (ed.), *Virginia Woolf in the Age of Mechanical Reproduction* (New York: Garland Publishing, 2000), 408–21.

[36] See Gillian Beer, 'Physics, Sound, and Substance: Later Woolf', *Virginia Woolf: The Common Ground. Essays by Gillian Beer* (Ann Arbor: University of Michigan Press, 1996), 112–24; and ' "Wireless": Popular Physics, Radio and Modernism', in Francis Spufford and Jenny Uglow (eds.), *Cultural Babbage: Technology, Time and Invention* (London: Faber & Faber, 1996), 149–66.

[37] Melba Cuddy-Keane, 'Virginia Woolf, Sound Technologies, and the New Aurality', in Caughie (ed.), *Woolf in the Age of Mechanical Reproduction*, 71.

[38] Ibid. 94. Cuddy-Keane helpfully proposes the term 'auscultation', the action of listening, as a narratological term to parallel focalization; she draws attention to the experimental quality of Woolf's use of 'ambient noise and environmental sound', which she links to her 'nonhierarchical, noncentered treatment of multiple voices', and she proposes, above all, that we pay attention to Woolf's 'heightened focus on sound as aural experience rather than intermediary for a nonaural signified'. Ibid. 71, 85, 90. Compare Edmund Carpenter's earlier suggestion of the term 'earpoint' as an alternative to 'viewpoint': Carpenter and Marshall McCluhan (eds.), *Explorations in Communication* (Boston: Beacon Press, 1960), 65–70.

nationality is over? All divisions are now rubbed out, or about to be." '[39] Yet Woolf's awareness of the potentially unifying effects of sound goes hand in hand with her registering the reverse of this process. Her inclusion of the occasional moment of sound that resists interpretation signifies a fear of the lack of communication that forms one of the central problems of social and political life in the 1930s. Thus in *The Years*, the strange ditty sung by the caretaker's children at Delia's party seems to symbolize an alarming gap: between classes, between generations, and—above all—between past and future. Rather than melody, however polyphonic, the sounds emitted by these youngsters suggest a jarring, and incomprehensible dissonance. In turn, Woolf's responsiveness to the different possibilities of sound demonstrates how her own deployment of its nature and effects resists any neat, harmonious reading. What she does do, however, is to make the reader continuously alert to the importance of attentive listening, rather than passive acceptance of ambient noise.

Woolf is well aware that consciously registered sounds can be the starting place for associative speculation. Thus Jacob, walking back into London after the bonfire party, hears his feet on the pavement, and—somewhat pretentiously—'it seemed to him that they were making the flagstones ring on the road to the Acropolis'.[40] For the more disturbed Septimus, the sparrows sing—as they had done for Woolf herself—piercingly in Greek. Moreover, one projects one's emotions into sounds, or one's feelings may be characterized as a kind of inner noise. When Orlando realizes that Sasha is not going to elope with him, the bells that toll in the deep, inky night of Elizabethan London 'see[m] to ring with the news of her deceit and derision'.[41] Clarissa registers her jealous hatred of Miss Kilman as primeval sound, a 'brutal monster' stirring within her, like twigs cracking and hooves being planted down 'in the depths of that leaf-encumbered forest, the soul'.[42] At other times, it can be curiously hard to determine who does the listening in Woolf's fiction. In *Jacob's Room*, when Florinda and Nick meet in a café on a hot afternoon, 'The door opened; in came the roar of Regent Street, the roar of traffic, impersonal, unpitying; and sunshine grained with dirt.'[43] Although these invasive noises are indubitably nuanced by Florinda's nervous, half-dulled, half-hysterical pregnant state, they are not overtly filtered through her aural attention: rather, they become assimilated into a composite impression of the oppressive atmosphere. As with the recording of the 'roar' of the shawl's loosening fold that disturbs the 'swaying mantle of silence' in the empty house in *To the Lighthouse*,[44] sounds, detached from those who might consciously register them, form part of Woolf's recurrent experiments with an impersonalized narrative voice.

[39] Virginia Woolf, *The Diary of Virginia Woolf*, ed. Anne Olivier Bell, asst. Andrew McNeillie, 5 vols. (London: Hogarth Press, 1979–84), iii. 145.

[40] Woolf, *Jacob's Room*, 123.

[41] Woolf, *Orlando*, 61.

[42] Woolf, *Mrs Dalloway*, 17.

[43] Woolf, *Jacob's Room*, 276.

[44] Virginia Woolf, *To the Lighthouse* (New York: Harcourt, Brace & Co., 1927), 196.

As we have seen, many of the complaints made against noise, whether in the later decades of the nineteenth century or in the twentieth century, inveighed against its invasive qualities, its capacity to force itself on an individual, so that she or he had no option but to be aware of it—as the sound of the aeroplane in *Mrs Dalloway* 'bored ominously into the ears of the crowd.'[45] But hearing, as has already been noted, has frequently been figured as a different thing from listening. 'Hearing,' wrote Roland Barthes, 'is a physiological phenomenon; *listening* is a psychological act. It is possible to describe the physical conditions of hearing (its mechanisms) by recourse to acoustics and the physiology of the ear; but listening cannot be defined only by its object or, one might say, by its goal.'[46] Indeed, to be overconscious of hearing—of the bodily activity, rather than the perceptual filtering and interpretation of impressions, may well be—Woolf suggests—to be vulnerable to being classified as neurotic, rather as Bartlett suggested. It certainly is symptomatic of the shell-shocked Septimus Smith's condition that he hears a nursemaid in the park, reading the sky-writing, spell out ' "Kay Arr" ' close to his ear, 'deeply, softly, like a mellow organ, but with a roughness in her voice like a grasshopper's, which rasped his spine deliciously and sent running up into his brain waves of sound which, concussing, broke.'[47]

Barthes distinguishes three types of listening: *alert*, which is shared by humans and animals (a wolf listens for the possible noises of its prey, a child or lover for footsteps which might be those of the mother or the beloved); *deciphering*, where the ear tries to interpret certain sounds, and that which involves the *development of intersubjective space*. Again, he draws a parallel between human and animal:

For the mammal, its territory is marked out by odors and sounds; for the human being—and this is a phenomenon often underestimated—the appropriation of space is also a matter of sound: domestic space, that of the house, the apartment—the approximate equivalent of animal territory—is a space of familiar, *recognized* noises whose ensemble forms a kind of household symphony.[48]

This description of the appropriation of familiar space through listening describes perfectly Clarissa Dalloway's aural alertness to her home environment, using the ambient domestic sounds as a means of assembling her own sense of individuation:

Strange, she thought, pausing on the landing, and assembling that diamond shape, that single person, strange how a mistress knows the very moment, the very temper of her house! Faint sounds rose in spirals up the well of the stairs; the swish of a mop; tapping; knocking; a loudness when the front door opened; a voice repeating a message in the basement; the chink of silver on a tray . . .[49]

[45] Woolf, *Mrs Dalloway*, 29.

[46] Roland Barthes, 'Listening', in *The Responsibility of Forms: Critical Essays on Music, Art, and Representation*, trans. from the French by Richard Howard (Oxford: Basil Blackwell, 1985), 245.

[47] Woolf, *Mrs Dalloway*, 32.

[48] Barthes, 'Listening', 246.

[49] Woolf, *Mrs Dalloway*, 56.

Clarissa listens here to 'the voice of the house—and all houses have voices':[50] an anthropomorphization that underscores yet further human filiation with one's environment.

For Barthes, it is important, if we are to practise listening at its best, that we should make 'no effort to concentrate the attention on anything in particular, but to maintain in regard to all that one hears the same measure of calm quiet attentiveness'.[51] This is a procedure which, in its deliberate avoidance of perceptual preselection, is one that he parallels to Freud's advice to analysts in his *Recommendations for Physicians on the Psychoanalytic Method of Treatment*:

> The analyst must bend his own unconscious like a receptive organ towards the emerging unconscious of the patient, must be as the receiver of the telephone to the disc. As the receiver transmutes the electric vibrations induced by the sound waves back again into sound waves, so is the physician's unconscious mind able to reconstruct the patient's unconscious which has directed his associations, from the communications derived from it.[52]

Freud found himself drawn to the language of new communications technology to describe the process of transmission and interpretation that must take place in attentive listening: the type of listening that Mrs Dalloway practises as the sounds of the house spiral up the stairwell. And it is precisely this idea of attentive listening, or aural alertness, that blurs the distinctions between hearing and listening that some commentators on the activity of the ear would like to establish, and which demands, moreover, that we understand the listening subject as someone who is in a constant process of interchange with their environment.

Thus in the longer run, rather than look to technological language to describe the effects that Woolf is striving after, recent terminology in anthropology may prove more enabling. Steven Feld, building on the conceptualizations of 'auditory space' that have emerged since the 1950s, has proposed the term 'acoustemology, by which I mean local conditions of acoustic sensation, knowledge and imagination embodied in the culturally particular sense of place'.[53] Such a term encompasses very usefully Woolf's understanding of the social shared aspects of an acoustic environment, whether the clock chimes that sound out across the city, or the particular combination of sound and space that produces the resonances of traffic sounds and hooting horns across the expanses of a London park, and the combination of this social knowledge with the physical and mental experiences of a given individual. For her, as for Feld,

[50] Woolf, 'The London Scene', 118.

[51] Barthes, 'Listening', 252.

[52] Sigmund Freud, *Recommendations for Physicians on the Psychoanalytic Method of Treatment* (1912), quoted in Barthes, 'Listening', 252.

[53] Steven Feld, 'Waterfalls of Song. An Acoustemology of Place Resounding in Bosavi, Papua New Guinea', in Steven Feld and Keith H. Basso (eds.), *Senses of Place* (Santa Fe: School of American Research Press, 1996), 91.

sound is central to making sense, to knowing, to experiential truth. This seems particularly relevant to understanding the interplay of sound and felt balance in the sense and sensuality of emplacement, of making place. For bodies are as potentially reverberant as they are reflective, and one's embodied experiences and memories of them may draw significantly on the interplay of that resoundingness and reflectiveness.[54]

A writer in *Macmillan's Magazine* for 1893 claimed that

The supremacy of one sense over all the others is now so completely established that the world of our waking moments is a world of sights, even as the world of our dreams is a world of visions. We are always looking, and but rarely listening; always attending to the shapes and colours before our eyes, seldom noticing the sounds which reach our ears. The visible has become the real, while the audible and the tangible appear but as casual properties of the visible.[55]

Sight, this author suggests, is the sense of the modern: a view endorsed by Woolf, one might say, in her 1927 essay 'Street Haunting', when she imagines the individual employing 'a central oyster of perceptiveness, an enormous eye':[56] the primary organ, in this piece, for assimilating and entering into the urban environment. Yet she admits, here, that the eye 'is not a miner, not a diver, not a seeker after buried treasure';[57] rather, it keeps one tied to the surface. The imagery she employs of the body's own surface, however, figuring it as 'the shell-like covering which our souls have excreted to house themselves',[58] suggests the potential for a more complex nacreous structure than that of the bivalve: the shape to which the folding passageways and chambers of the ear have frequently been compared. The writer of the *Macmillan's* article goes on to hypothesize that 'The present complete ascendancy of sight prevents us from realising that there might have been, and probably was, a time in the past history of man when sounds were of far more importance relatively to sights than they are at present.'[59] In Woolf's work, we see this importance reasserted. The attention she pays to the acoustic world within her fiction distinguishes her from many of her contemporary commentators on urban life, who register city noise in terms of damaging invasiveness. Whilst Woolf's representation of noise does not have the onomatopoeic directness of Joyce's newspaper presses juddering away in *Ulysses*, its place in both individual and collective consciousness is readily acknowledged. Indeed, so prominent a place does Woolf accord listening in our perception; giving due weight to its associative and connective powers, that in addition to that 'enormous eye', she demands that we pay attention with, and to, the organ which Lucy Swithin imagines in *Between the Acts*: a 'gigantic ear attached to a gigantic head'.[60] This instrument, as Mrs Swithin would have it, is capable of producing harmony even out of discordant noise—although such harmony is not always

[54] Feld, 'Waterfalls of Song', 97.
[55] J.B.C., 'In the Realm of Sound', *Macmillan's Magazine* 67 (1893), 438.
[56] Virginia Woolf, 'Street Haunting: A London Adventure' (1927), in Bowlby, *Crowded Dance*, 71.
[57] Ibid. 71.
[58] Ibid. 71.
[59] J.B.C, 'In the Realm of Sound', 438.
[60] Virginia Woolf, *Between the Acts* (New York: Harcourt, Brace & Co., 1941), 175.

achievable by those with a less optimistic world-view. Individual subjectivity inevitably conditions how we register noise, which, for Woolf, is more internalized affect than an objectively quantifiable property of the external world. This internalization is quite literal, since the ear is the instrument, too, which has the potential to register the reverberations of this outer environment as reverberations within the body's own organs. It is thus ideally poised to reveal the connections that exist between people in the urban environment: connections that, like noise itself, may be welcomed or rebuffed, but which are, none the less, an unalterable fact of Woolf's perception of the modern world.

CHAPTER 12

'Chloe Liked Olivia': The Woman Scientist, Sex, and Suffrage

MAROULA JOANNOU

'Chloe liked Olivia. They shared a laboratory.'[1] The well-known line from Virginia Woolf's A *Room of One's Own* (1928) evokes the *frisson* of attraction between women engaged in scientific work in a medical laboratory. The two women are mincing liver, apparently a cure for pernicious anaemia. But who were Chloe and Olivia? Women workers in early twentieth-century science laboratories were few and far between. And what if, instead of sharing their laboratory amicably as women scientists, Chloe and Olivia found themselves the objects of unwanted sexual interest and rivalry on the part of male scientists?

This chapter explores the representation of the woman scientist in three early twentieth-century novels: H. G. Wells's *Ann Veronica* (1909) and what I take to be two feminist responses to it some fifteen years on: Edith Ayrton Zangwill's *The Call* (1924) and Charlotte Haldane's *Man's World* (1926).[2] Wells's novel, the best remembered of the three works, established a powerful and provocative typology for the representation of the female scientist—a typology that Zangwill and Haldane both endorsed in several important respects but that they were also determined to resist. Wells took as his protagonist a gifted young woman scientist who chooses to become a militant suffragette, and who finds both her work and her consequent political commitments bringing her into conflict with the male scientist she loves. *Ann Veronica* thus offered to later novelists a multi-plot model for thinking about female science: one in which the narrative of scientific interest and discovery was in tension both with the narrative of heterosexual romance and with a developing narrative of feminist activism.

Ann Veronica also proved predictive in its strong reliance on the author's own experience of women scientists. Wells modelled his heroine, Ann Veronica, on the young Amber Reeves, whom the married Wells had seduced when she was a student at Newnham College, Cambridge. But the novel reverberates, too, with an earlier illicit love affair with Amy Catherine Robbins, a New Woman for whom Wells left his first wife, Isabel, in 1894 and whom he later married. They met when

[1] Virginia Woolf, *A Room of One's Own* (London: Hogarth Press, 1928), 181.

[2] H. G. Wells, *Ann Veronica* (London: Mills & Boon, 1909), Edith Zangwill, *The Call* (London: Allen & Unwin, 1924), Charlotte Haldane, *Man's World* (London: Chatto & Windus, 1926). All quotations are from the first editions and given parenthetically.

he was a tutor of biology at the College of Preceptors and she was his student: 'We were the most desperate of lovers; we launched ourselves upon our life together with less than fifty pounds between us and absolute disaster.'[3]

Edith Ayrton Zangwill's *The Call* also dramatizes the sexual competition and sexual attraction between men and women who meet through their interest in scientific work, and whose conflict in the laboratory leads the heroine to direct involvement in suffragette politics. *The Call* is loosely based on the life of Zangwill's stepmother, the renowned physicist and suffragette, Hertha Ayrton, who died the year before publication. Zangwill takes key motifs from *Ann Veronica*, such as a mature male scientist's infatuation with a beautiful young student, and gives them a feminist inflection. Charlotte Haldane's *Man's World* (1926) was inspired by Haldane's love affair with her scientific mentor and husband-to-be, the geneticist J. B. S. Haldane (a relationship described in her 1949 autobiography, *Truth Will Out*). Set in the future, *Man's World* draws on his notorious 1923 'Daedalus', which predicted, among other things, 'the growing of a human foetus in the laboratory'.[4] Unlike *The Call*, *Man's World* relies on the standard tropes of romantic fiction,[5] taking its lead from J. B. S. Haldane's romanticization of the male scientist. ('Daedalus' concludes with the assertion that 'the biologist is the most romantic figure on earth.'[6]) But Charlotte Haldane also cites Wells as a key influence on *Man's World*—and in her view, unlike Zangwill's, *Ann Veronica* stands as an exemplary feminist work: '*Ann Veronica* and *Joan and Peter* were my contemporaries . . . Wells's girls were, like myself, sincere, honest, puzzled, and determined to be worthy of their noble feminist ancestresses.'[7]

*

Ann Veronica is probably the first New Woman novel whose heroine is also a New Scientist. Ann Veronica attends anatomy classes at the Tredgold Women's College, where she chafes at the intellectual limitations of the 'lady instructor' who 'retail[s] a store of faded learning' and is 'hopelessly wrong and foggy' about the composition of the skull (5). Her own work is 'high above the normal female standard' (133). (Wells is almost certainly jibing at the reputedly low standards of teaching in Bedford College, the subject of a heated correspondence initiated by Karl Pearson in the *Pall Mall Gazette*.[8]) Having escaped from the strict control of her father and the watchful eyes of her Victorian maiden aunt, Ann Veronica

[3] H. G. Wells, *The Book of Catherine Wells* (London: Chatto & Windus, 1928), 11.

[4] *Truth Will Out* (London: Weidenfeld & Nicolson, 1949), 17.

[5] Susan Merrill Squier, *Babies in Bottles: Twentieth-Century Visions of Reproductive Technology* (New York: Rutgers University Press, 1994), 119.

[6] Quoted in Judith Adamson, *Charlotte Haldane: Woman Writer in a Man's World* (London: Macmillan, 1998), 53.

[7] *Truth Will Out*, 15.

[8] Rosaleen Love, ' "Alice in Eugenics Land": Feminism and Eugenics in the Scientific Careers of Alice Lee and Ethel Elderton', *Annals of Science* 36 (1970), 145–58 (146).

enrols at Imperial College, home of 'the infidel Russell' (24). Wells shifts the location from Cromwell Road in South Kensington to the area between Euston Road and Great Portland Street. Imperial College had grown out of the Normal School of Science where Wells had attended Huxley's lectures, securing a first-class degree in zoology in 1890. The influence of Huxley's work on evolution for his subsequent intellectual development was profound. In his autobiography Wells would pay tribute to the 'fundamental magnificence of Darwin and Huxley's achievement' in putting the 'fact of organic evolution upon an impregnable base of proof and demonstration'.[9]

Ann Veronica expresses Wells's view of biology as the lynchpin on which social behaviour revolves. The *dramatis personae* reflects debates between the 'new' and the traditional in the field. The Cambridge Mendelians, mentioned early on, were opposed to 'the big names of the 1890s' (24). The rediscovery of Mendelian genetics in 1900 had illuminated the enigma of heredity. Mendel argued that children acquired 'free' genetic substances from both parents and that the physical attributes of each partner therefore mattered. Affirming the importance of sexual selection in human mating, and opposing Victorian prudery, this modern science appealed to socialist eugenicists. Its attraction for Wells was that 'practically all its questions and phenomena lie within the scope of normal experience'.[10] Of all the branches of science biology made the most 'vivid, sustained attempt to see life clearly and to see it whole'.[11] The key insight the novel affords is that the New Woman is herself the end of a long evolutionary process. Although Wells was 'far from endorsing the Social Darwinists' view of a moral order entirely compatible with, because exactly analogous to, the natural scheme of conflict in a brute struggle for existence', he did believe that human beings were still 'subject to instinctual drives derived from ape-like ancestors'.[12] By a process of extension and concatenation, biology—with its tentacular generalizations—becomes the novel's dominant source of metaphors whereby the reader is invited to understand a number of disparate contemporary social phenomena, including the sexual behaviour of men and women. But if *Ann Veronica* pinpoints the tensions and contradictions inherent in biological definitions of women, it also compounds the confusions inherent in attempts to explain social and political relations by means of biological discourse.

Biology, in *Ann Veronica*, is an 'extraordinarily *digestive* science. It throws out a number of broad experimental generalisations, and then sets out to bring into harmony or relation with these an infinitely multifarious collection of phenomena . . . stretching out further and further into a world of interests that [lie] altogether outside their legitimate bounds' (134). In the pursuance of her scientific interests it dawns upon Ann Veronica that

9 *Experiment in Autobiography: Discoveries and Conclusions of a Very Ordinary Brain (since 1866)*, 2 vols. (London: V. Gollancz; Crescent Press, 1934), ii. 203.

10 Ibid. 203.

11 Ibid. 210.

12 See Robert M. Philmus and David Y. Hughes (eds.), *H. G. Wells: Early Writings in Science and Science Fiction* (Berkeley: University of California Press, 1975), 179–80.

this slowly elaborating biological scheme had something more than an academic interest for herself. And not only so, but that it was, after all, a more systematic and particular method of examining just the same questions that underlay the discussions of the Fabian Society, the talk of the West Central Arts Club, the chatter of the studio, and the deep, bottomless discussions of the simple-life homes. It was the same Bios whose nature and drift and ways and methods and aspects engaged them all. (134)

Much of *Ann Veronica* takes place in the modern biological laboratories where all animal life comes to be viewed as 'pairing and breeding and selection, and again pairing and breeding' and where Ann Veronica's own desire for love comes to seem 'only a translated generalisation of that assertion' (142). Here she learns to understand herself primarily as a biological being: 'she, in her own person too, was the eternal Bios, beginning again its recurrent journey to selection and multiplication and failure or survival' (134). She enquires how her own notions of beauty fit into the scheme of the survival of the fittest. Was it that the 'struggle of things to survive produced as a sort of necessary by-product these intense preferences and appreciations?' (146). She is continually reminded of her own affinities to the animal world. Examining the fine hair on her arm she remarks, 'etherealised monkey' (148). On a visit to the Zoological Gardens she admires the gentle eyes of the chimpanzees 'so much more human than human beings' (223). The suffragettes in prison with their 'barkings, yappings, roarings, pelican chattering, and feline yowlings, interspersed with shrieks of hysterical laughter' will be likened by the unsympathetic narrative voice to carnivora (205).

More specific than its debt to Wells's interest in biology is the testimony *Ann Veronica* makes to his belief in eugenics. Like many other literary figures, including Grant Allen and George Bernard Shaw, Wells saw conventional sexual morality as an obstacle to better breeding and racial regeneration. He would readily have agreed with the view expressed in Havelock Ellis's 1906 essay 'Eugenics and St. Valentine', in which Ellis argued that a modern St Valentine would be a saint of science and that sexual attraction was essentially eugenic in its nature.[13] This interest in eugenics lends a particular twist to Wells's representation of women in science. More and more women were entering the sciences in the early decades of the twentieth century, and in the main they met with resistance. Dorothy Needham recollected that Gowland Hopkins' Biochemistry laboratories were the only place in Cambridge University where women were accepted when she began her scientific career there in 1920.[14] However, five of the fourteen researchers working in the Eugenics Laboratory at University College, London in the first decade of the twentieth century were women.[15] There, at least, eugenics offered women something approaching equivalent status to that of male scientific

[13] *Nineteenth Century and After* 49 (May 1906), 779–87.

[14] Dorothy Needham, 'Women in Cambridge: Biochemistry', in Derek Richter (ed.), *Women Scientists: The Road to Liberation* (London: Macmillan, 1982), 161.

[15] Richard Allen Soloway, 'Feminism, Fertility and Eugenics in Edwardian Britain', in Seymour Drescher *et al.* (eds.), *Political Symbolism in Modern Europe* (New Brunswick: Transaction Books, 1982), 121–46 (121).

researchers. But eugenics also did much to lend force to arguments for keeping women out of the sciences. A major practical obstacle to women who were working for the vote, or breaking the mould through new types of work, was the pervasive belief in biological imperatives to which eugenics lent new authority. The problem was experienced particularly keenly by educated New Women in the sciences because many prominent eugenicists, like Galton, who coined the term 'eugenics' in 1880, perceived eugenics essentially as an 'extension of nineteenth-century social Darwinism'. For such men, the resulting politics of science were highly conservative,[16] with little sympathy shown for the collectivist proposals of social reformers who urged the primacy of nurture over nature.

Although eugenic ideals have retrospectively come to be associated with the right of the political spectrum, and the pronatal policies of the Nazis in particular, they also had a strong influence on the left in the early part of the twentieth century. Sylvia Pankhurst, for example, a socialist, suffragist, feminist, and single parent by choice, spoke of her pride in her 'eugenic' baby: 'It is good eugenics, I believe if one desires parenthood, to consider if one is of sufficient general intelligence, bodily health and strength, and freedom from hereditary diseases to produce an intelligent and healthy child. I believe that of myself. I also believe that of my baby's father.'[17] A major area of contention between eugenicists of the left and right was therefore the relative importance to be attached to environmental rather than hereditary factors in determining human behaviour. It was possible for leading eugenicists to disagree fundamentally over the role of women in the workplace—including, of course, science. Karl Pearson, the most prominent of the socialist eugenicists, wrote to Galton in defence of the women research scientists at UCL: 'They are women who in many cases have taken higher academic honours than men and are intellectually their peers. They were a little tried therefore when your name appeared on the Committee of the Anti-Suffrage Society.'[18]

Wells's novel never strays far from the biological imperative of race-motherhood[19] by which eugenicists defined the modern woman in an evolving society. In *Ann Veronica* 'Endowment of Motherhood' is written in letters of light across the 'cloud paradise of an altered world in which [. . .] the Fabians and reforming people believed' (184). As Lesley Hall has noted, the key problem with eugenics for many women was the passive role that it ascribed to them. Women were either genetically good, fit, stock, 'in which case (provided that they were married) they were supposed to have as many children as they could, to replenish the nation, or, if they were of unsound stock, they were to refrain from breeding'.[20] Ann Veronica falls into the former category and enthusiastically embraces the prospect

[16] Diane B. Paul, 'Eugenics and the Left', *Journal of the History of Ideas* 45/4 (1984), 567–89 (568).

[17] Sylvia Pankhurst, *The News of the World*, 8 April 1928. Quoted in Patricia W. Romero, *E. Sylvia Pankhurst: Portrait of a Radical* (New Haven: Yale University Press, 1987), 169.

[18] Quoted in Love, ' "Alice in Eugenics Land" ', 146.

[19] For a discussion of race-motherhood see Anna Davin, 'Imperialism and Motherhood', *History Workshop Journal* 5 (1978), 9–66.

[20] Lesley Hall, 'Women, Feminism and Eugenics', in Robert A. Peel (ed.), *Essays in the History of Eugenics* (London: The Galton Institute, 1998), 36–52 (37).

of maternity. As a student, she forms a romantic attachment to Capes, a married scientific demonstrator whom she meets, appropriately enough, when she is dissecting a dogfish. Ann Veronica and Capes are cast as the defenders of those aspects of Modern Science which will free humanity from sexual repression. Through their eugenic desire for each other, the novel expresses Wells's deep-rooted objection to the social practices and prejudices that place impediments in the way of better breeding. As he later wrote in *The New Machiavelli*, 'It is not so much moral decadence that will destroy us as moral inadaptability. The old code fails under the new needs.'[21] Ann Veronica speculates more tentatively that if 'in some complex yet conceivable way women were endowed, were no longer economically and socially dependent on men' (184) this would give her the freedom to go to Capes without burdening him with obligations.

Ann Veronica's adventures are those of the classic New Woman struggling to exert her right to her autonomy and freedom. Had she lived twenty years earlier, Wells suggests, she 'would have been called a Young Person' whose chief duty in life would have been 'not to know, never to have heard of, and never to understand' (62). But as a vulnerable young woman trekking the streets in search of work Ann Veronica comes to recognize that nearly all the things that a woman can do to support herself will demand all her time and energy and provide only subsistence wages. Moreover, she is sexually molested, chafes against the almost unavoidable obligation to some individual man, and has to resist entreaties from her brother Rodney to return home: 'the only possible trade for a girl that isn't sweated is to get hold of a man and make him do it for her . . . it's providence. That's how things are; and that's the order of the world' (100). All this persuades her of the justice of the woman's cause and propels her into the arms of the militant suffragettes. She participates in a raid on the House of Commons—the suffragettes had organized the 'pantechnicon raid' on parliament, on which the fictional one is based, in February 1908—and becomes a prisoner in Canongate. But Wells's purpose in consigning his protagonist to prison appears to be to reinforce her dislike of the other women and to confirm her liking for men: 'I've got no feminine class feeling. I don't want any laws or freedoms to protect me from a man like Mr Capes' (206). Her final conclusion is that 'A woman wants a proper alliance with a man, a man who is better stuff than herself . . . She wants to be free—she wants to be legally and economically free, so as not to be subject to the wrong man; but only God, who made the world, can alter things to prevent her being slave to the right one' (206).

Wells was not opposed to women's suffrage but was deeply hostile to the social purity and anti-male strands within the organized feminist movement which culminated in the slogan of the Women's Social and Political Union (the militant wing of the women's suffrage movement) 'Votes for Women and Chastity for Men'. He believed that Free Love 'and not any petty political enfranchisement must surely constitute the real Magna Carta of women'. But because the notion

[21] H. G. Wells, *The New Machiavelli* (London: John Lane, 1911), 413.

that 'feminism had anything to do with sexual health and happiness was repudiated by these ladies with flushed indignation',[22] Wells and the suffragettes found themselves at cross purposes. This antagonism is a continual theme in the novel. A thinly disguised Christabel Pankhurst appears as Kitty Brett, caricatured for being 'as capable of intelligent argument as a runaway steam roller' (188). In fact, Christabel Pankhurst was famed for her trained, logical mind, having been the only woman in her year to study law at Manchester University from where she had graduated with a first-class degree. Kitty Brett's role is to plead with a questioning Ann Veronica not to lose herself in a 'wilderness of secondary considerations' (189–90) when the latter ventures to advocate (as Wells did himself) that the roots of women's discontents might to some extent be economic. The 'enthusiastic continence'[23] of the social purity suffragettes which promised an end to the social evils associated with sex is similarly ridiculed through Nettie Miniver: ' "Bodies! bodies! Horrible things! We are souls. Love lives on a higher plane. We are not animals. If I ever did meet a man I could love, I should love him"—her voice dropped again—"Platonically" ' (44).

Wells's preferred ideal of women making themselves sexually available to men without economic dependency is clearly wishful thinking. As Jane Eldridge Miller has pointed out, his support for women's sexual freedom was premissed largely on a recognition of his own sexual and emotional needs.[24] Whatever qualities he might admire in women such as Ann Veronica and her counterparts, such women existed for him primarily as potential sexual partners.[25] And like so many other eugenicists, Wells found it difficult, even impossible, to reconcile sexual desire, the racial biological imperative, and the goal of economic and social freedom for women. Eugenicist attitudes to womanhood at this time eulogized the *function* of the 'superior' woman as wife and mother. Maternity did not merely require of a woman that she should give birth but that she should relinquish her own ambitions outside the home. Because the willing co-operation of the mother was essential to the nurturing and education of the young it was necessary to *convert* 'superior' women to the task of 'race-preservation' by convincing them of the unrivalled importance of their ordained work as mothers and nurses of children.[26] As Lucy Bland has argued, the fact that middle-class women were ' "shirking" their "racial" duty to breed' was attributed to 'the invidious effects of feminism'.[27] If, as eugenicists argued, the only way in which the prospect of racial

[22] *Experiment in Autobiography*, ii. 483, 484.

[23] The phrase is used by Claudia Nelson and Ann Sumner Holmes in *Maternal Instincts: Visions of Motherhood and Sexuality in Britain, 1875–1925* (Basingstoke: Macmillan, 1997), 2.

[24] *Rebel Women: Feminism, Modernism and the Edwardian Novel* (London: Virago, 1994), 170–1.

[25] See Patricia Stubbs, *Women and Fiction: Feminism and the Novel 1880–1920* (London: Methuen, 1979), 193.

[26] See Jean Eason, 'The Eugenics Revolution: An Inquiry into the Relation of Eugenics to Ideologies of Gender and the Role of Women', MA Dissertation in Women's Studies, Anglia Polytechnic University (1994), 55.

[27] *Banishing the Beast: English Feminism and Sexual Morality, 1885–1914* (London: Penguin, 1995), 226.

degeneration could be halted was through the adjustment of the birth rate to favour those with 'superior' heredity, 'superior' women like Ann Veronica would be required to give up their aspirations towards education and independence.

This is the catch 22 expressed by the narrative of *Ann Veronica*. In order to fulfil a biological imperative which both reason and desire endorse—that is, live with Capes and start a family—Ann Veronica must relinquish her scientific career. With a pregnant Ann Veronica reunited with the respectable middle-class family from which she had escaped, into which an unrepentant Capes is somewhat improbably welcomed as a new son-in-law, the epicene, independent New Woman of the novel's opening is reformulated into the anodyne New Mother thus counteracting any threat that the scandalous New Woman scientist may have posed to the stability of social formation. Ann Veronica's pregnancy forms the bridge between traditional sexual and marital morality and the eugenically blessed long-term partnership that sanctions her initial defiance of social codes of respectable behaviour. Through this pregnancy Wells illustrates his own conviction that physical desire between superior people should rarely be separated from procreation. As Richard Remington, a character in *The New Machiavelli*, observes, 'Physical love without children is a little weak, timorous, more than a little shameful. With imaginative people there comes a time when it is impossible for that to go on' (413).

For many readers *Ann Veronica* is most revealing of the gender-political and, in effect, intellectual bias behind Wells's representation of women in science when read as a continuation of his vendetta against the Fabians, also pursued in *The New Machiavelli* (in which Beatrice Webb is pilloried as Altiora Bailey). The Fabian Society was the major forum in which intellectuals of Wells's day met to discuss radical ideas.[28] Wells presented his projects on eugenics and the endowment of motherhood to the society in various guises until the scandal caused by his affair with Amber Reeves, the daughter of two prominent Fabians, forced him to resign. The Fabians did publish a tract on the endowment of motherhood, but it was a 'pallid, actuarial treatment that did not mention Wells, even in the bibliography'.[29] Half the meeting of the Fabian Society in *Ann Veronica* consists of a 'great variety of Goopes-like types' (Goopes is an eccentric fructarian) (115). Moreover, the agenda is largely made up of high-minded and rather pointless speculation. As Samuel Hynes has put it, Wells's motive in all this was 'simply spleen. Wells used his fiction to revenge himself upon his enemies. Having broken with the Fabians, he promptly turned them into fiction.'[30]

There is a particularly marked contrast between the intellectual strength of the real Fabian Society women and their fictional representatives in *Ann Veronica*. The Fabian Women's group, founded in 1908, included distinguished trade unionists such as Margaret Bondfield and Mary McArthur, as well as Maud

[28] *Experiment in Autobiography*, ii. 247.
[29] Samuel Hynes, *The Edwardian Turn of Mind* (London: Oxford University Press, 1968), 117.
[30] Ibid. 20.

Pember Reeves, the mother of Amber Reeves. The group had thrown its formidable weight behind the attempt to include formal equality between the sexes as part of the society's aims. The publication of two major studies, *Round About a Pound a Week* (1909–13), Maud Pember Reeves's report on how working-class women in Lambeth managed their day-to-day lives, and Beatrice Webb's *Minority Commission on the Poor Law* (1909), published the same year as *Ann Veronica*, demonstrated that the Fabian Society women had a far stronger investment in practical political projects than Wells, whose views on eugenics and reproductive issues were considerably less likely to find general acceptance than the projects of his fictional Fabians on which he poured derision.

The novel also evokes anthropological and other contemporary controversies over the status of women that had riven the intellectual world of Wells and his circle. Capes has launched in *The Nineteenth Century* (one of the leading liberal journals of the day) an attack on Lester Ward's case for the existence of a primitive matriarchate. Ward was a pioneering American sociologist whose controversial gynocentric theories about the innate superiority of the female species had been expounded in *Pure Sociology* (1903).[31] A number of contemporary British feminist intellectuals including Frances Swiney and Catherine Hartley[32] had developed variations on the theme that women had reached a higher stage of evolutionary development than men and had argued that women were charged with a special responsibility to regenerate the race.[33] In *Ann Veronica* such views are ostensibly endorsed, but in practice mocked. Capes is initially dismissive of Ward's work suspecting that theories which emphasized women's innate moral, spiritual, or physical superiority to men would be put to use to justify women's sexual repression and ignorance. He changes his mind because he recognizes Ann Veronica as a High Priestess to be placed on a pedestal and worshipped—'You have converted me to Lester Ward' (284)—but the language of love blocks the question of where serious belief might begin. When Ward's views are espoused by the character of Nettie Miniver, a militant suffragette who studies science at Imperial College, the satire is transparent: 'The Matriarchate! The Lords of Creation just ran about and did what they were told.' Moreover, among the first animals there 'were no males, none at all' (31).

Here, as at so many other points, *Ann Veronica* is transparently a mouthpiece for Wells's own biological essentialism—his fetishization of relationships that ideology presents as natural and not subject to change. Women are to be freed from economic wrongs but only so that they may arrive at eugenically proper unions. Moreover, such alliances must be predicated on a woman's submissiveness to a man who, in the evolutionary discourse that permeates the whole novel, is 'better stuff' than herself. With her marriage to Capes, albeit a union determined

31 Lester F. Ward, *Pure Sociology: A Treatise on the Origin and Spontaneous Development of a Society* (London: Macmillan, 1903).

32 Catherine Gasquoine Hartley, *The Truth about Woman* (London: Eveleigh Nash, 1913).

33 See George Robb, 'Eugenics, Spirituality and Sex Differentiation in Edwardian England: The Case of Frances Swiney', *Journal of Women's History* 10/3 (1998), 97–117.

by the sexual attraction of the two parties rather than social convention, Ann Veronica returns voluntarily to the very state of domesticity that she had fled her suburban home to escape when she sought a career as a woman scientist. As Ann Veronica contemplates her future with Capes the reader is presented with the prospect of evolution in reverse: 'Modern indeed! She was going to be primordial as chipped flint!' (260).

*

Edith Ayrton Zangwill's novel *The Call*, published before the extension of the vote to women over 21 in 1928, is dedicated 'admiringly and affectionately' to 'all those who fought for the FREEDOM OF WOMEN'. *The Call*, which dramatizes the profound difficulties experienced by women as holders of scientific authority, and charts the processes of their intellectual marginalization, is loosely based on the life of her stepmother, Hertha Ayrton, the first woman to be elected a full member of the Institute of Electrical Engineers. Like the Nobel prize-winning chemist, Dorothy Hodgkin, Hertha Ayrton (née Marks) was not only one of the most distinguished scientists of her day but also a committed socialist and internationalist. She met her husband, the physicist William Ayrton, when she attended his evening classes in electricity at Finsbury Technical College in 1884 and accepted responsibility for bringing up Edith, his daughter from his first marriage to Mathilda Chaplin, herself a pioneering doctor and a suffragist. Hertha Ayrton was awarded the Royal Society's Hughes gold medal in 1906 for original discovery in the Physical Sciences, in recognition of her work on the electric arc and on the formation of sand ripples by the oscillation of water. The Royal Society award was widely acclaimed as a public acknowledgement that it was possible for women as well as men to excel in the hitherto male preserve of pure science. As the Mistress of Girton College, Cambridge observed, it was taken as having been conferred also upon 'the college and the cause of women generally'.[34] The Royal Society statutes nevertheless prohibited Ayrton from ever becoming a Fellow.

Following in the footsteps of another prominent feminist, the mathematician and physical scientist, Mary Somerville, whose signature had been the first collected on John Stuart Mill's petition for women's suffrage, Hertha Ayrton became a role model in feminist and scientific circles. Her studies at Girton had, indeed, been partially financed by the pioneering Victorian feminist, Barbara Boudichon, after whom her own daughter, Barbara was later affectionately named. At the time that Ayrton joined the WSPU, she was already one of the most respected scientists in the country. Until that point she had argued that the disabilities of scientific women 'did not differ sufficiently from the general disabilities of voteless women to demand separate treatment'.[35] As her biographer

[34] Robb, 'Eugenics, Spirituality and Sex Differentiation in Edwardian England', 184.

[35] Quoted in Evelyn Sharp, *Hertha Ayrton, 1854–1923: A Memoir* (London: E. Arnold, 1926), 196.

Evelyn Sharp put it, 'to many who were lukewarm suffragists or definite opponents—and nearly everyone in the particular circle in which she moved was one or the other—her championship of the less "respectable" section of the suffrage movement, when it did not appear merely incomprehensible, seemed to them suicidal from the standpoint of her career'.[36] Hertha Ayrton, who gave lectures to women on electricity anticipating its domestic uses and presided over the physical science section of the International Congress of Women in 1899 (at which she called for the employment of women in the electricity industry), was soon much in demand to preside at suffrage functions such as the opening of the suffrage exhibition in 1909. When early issues of the radical journal *The Freewoman* launched attacks on WSPU militancy, both Hertha Ayrton and Edith Ayrton Zangwill replied with impassioned defences of the Pankhursts. Zangwill, who had contributed a short story to the first issue of *The Freewoman*, commented that a 'public and acrimonious attack on a prominent militant leader is ill-judged and likely to harm the suffrage cause'.[37] Her stepmother expressed 'utter detestation' of the attack made upon Christabel Pankhurst: 'My promise of a subscription was made under the impression that the editors were still loyal friends to the cause of Woman's Suffrage and to the Women's Social and Political Union.'[38]

Hertha Ayrton wished to believe that, in general, she could help 'the cause' most by pursuing her own researches: being called to make speeches about suffrage interrupted her scientific work more disastrously than anything else. But because absenting herself from the militant campaign caused her acute ethical dilemmas much time was taken up with suffrage to the detriment of her scientific activity. Her participation in the militant demonstrations in November 1910 led to the postponement of an important scientific paper to the Royal Society as she fully expected to be in prison.[39] She walked in every major suffrage procession including the 'Mud March' in February 1907 (where Edith Ayrton Zangwill joined her) and the suffrage procession of January 1911 in which she headed the section on women in the sciences. She regularly filled her house with suffragettes from the provinces, opened her doors to supporters of the Women's Freedom League protesting against the census in 1911, and made her home a refuge for Emmeline Pankhurst and other prisoners who had been forcibly fed and needed devoted nursing back to health.

Moreover, Ayrton was one of the WSPU's most generous financial contributors, giving over £1,000 each year in 1909–10, 1910–11, and 1912–13[40] and laundering the WSPU's money through her personal bank account when it appeared the union's funds were about to be sequestered in March 1912.[41] She did much to secure support for the suffragettes from the international scientific community which included her friend, Marie Curie—'J'ai été très touchée', Curie wrote, 'par tout ce que vous me dîtes de la lutte des femmes anglaises pour leurs droits; je les

[36] Ibid. 194.
[37] *The Freewoman*, 20 November 1911, 30.
[38] Ibid. 7 December 1911, 51.
[39] Sharp, *Hertha Ayrton*, 225.
[40] Andrew Rosen, *Rise Up, Women! The Militant Campaign of the Women's Social and Political Union 1903–1914* (London: Routledge & Kegan Paul, 1974), 224.
[41] Sharp, *Hertha Ayrton*, 234.

estime beaucoup et je souhaite leur succès.'[42] Ayrton championed Curie when the discovery of radium was attributed to her husband, writing in *The Westminster Gazette* in 1909 that 'Errors are notoriously hard to kill, but an error that ascribes to a man what was actually the work of a woman has more lives than a cat.'[43] One of the many scientists to whom she wrote about votes for women was the much less sympathetic Francis Galton who replied unhelpfully, 'You have asked a Balaam to bless your cause!'[44]

In contrast to *Ann Veronica* and *Man's World*, *The Call* is not directly influenced by eugenics, although it explores issues that were of concern to eugenicists of the day. The protagonist is a woman of exceptionally high intelligence and the novel is concerned with her attempt to gain membership of an exclusively male scientific elite. Ursula Wingfield is a brilliant young researcher in chemistry whose unhappy experiences with the British Association are reminiscent of Hertha Ayrton's experiences with the Royal Society. In *The Call* it is only through the patronage of a former president, Vernon Smee, that Ursula is able to attend their meetings. Yet she knows that her work is as good as any man's. Ursula reflects bitterly on the injustice that inexperienced young men are permitted to become Fellows while she, with all her experience, is excluded.[45] She becomes the first woman to speak at the Chemical Society after a distinguished guest speaker has commended her paper, wrongly assuming it to have been written by a man. Smee has unsuccessfully attempted to change the charter of the society to admit ladies and is regarded as 'a bit of a traitor—one of a beleaguered garrison who had helped the enemy to scale the wall' (26). (Ayrton's fictional version of the Chemical Society is thus significantly different from the real Society which had admitted women to membership in 1920.[46])

The idealized relationship between a mature male scientist and his adoring pupil in *Ann Veronica* is here replaced by competition and resentment. Ursula's mentor and champion, like Ann Veronica's, is an unhappily married middle-aged man but his romantic attachment to his protegée is not reciprocated, and Ursula in due course becomes engaged to a man of her own age. Moreover, her change of attitude to Smee is prompted by sympathy for his wife who has had five miscarriages although her husband had been told that further pregnancies would endanger her life. (Capes's wife is a structuring absence in *Ann Veronica.*) It is also suggested that Ursula may be the better scientist of the two. Smee is a charismatic figure but his scientific work is thought to be merely the 'skilful verbal and liter-

[42] Sharp, *Hertha Ayrton*, 237.

[43] Quoted in Joan Mason, 'Hertha Ayrton: A Scientist of Spirit', in Gill Kirkup and Laurie Smith Keller (eds.), *Inventing Women: Science, Technology and Gender* (Cambridge: Polity, in association with The Open University Press, 1992), 168–77 (172).

[44] Sharp, *Hertha Ayrton*, 213. There is some ambiguity here because Balaam did give his blessing. Galton may have meant that the angel of the Lord might extract this from him.

[45] Francis Galton resigned from the Royal Geographical Society in 1893 when it refused to elect fifteen well-qualified women.

[46] In 1927 eighty-four out of a total membership of 4,101 were women. See Charlotte Haldane, *Motherhood and its Enemies* (London: Chatto & Windus, 1927), 109.

ary presentation of other men's ideas' (230). When her paper to the British Association is not well received Ursula feels a 'distinct current of hostility' and speculates that 'these male chemists may have unconsciously resented a woman's claim to discovery' (155). Zangwill also draws directly on the difficulty Hertha Ayrton had in persuading the scientific and political establishment to make use of her scientific expertise. Ayrton had invented a fan to dispel gas from the trenches in the 1914–18 war, but met with obfuscation, delays, and incomprehension when trying to persuade the authorities of the importance of the technology she had developed. The War Office adopted the 'Ayrton fan' tardily, and it appears to have saved thousands of soldiers from gas poisoning (some soldiers claimed to have found it of little use, but Ayrton always insisted that this was because they had not been properly trained). In *The Call* the delays in securing official approval for the new device are specifically attributed to the highly placed Vernon Smee, whose own invention is directly in competition with Ursula's.

But Zangwill's most direct challenge to the model of the woman scientist Wells had put forward in *Ann Veronica* comes with her revaluation of the heroine's association with the suffragette cause. Where Wells had charted his intelligent protagonist's disillusionment with militancy, and shown love to be a better choice for an intelligent young woman than either science or suffrage, Zangwill does the very opposite. Ursula Wingfield at first disapproves of the rowdy behaviour of the suffragettes but is converted after witnessing a man convicted of child sexual abuse being given a derisory sentence by the courts. Like Barbara Ayrton Gould (Edith Ayrton Zangwill's stepsister) who had joined the WSPU as a student of chemistry at University College, London, Ursula suspends her postgraduate research in order to become a paid full-time organizer for the suffragettes.

The Call is one the finest of the forgotten suffrage novels and vividly depicts Ursula's work for the cause across the country and her imprisonment. Instead of giving up her suffrage work for love, Ursula, at least temporarily, gives up love for the suffragette cause. Her activism puts her directly at odds with her uncomprehending fiancé, and the impasse between them is ended only after women are given the vote in 1918. In *Ann Veronica* the noises made by women in prison produced in the heroine a 'violent reaction against the suffrage movement' (204) and prompted her to call her fellow suffragettes 'Intolerable idiots!' and to declaim that she has 'no feminine class feeling' (206). But Ursula's despondent sense of isolation in prison is lifted by the communal singing of the Women's *Marseillaise* to which she is happy to add her voice. Nothing remains in that scene of the reactionary biologism with which Wells dismissed female militancy as a regression to the animal state. But in one respect Zangwill stays pointedly close to Wells. When this retelling of the New Woman scientist's story finds itself obliged to arbitrate between three claims—love, science, and politics—science again is a loser. One can, it seems, be a New Woman activist or a lover or a scientist but not all three, and not even two of the three. Zangwill's priority marks her difference from Wells, but the structure of choice remains the same, fifteen years on.

*

Charlotte Burghes Haldane (née Franken) was a novelist and scientific journalist whose articles in the popular press brought early forms of reproductive technology (such as amniocentesis) to the attention of a wide non-scientific readership.[47] *Man's World*, her futuristic novel of 1926, needs to be understood partly as a forerunner to what would prove her most controversial legacy for later twentieth-century feminism, her enthusiastic advocacy of vocational motherhood in *Motherhood and its Enemies* (1927). *Man's World* has a somewhat more ambivalent attitude to maternity, but in both works Haldane's ideas resist simple categorization and can be disturbing to the sensibilities of feminists and anti-feminists alike. Her claim in *Motherhood and its Enemies* is that all women are born with maternal instincts upon which their personal fulfilment depends. In this she does not seem to have been saying anything very different from other women such as Marie Stopes in such works as *Radiant Motherhood* (1920).[48] But her idealization of motherhood reinforced the widespread prejudice against the two million 'superfluous women' in the 1920s and may also have played into the hands of the European Fascists who venerated motherhood for their own purposes.

In *Truth Will Out*, her 1949 autobiography, Haldane writes that she had wanted, in *Man's World*, to ask 'what would be the effect on society if the human race could determine in advance the sex of its children. But to deal with this theme in fiction would mean taking a large imaginative stride into the future, since present biological research had not yet made a society built on such a postulate a practical possibility.'[49] Charlotte Haldane's mentor, the geneticist J. B. S. Haldane, who advised her about many of the scientific ideas underpinning *Man's World*, was particularly interested in ectogenesis (the development of a baby outside its mother's body) and 'had an unusual enthusiasm for the popularising of science, and a firm conviction that this was both desirable and necessary'.[50] His own (in its day much talked about) prediction for the development of science 'Daedalus, or Science and the Future', was first delivered as a lecture before the Heretics Society at Cambridge, on 4 February 1923, and published by Kegan Paul in London in 1924. In it he foresaw, among other things, a time when babies would be made in the laboratory, thus taking the decision about the future of the race out of the fickle sphere of sexual choice altogether. Like Wells, J. B. S. Haldane refused to see science and ethics as distinct or separate from one another. *Daedalus* argues that the future should be determined by the socially responsible scientist who will use his knowledge of biology to reshape society.

For many left eugenicists, including Charlotte Haldane and H. G. Wells, socialism and science were inextricably linked. They welcomed an extension of

[47] Charlotte Burghes [Haldane], 'The Sex of Your Child', *The Daily Express*, 18 May 1824, 7.

[48] Marie Carmichael Stopes, *Radiant Motherhood: A Book for Those who are Creating the Future* (London: Putnam, 1920).

[49] Haldane, *Truth Will Out*, 15.

[50] Ibid. 21.

state authority and power in order to facilitate rational planning. The idea of a programme of socially responsible eugenics also appealed strongly because it reflected their interest in a planned society and their demands that the state concern itself seriously with issues of motherhood and reproduction. *Man's World* is the locus of some of these tensions and contradictions. Its setting is Nucleus, a white racist, male-dominated, eugenically organized society that is experimenting with the Perrier method whereby the sex of babies can be determined in advance. The advantage of the Perrier method (prenatal exercises performed by the mother) is that it produces male children on demand: the nations will have an unlimited supply of 'Man Power' (36) as well as ending the problem of 'surplus women' (44). Nucleus is supposedly a 'socialist society' which extends over the white world, Australia, the Americas, and Europe. Its inhabitants are of 'varying branches of the white race' (8) and collectively make obeisance to the male master figure of the scientist who founded the state. There are no women scientists in this community. Man's World is precisely that.

The demand, voiced by many early twentieth-century eugenicists, that the state concern itself with motherhood and reproduction is taken to its logical extreme here. Motherhood is organized entirely on eugenic principles. The state 'wish[es] to breed from the best available material' (165). 'The race is the ideal, ultimately' (65). The sexual behaviour of women, although notionally controlled by women themselves through motherhood councils, is tightly regulated in the interests of the social body as a whole. Individual desires are subjugated to the ultimate goal of race-motherhood. Women are 'perfect vessels singled out for the propagation of our race' (51). They are reduced to their biology which gives them a social identity, place, and purpose. All women either 'choose' to become vocational mothers or become neuters and are sterilized and turned into 'entertainers'. The entertainers are 'perfectly trained and fashioned to bring beauty to all the world . . . They smil[e] perpetually' (130).

The fantasy of Nucleus can be seen as Haldane's thinking through of the widespread belief among many intellectuals in the 1920s that the 1914 war was a watershed and that post-war reconstruction must be based on new scientific lines. To a large extent the fantasy world it imagines can be read as an endorsement of science. From its inception Nucleus is a self-governing community 'created by the scientific mastery of man's instincts to fight and to propagate his species' (10). The solution to the problems that humanity faces lies with the scientist who is 'the new director, the inevitable successor to the priest and the politician' (4) in this brave new world. And, as in *Ann Veronica*, the social and political in *Man's World* find their most pointed expression in biological discourse: 'When the propagandists had provided the imagery which would make the masses submit to the works of the Patrol, they had translated the terms of the social organisation into those of the body. The body then stood symbolically for the entire white race' (63).

But at the same time that *Man's World* dramatizes the ultimate triumph of scientific principles, it testifies to the fears that Charlotte Haldane expressed about their abuse in the hands of racists:

Let the class-conscious or race-proud individual . . . with a mere smattering of scientific knowledge, attain any influence in this matter, and those whom he fears or hates (the same thing) will fare hardly. One would require a certificate of psychological purity even in the case of certain scientists before one would entrust them with so dangerous a profession as that of human geneticist.[51]

As in a much later novel, Margaret Atwood's *The Handmaid's Tale* (1984), the state supposedly celebrates maternal agency, but in practice denies women any but the most rudimentary biological definition of agency by its policing of pregnancy and childbirth. For today's readers, *Man's World* startlingly anticipates that dystopian vision, with its exploration of the implications for women of the control of reproduction by male scientists. Susan Merrill Squire has argued that it 'offers a prescient analysis of how reproductive technology can produce power/knowledge for a patriarchal state through control of the (female body)'.[52] *Man's World* is essentially an analysis of gender distinctions, according to which men are seen to be denied the possibility of achieving wholeness because the feminine principle is withheld from them.[53] As Jessica Benjamin has noted, 'although the ideal may be the structure of gender polarity, which upholds masculine rationality and autonomy in the public world and honours feminine nurturance in the home, the masculine principle cannot be contained in public life. It inevitably threatens to exceed these limits and devalue the threatened haven of the home.'[54]

Although the novel is clearly critical of male control of women's fertility it remains contradictory in that it reflects Haldane's fascination with the idea of male genius and her adulation of the figure of the strong male scientist. She conceded in her autobiography that she had an 'emotional need for a religion, or a substitute for one'.[55] As Judith Adamson points out, Charlotte Haldane had 'little more than her feminist good sense to counterbalance the "passionate adulation of JBS and his world" ' that permeates the novel.[56] *Man's World* consequently illustrates how Charlotte Haldane is torn between the scientific ideal of race-motherhood and a romantic individualism that runs counter to the scientific rationalism of the state. The latter asserts a woman's right to follow her instincts and choose the father of her child, or indeed choose not to be a mother at all; the former insists upon her social and ethical responsibility rather to follow reason and science and embrace eugenic motherhood. The predicament is expressed through the character of Nicolette, whom everyone believes will eventually choose the vocation of motherhood. But her 'training for motherhood did not

[51] Haldane, *Motherhood and Its Enemies*, 238.

[52] Squire, *Babies in Bottles*, 121.

[53] Elizabeth Russell, 'The Loss of the Feminine Principle in Charlotte Haldane's *Man's World* and Katherine Burdekin's *Swastika Night*', in Lucie Armitt (ed.), *Where No Man Has Gone Before: Women and Science Fiction* (London: Routledge, 1991), 1–15.

[54] Jessica Benjamin, *The Bonds of Love: Psychoanalysis, Feminism and the Problem of Domination* (London: Virago, 1990), 99–200.

[55] Haldane, *Truth Will Out*, 31.

[56] Adamson, *Charlotte Haldane*, 53.

involve the clear interpretation of very obscure emotions' (133). Instead, Nicolette falls in love romantically with Bruce Wayland, a member of the ruling scientific elite and clearly modelled on J. B. S. Haldane. Nicolette's dilemma pinpoints the quandary of the rational person with an attachment to the affective realm for which the course advanced by reason (rational family planning in a socialist state) appeared to leave no space.[57]

If Haldane fudges the question of reason's compatibility with emotion by allowing her heroine to marry a eugenically desirable scientist with whom she is—opportunely—deeply in love, she gives up entirely on the other dilemmas at the heart of the woman scientist story as Wells had formatively imagined it. For all her own involvement in scientific journalism and her desire to tell a story about women and modern science, Haldane chooses a heroine who is neither a scientist nor even a promoter of the public understanding of science. By setting the novel in a future that is ambiguously both dystopian and potentially utopian, she also sidesteps a direct imaginative engagement with the early twentieth-century reform efforts of the suffragette movement. Both Wells and Zangwill had recognized those efforts as closely interwoven with the slow but gradually increasing acceptance of women as workers in research laboratories. Haldane seems to have seen them as the most readily dispensable element in a narrative that already had more strands at work in it than could easily be reconciled.

It would be possible to read Wells's *Ann Veronica* as a representation of the woman scientist so biased against its heroine's success *as* a scientist that it could not but stimulate later scientifically educated women to defensive rebuttal. Zangwill's *The Call*, certainly, could be read in that light. But my argument here has been that the most important legacy of the first New Woman scientist novel was not so much its patent ideological prejudices as its recognition of deep, perhaps irreconcilable conflicts of desire at work when imagining what it might mean to be a woman scientist in early twentieth-century English society. If there was a narrative waiting to be told about Chloe and Olivia companionably at work in their laboratory, there were also other narratives to be told, and to be somehow made compatible with that narrative, about how Chloe and Olivia liked and loved, and how their liking and loving affected their ability to find acceptance as scientists. If scientific and personal aspirations could not easily be imagined in harmony, as for H. G. Wells they clearly could not, what other, more explicitly political narratives might then have to be told about the need for continuing change?

[57] This was, of course, no uncommon predicament for men and women on the left. As Sydney Webb once put it, 'No consistent eugenist can be a "*Laissez Faire*" individualist unless he throws up the game in despair.' See 'Eugenics and the Poor Law: *The Minority Report*', *Eugenics Review* 2 (1910–11), 233–41 (237).

CHAPTER 13

The Chemistry of Truth and the Literature of Dystopia

ALISON WINTER

> Nothing [is] your own except the few cubic centimeters inside your skull.
> (George Orwell, *Nineteen Eighty-Four* (1949))[1]

> You must realize how important a discovery it is; from now on no criminal can deny the truth. Not even our innermost thoughts are our own . . .
> (Karin Boye, *Kallocain* (1940))[2]

The idea that memories lie within us, almost as physical entities within a container, has an intuitive appeal. No less alluring is its corollary: that such entities can be accessed by putting the body or mind into a controllable state of openness. Efforts to do just this have a long and consequential history in the modern era. In particular, during the twentieth century several discrete research programmes explicitly pursued this end via the use of pharmaceuticals. The period also saw intensive literary reflection on the possibilities of 'truth serum' drugs and technologies. Despite scepticism—and often downright dismissal—of the feasibility of such 'truth techniques', they have had a stubbornly recurrent, indeed almost irrepressible, appeal. Newspaper reports that truth drugs may have been used on terrorist suspects at Camp X-Ray are only the most recent call for our attention to their problematic promise. My chapter will explain how and why people first sought to develop a 'truth serum', and will explore how their researches fostered reflection on the implications of such a technique for society and the individual's imperilled place within it. That reflection took shape largely within the fictional literature of dystopia, drawing on the genre's long association with political and philosophical dissent and recasting its common, though not inevitable, hostility to the intervention of the chemical sciences in the domain of psychology.

There exists to date no sustained history of 'truth serum' *per se*. But there is a well-developed scholarly literature on the history of forensics and of lie detection.

[1] *Complete Works of George Orwell*, ix: *Nineteen Eighty-Four* (1949), ed. P. Davison (London: Secker & Warburg, 1986–7 (rpt. 1997)), 29.

[2] Trans. Gustaf Lannestock, ed. R. B. Vowles (Madison, Wis.: University of Wisconsin Press, 1966), 13.

This literature tends to portray innovators as wishing to avoid the messy terrain of the human mind altogether by resorting to the body and physical evidence—their assumption being that, in the words of one recent historian, 'while the mind can lie, the body is honest'.[3] Examining the history of truth serum itself offers the possibility of understanding the development of truth technologies in a somewhat different light. It suggests that innovators assumed that the body could indeed lie, since it was the conduit for all communication. The challenge of obtaining truth from an individual could be met only by studying exactly *how* the body conveyed or concealed information. These researchers shared a hope that personal evidence could be validated by what one might call 'technologically situating' human testimony.

The notion that science might be able both to validate existing testimony and to procure truthful statements from unwilling subjects was inspired by reflection upon the question of what it would mean to have a more 'truthful' society—and it in turn provoked new forms of attention to that question. The allure of truth-telling techniques for their developers lay, both narrowly and broadly, in this: the prospect that science could supply a means to purge from a given arena the problems posed by the ability of human agents to conceal the truth and to speak or act in bad faith. The arena could be very constrained or all-encompassing—the interrogation room, the criminal justice system, or even society in general. One of the recurrent themes in the history of truth techniques became the question of what would happen to the dynamics of society as a whole if lying could be scientifically expunged. The possibility excited surprisingly conflicting responses, and I want to identify some of them here.

Truth techniques were mooted and first developed at a time of much angst about the social consequences of dishonesty, ranging from the criminal behaviour of individuals to corruption within government structures—in particular, the police forces, the judiciary, and the criminal justice system more generally. America's role in this history was pivotal, and, given the constraints of space, I shall stay with the American story in charting the origins of these techniques; but the pharmaceuticals developed in the United States, and the applications to which they were put, were publicized and widely emulated across Britain and Europe in the 1920s and 1930s. The first major forensic laboratory in America, the Scientific Crime Detection Laboratory, was established at Northwestern University rather than within a governmental agency because its private founders did not trust government with such potentially powerful tools. They hoped, in fact, that the new techniques of scientific crime detection would become instruments against government corruption. They would slice through webs of covert alliances and dishonest practices, identifying lies and establishing a method by which true information could be made to flow. This in turn was expected to help with broader reforms that were intended to produce a 'truer' society. The harder it was

[3] Ken Alder, 'To Tell the Truth: The Polygraph Exam and the Marketing of American Expertise', *Historical Reflections* 24/3 (1988), 487–524 (488).

to lie, they thought, the healthier society would become. Of course, this was not the only opinion one could hold about truth techniques. Forensic innovators were regarded as dangerously naïve by sceptics, who anticipated that successful techniques would be more likely to become tools of a corrupt government than forces for removing corruption. These warnings were most vividly elaborated in the 1940s literature of dystopia—especially in the work of Karin Boye and George Orwell. In their futuristic fantasies, the ability to conceal truths is held out as a crucial feature of private and civic life, its removal imperilling the very bases of Western society.

Another question emerges out of this history and literature that may be deeper than, or prior to, the other themes outlined so far. This is the question of the definition of truth itself. Over the course of the period and within the different projects addressed by this chapter, different and conflicting definitions of truth were in play. What these different definitions were, and how they related to or conflicted with one another, will emerge case by case, and then will be addressed more directly in the conclusion.

First, however, we need to examine how truth serum brought to centre-stage problems of self and society, narrative and truth, and subjectivity and objectivity. Understanding this will require the kind of sensitivity to confluences and intersections between fields that has long been a hallmark of Gillian Beer's work, and that, through her work, has come to have a central place in both the history of science and the historical study of literature and culture: namely, an attention to the mutual construction of cultural meaning in scientific work and in the composition and consumption of literature, and an analysis of both that is contextualized in relation to the social, cultural, and scientific concerns of the day.

THE PHYSIOLOGY OF SELF-EXPRESSION

In one sense, the notion of a relationship between drugs and honesty is ancient. Alcohol is the outstanding example, as in Pliny's famous saying, 'in vino veritas'. But although the proverbial relationship was well known and of long standing, it was only over the course of the nineteenth century that a notion emerged that there might be distinct physiological states coextensive with degrees and manners of truth-telling. At that time a number of experiments and debates explored how human thought and communication could be managed by scientific techniques.[4] The widespread proliferation first of mesmerism from the 1820s, then of hypnotism and chemical anaesthesia from the 1840s, and finally psychoanalysis at the close of the century further developed the notion. By the century's end it was widely accepted that there was an as yet ambiguous, but nevertheless potentially knowable, relationship between a precisely calculated and controlled, artificially produced state of mind

[4] e.g. Terry Parssinen, *Secret Passions, Secret Remedies: Narcotic Drugs in British Society 1820–1930* (Manchester: Manchester University Press, 1983); Lloyd Stevenson, 'Suspended Animation and the History of Anaesthesia', *Bulletin for the History of Medicine* 49 (1975), 482–511; Alison Winter, *Mesmerized: Powers of Mind in Victorian Britain* (Chicago: Chicago University Press, 1998), ch. 7.

and the meaning of what an individual said while in that state. The link between physiological state of mind and truth status had become a commonplace.

More generally, by this time enormous optimism obtained about the powers of 'mental science' to 'lay bare the mind'.[5] The period saw the emergence of an evidential paradigm framing the relation of certain new kinds of evidence to the understanding of human beings. For instance, as Carlo Ginzburg has famously argued, the close of the nineteenth century saw the triumph of the concept of a peripheral 'trace'—such as fingerprints or human fibres—that bore a central and unique relation to the individual who made it.[6] Early twentieth-century truth techniques participated in this trend as well as descending from earlier, Victorian conventions. The physiology of mental reflexes is an example. This was developed in Germany by Wilhelm Wundt and in England by William Carpenter and associated physiologists.[7] Although Carpenter's work was first published in the 1850s and 1860s, the mental physiology part of his general physiology was greatly expanded over the following years, and had its major impact in America towards the end of the century. A central tenet of this physiology was that the individual contained an extensive set of reflexes ranging from the most basic in the extended nervous system to an elaborate mechanism within higher brain function, all of which operated in semi-autonomy from the conscious mind. Carpenter thought that we all build our own personal 'automata', which act for us both when they carry out routine actions to which we do not pay attention and, in special circumstances, when our conscious minds are impaired from their ordinary vigilance. It was when the proliferating phenomena of altered mental states were coupled with such reflex physiologies that the idea arose of techniques—or drugs—being coded for their effects on memory and communication.

The early years of the twentieth century saw a number of lie-detection techniques emerge out of reflex physiology, and more immediately out of the sciences of emotion that were then flourishing.[8] These techniques were based on the premiss that emotion was a physiological state whose changes could be measured by scientific instruments. If emotions were affected by the truth status of what an individual was saying, then such instruments became, in effect, truth detectors. New truth technologies such as the polygraph, sphygmograph, and others were therefore championed during the mid-1910s as potential mechanical adjudicators. Popular articles represented them as having, in the words of one historian, an almost 'mythical' power, and sufficient potency to make the body 'betray the mind'.[9]

[5] Anon, 'Recent Discoveries in Mental Science Lay Bare the Mind of the Criminal to the Psychic Expert', *New York Times*, 22 September 1907, 9.

[6] Carlo Ginsburg, *Myths, Emblems, Clues* (London: Radius 1990).

[7] Winter, *Mesmerized*, chs. 11–12.

[8] Otniel E. Dror, 'The Affect of Experiment: The Turn to Emotions in Anglo-American Physiology, 1900–1940', *Isis* 90/2 (1999), 205–37.

[9] Otniel E. Dror, 'The Scientific Image of Emotion: Experience and Technologies of Inscription', *Configurations* 7/3 (1999), 355–401, 356–8. See also Anon., 'How the Body Betrays the Mind', *Literary Digest* 48 (1914), 153.

Promoting these techniques were a number of scientists and doctors who were more generally committed to a scientific revolution in the regulation of crime. Their campaign was based on a new technological claim to the knowability of the mind—one that promised not just understanding but access. The first forensic laboratories in America were founded during these years, touted as having the promise to realize a Holmesian programme for tracing and identifying criminals. The new interrogation techniques made possible by 'lie detectors' enhanced the potential power of such programmes because they brought technology to bear on the interpretation of narrative, as opposed to merely physical, evidence. They raised expectations that the most ephemeral of human phenomena—tiny physiological traces invisible except via the mediation of scientific techniques—were in principle accessible to inquiry, and in situations where they might hold the greatest consequence.

It is worth noting that for all the emphasis that has been placed on the instrumental character of the new forensic tools of the late-nineteenth and early twentieth centuries—and on the notion that these tools were supposed to remove the old reliance on human evidence, imagination, and testimony—the core element and the final result of forensic science was a *story*. Every clue led to or was already part of a plot. A narrative of human experience was the only fully satisfying end result of the investigation of a crime. Moreover, the self-image of early twentieth-century forensics and of the forensic scientist was more indebted to literature than vice versa. The forensic authors of the period represented themselves as realizing the promise of Sherlock Holmes.[10] Forensic scientists were both the storytellers and characters, their clues the catalysts for the unfolding of scientific plots.

HOUSE'S PLOTS

The drug that inspired the term 'truth serum' was scopolamine. It was synthesized in Germany in the 1890s and first widely used as an obstetric anaesthetic to produce so-called 'twilight sleep'. Twilight sleep became particularly well known for its tendency to suspend the memory of surgical events (including that of the pain of the knife, although the patient was aware of that pain during the surgery itself). To attain the desired state, many obstetricians used the patient's memory as a calibration tool, and it was through this use of memory as an instrument that scopolamine came to be used as an interrogation technique.

To calibrate the dosage of scopolamine, an obstetrician would commonly show an object (for instance a cake of soap) to the patient before administering the drug. After each of a succession of doses the patient was asked to recall the object. When she became unable to do so, the correct depth of trance had been reached. Scopolamine's efficacy in twilight sleep was thus calibrated through the

[10] Ronald R. Thomas, *Detective Fiction and the Rise of Forensic Science* (Cambridge: Cambridge University Press, 1999).

erasure of short-term memory. The physician who has been credited with the discovery of 'truth serum' was a Texan, R. E. House, who used scopolamine in his own obstetric practice in exactly this fashion. House's story of how he thought of the notion ran thus: in 1916, as he delivered a baby at a patient's home, he asked the husband for the scales and was told that they could not be found. The patient—still deep in twilight sleep—gave precise and articulate instructions about their location. House was surprised by this, perhaps because (although he himself did not explain his surprise) the prominence of memory loss in the twilight-sleep procedure would have led one to assume that the drug blocked memory and communication in general. He claimed that this experience gave him the idea that scopolamine could be used to extract truthful statements from individuals who might otherwise lie.[11] House published his initial findings in 1922, and followed this paper with a series of others. Legal and police journals soon hailed his work as promising and suggestive;[12] the popular magazines crowed that it was 'revolutionary', and coined the term 'truth serum'.[13] This term subsequently proved irresistible even to doctors who warned that scopolamine and similarly acting drugs were neither 'sera' nor straightforward extractors of 'truth'.

One striking aspect of this incident is how underdetermined it was. House might have drawn a great many different conclusions. For instance, the woman's speech, as he reported it, was in no obvious sense a confession or revelation. What was impressive was her ability to communicate at all, and to impart reliable information (reliable being the operative word, rather than 'true', because she had no evident incentive to lie). She did not relate information that she would otherwise have wished to conceal, or that she, in her sober state, had forgotten. So why did House move to this conclusion? One reason may have been the very prevalence of new forensic techniques in the early twentieth century, and the intensity of publicity surrounding them in the mid-1910s. House's interpretation probably arose from a combination of the promise of forensic science and a well-entrenched physiology of communication—along, perhaps, with a role for doctors in the criminal justice system that had long been substantial. But an additional possibility is that this patient did in fact reveal something during her anaesthetized state that House (remembered in his home town as the 'perfect gentleman' even today by those few remaining citizens of Ferris, Texas, who knew him personally[14]) could not bring himself to repeat in print.

The explanation House gave for his technique was drawn from the mid- and late-Victorian notion of automatism, onto which he grafted a loosely psychoanalytic

[11] R. E. House, 'The Use of Scopolamine in Criminology', *Texas State Journal of Medicine* 18 (1922), 256–63.

[12] C. T. McCormic, 'Deception Tests and the Law of Evidence', *California Law Review* 15 (1926–7), 484–504; similarly, C. T. McCormic, 'Deception Tests and the Law of Evidence', *American Journal of Police Science* 2 (1931), 388–400 (397–8).

[13] Ostensibly by the Los Angeles *Record* in 1922, according to multiple sources of the 1940s and 1950s. I have yet to locate this article.

[14] Personal communication, May 2001, Virginia Leys of Ferris, Texas.

notion of the unconscious. He defined the phenomenon as 'a process by which memory can be extracted against the will from an individual's subconscious mind'.[15] For instance, acting as an expert witness for the defence in a 1920s rape case, House gave the following account of how a query would produce 'automatic' information from an anaesthetized patient:

> When you ask an individual a question, the sound waves hit the drum of the ear, and the nerve of hearing, like a telephone wire carries the sound waves to the center of hearing. The only function of the center of hearing is to evoke memory by sending the impulse to that part of the brain where the answer is stored for future use, and in a fraction of a second the brain sends the answer to the nerve of the tongue. To illustrate—If I ask a person 'What is your name?' he cannot by the will power or by any other function of the brain keep from hearing the question or thinking the answer, but his will power can prevent the tongue from articulating the name, and the power of reason can also take the answer and by calling on the imagination make the tongue tell a lie, but when the will power and the power to reason are removed, the replies are automatic. Hence the truth. When the will and the power to reason are nonexistent, then man is too unconscious and too helpless to prevent himself from inventing replies to questions propounded.[16]

House himself gave no more sophisticated account than this of 'automatic' mental actions and the nature of memory, volition, or self-expression. But this may have been an advantage: his claim drew on assumptions about the will and memory that were powerful because familiar, and his explanations were sufficiently simple to be quoted in the many local newspapers that followed his proselytizing efforts. He assured readers straightforwardly that 'Under the influence of the drug, there is no imagination.' It was impossible to lie, he said, because drugged individuals were merely conduits for information, with 'no power to think or reason'.

For House, and for many of the journalists who publicized his work, truth serum had profound implications. First of all, there was the deplorable state of the criminal justice system, currently the focus of much public concern for the corruption of its administrators and for its brutal techniques of interrogation. The so-called 'third degree' had become almost synonymous with police interrogation by the mid-1920s, and House himself used the resulting controversy to assail the system for seeking confessions without concern for their truth status. This was destructive in two ways, he said: '[T]he zealous peace officer . . . goes to such brutal and unwarranted lengths to get confessions, as to destroy the value of what he seeks so hard to obtain. "They go after confessions and get them, sometimes true, sometimes false; but they always get confessions" ' (143). By now this was so well known to the general public that a reference by defence counsel to 'the third degree' had become a staple technique for helping to secure an acquittal. Hence the double destructiveness of the third degree itself: the innocent were

[15] House, speech to law enforcement officers in 1924, quoted by A. A. Moenssens, 'Narcoanalysis in Law Enforcement', *Journal of Criminal Law and Criminology* 52 (1961), 453–8.

[16] People *v.* Hudson (manuscript), Missouri Supreme Ct. Archives, 162.

forced to produce false confessions, and the guilty could represent their own true confessions as involuntary fabrications. House represented truth serum as a humane, 'harmless' form of the third degree, by which he meant that it fulfilled the ideal purpose of the third degree (namely, securing involuntary confessions) without the fallibility and destructive by-products.

But there was a wider implication to the new truth technique, related to the systemic effects of crime and dishonesty on society. For House, the current levels of crime—'Three hundred thousand youths sent every year to our penal institutions'—showed society to be in a deeply unhealthy state: 'The wealth of any nation is not in its undeveloped resources but in the undeveloped "character" of its children . . . Society cannot exist unless the criminal can be controlled; the criminal cannot be controlled unless he is understood; and the criminal will never be understood until society demands that the crime problem be mastered' (138). The key to this mastery lay in the freeing and disseminating of truth—in understanding the nature of criminal psychology and behaviour, and in unlocking the memories of all suspects, thereby protecting them and the general public from lies and falsehoods of all kinds. For House, then, the ability to conceal the truth was the germ of social disorder and disease; truth serum supplied the vaccine.

House followed his initial trials with other experiments, and he inspired others to make their own. Throughout the 1920s he toured the country carrying out 'scopolamine interviews' on criminal suspects and convicts. In a few very celebrated cases, individuals interviewed whilst under the influence of scopolamine incriminated themselves dramatically. One of the earliest was a series of nearly two dozen axe murders in Birmingham, Alabama, in the early 1920s. According to newspaper reports, five people were subjected to scopolamine interviews, during which they confessed to the murders and gave corroborating evidence to the investigators after the effects of the drugs had worn off.[17] Another case, about which more details are known, is that of a man accused in the murder of his lover's husband in 1934. As one of the many medical accounts of the case told it,

> Suspicion was directed at both the wife and a person identified as her paramour. The latter was so insistent in professing his innocence of the crime that he consented to take a scopolamine test. While under the influence [he was asked what he did with the pistol after the deceased had been shot; he replied that] he threw it into a river. At this point the investigators were in a quandary, because obviously nothing like this could have taken place in the instant case (the body having been found in bed with a pistol at its side). For that reason the questions were repeated. The subject then answered that he hid the gun in a patch of heather in a town in Ontario, Canada. [The subject also referred to covering the body with branches.] Concerning the present crime, however, he continued to profess his innocence.[18]

[17] *Birmingham Age Herald*, 7 January 1924, 1; 8 January 1924, 1; *New York Times*, 8 January 1924, 1. The case is also cited in John Henry Wigmore, *Principles of Judicial Proof, or, The Process of Proof: As Given by Logic, Psychology and General Experience, and Illustrated in Juidical Trials*, 2nd edn. (Boston: Little Brown, 1931), 610.

[18] 'Scientific Evidence in Criminal Cases', *Journal of Criminal Law and Criminology* 24/6 (1934),

The police subsequently checked with officials in Canada, whereupon they learned that he was wanted for several murders.

The truth serum drugs were widely discussed in popular journals, and seem to have been growing in popularity with police for interrogations. However, there were relatively few efforts to introduce scopolamine interviews as court evidence, and those few attempts that were made were unsuccessful. One judge, ruling on an early attempt, summed up in unusually colourful terms the general opinion of jurists: 'We are not told from what well this serum is drawn nor in what alembic its alleged truth compelling powers are distilled. Its origin is as nebulous as its effect is uncertain.'[19] Even those who were willing in principle to grant the validity of a 'truth technique' did not necessarily buy the idea that this was, as House claimed, a 'harmless' practice. George W. Kirchway, for example, the one-time dean of Columbia Law School, bluntly described its use as 'a mild form of torture', and thought that if it did secure truthful statements its use would violate two fundamental rules of common law: one, 'that the accused person may not be questioned after arrest or during a magisterial hearing', and another, 'that torture may not be used to extort the truth'.[20]

Yet the intense interest that House's claims produced among many different constituencies was enough to mitigate the significant scepticism that they also provoked, and the legal blockades that had been erected against them. Calls came forth from leaders within the legal and law enforcement communities for House to be given an institutional base and funds with which to develop scopolamine's potential—to see whether its truth-producing powers, at least, could be established, leaving aside the legal hurdles to its forensic use. But before this nascent campaign could bear fruit, House died suddenly of a massive stroke. He had been the leading proponent of scopolamine as a 'truth serum', and his death in 1930, coupled with the rise of the newly synthesized barbiturates, marked the decline of scopolamine as a forensic agent.

Throughout the 1920s and 1930s, scopolamine, and then the new barbiturates sodium amytal and sodium pentothal, were tested as 'truth sera'.[21] All three drew national publicity,[22] and psychiatric and legal societies debated their potential.[23]

1140–58. See also Fred Inbau, 'Scientific Detection of Crime', *Minnesota Law Review* 177 (1933), 602–3. This case was handled by the Scientific Crime Detection Laboratory at Northwestern University. The centre had been founded in 1929 to further the development of forensic techniques for studying traces, and was becoming the flagship institution in the application of laboratory medicine and 'trace' analysis to the creation of a police science.

[19] State *v.* Hudson, 31 4 Mo (1926), 599; Moenssens, 'Narcoanalysis in Law Enforcement', 455.

[20] Anon., 'Does Scopolamine Make Criminals Tell the Truth, as Declared?', *Current Opinion* 75 (September 1923), 344–5 (345). For similar worries about suspects' rights to privacy see *Harvard Law Review* 44 (1930), 842–6.

[21] Sodium amytal is a barbiturate compound first synthesized by the Lily Company around 1927; sodium pentothal was first synthesized around 1929 by Abbott Laboratories. M. Naples and T. P. Hackett, 'The Amytal Interview: History and Current Uses', *Psychosomatics* 19 (198), 98–105.

[22] C. Goddard, 'How Science Solves Crime III: "Truth Serum", or Scopolamine, in the Interrogation of Criminal Suspects', *Hygeia* 10 (1932), 337–40.

[23] Lorenz, 'Criminal Confessions under Narcosis', *Archives of Neurology and Psychiatry* (1932),

They were generally associated with each other during these early years. In fact amytal and pentothal rapidly eclipsed scopolamine as 'truth' drugs.[24] But the understanding of truth and falsehood associated with House's original 'truth serum' campaign was the one that ultimately prevailed in the public understanding—the one popularized in magazines of the 1930s and 1940s, and in films for the remainder of the century.

House had described his drug as a facilitator. It had the potential to create an environment in which an effective interview could take place, as long as it was dispensed by a specialist trained in the administration of a chemical agent whose effects were highly volatile, and as long as the interview was carried out by someone skilled in addressing questions to a consciousness that was also in a volatile state. Thus the drug itself was understood to reach into the subject's mind and secure a truthful narrative. This was one component of the 'truth drug' that remained consistent over the next several decades of popular representation: its capacity to exert an overwhelming power over the mind of the subject. In the Hammer film noir *Kiss Me Deadly* (1955) for instance, an interrogator injects the detective with 'truth serum', then does not even bother to ask him questions; the drugged man is shown, half-visible in the murky light of a darkened room, thrashing in a nightmarish drugged sleep, alone. Observers walk in and out intermittently, noting his behaviour and presumably transcribing his speech. A similar example may be found in James Cameron's *True Lies* (1994) where a single injection reduces Arnold Schwarzenegger to a pliant interviewee who gives a stream of reliable information in reply to every question put to him. As the next section will explain, this image of the drug as straightforwardly commandeering the mind was operationally inapplicable, because in the next generation uses of sera actually depended on the possibility that a subject might be faking the telling of truths. House's most lasting accomplishment, then, may have been to catalyse a public reputation for the serum as an agent that by definition reached into the subject's mind and pulled out 'the truth' regardless of anything the subject could do to resist.

THE PERSONAL TRUTHS OF SODIUM AMYTAL

The clinical and forensic future of truth serum was assured by the new barbiturates sodium amytal and pentothal. Yet they made their names not initially (or

1221–2; Anon., 'Selections: Sodium Amytal in Criminal Investigations under Narcosis', *Medico-Legal Journal* 48–50 (1931–3), 66–8.

[24] M. W. Thorner, 'The Psychopharmacology of Sodium Amytal', *Journal of Nervous and Mental Disease* 82 (1935), 299–303; David Lamar Mayer, 'The Effects of Sodium Amytal upon the Affective Behavior of the Normal Adult'. Unpublished Master's thesis presented to the Graduate School of Southern Methodist University (1933); E. Lindemann, 'Psychological Changes in Normal and Abnormal Individuals under the Influence of Sodium Amytal', *American Journal of Psychiatry* 11 (1932), 1083–91; M. Herman, 'The Use of Intravenous Sodium Amytal in Psychogenic Amnestic States', *Psychiatric Quarterly* 12 (1938), 738–42; C. P. Wagner, 'Pharmacological Actions of Barbiturates. Their Use in Neuropsychiatric Conditions', *JAMA* 101 (1933), 1787–92.

primarily) in House's forensic context, but in a psychiatric one. Hospital psychiatrists' understandings of these agents and their effects differed in significant ways from their forensic predecessors'. They had different perceptions as to the nature of the psychical states induced by the drugs, and of the statements subjects made in these states. In their own way these therapists were also using these drugs to hunt for truth. They were seeking the true nature of their patients' psychical disorders, and to gain access to the content of patients' distorted understandings and hallucinations. In some cases they were even trying to restore what one might call an authentic element of their patients' identities—something submerged, displaced, or distorted by the disorders from which they were suffering. But they were not particularly concerned with the factual truth of the narratives thus recovered from their patients. While some of them did see forensic potential in the drugs they were using for clinical purposes, it was the clinical context that loomed largest in their accounts. The truths their sera revealed were still narrative, but they were internal and psychological, not external and juridical.

In the late 1920s, a hospital psychiatrist working within a broadly psychoanalytic framework began using the recently synthesized barbiturates, particularly sodium amytal, to treat catatonic patients diagnosed with schizophrenia. W. J. Bleckwenn of the Wisconsin Psychiatric Institute in Madison induced a deep state of unconsciousness in these patients—deep enough to remove many basic reflexes—and let them sleep for hours. As they began to recover from the effects of the drug, they emerged (usually temporarily) from their catatonic states. These patients were long-term cases of catatonia—some had been on feeding tubes for months or years. As they emerged from the deep narcotic coma, they also awoke from their catatonic state: they walked, talked, ate, drank, and demonstrated that they could read or write. Bleckwenn found that he could get them to talk about the content of paranoid delusions that they had hitherto refused to discuss with him. He recorded his experiments in a dramatic film of 1934, in which his patients are shown in 'before' and 'after' shots. (The film itself is strikingly reminiscent, for the modern viewer, of Penny Marshall's 1990 film adaptation of Oliver Sacks's *Awakenings*, in which patients similarly seem to come briefly back to life as recognizably 'ordinary' and fully functioning human beings.) I have not been able to locate responses to Bleckwenn's film on the part of contemporary psychiatrists, but to my eyes at least they seem calculated to display a movement towards the recovery of some lost selfhood or authentic personal identity. The patients go from anonymous, depersonalized bodies to the real people one imagines they once were. For Bleckwenn, the 'truths' amytal could get for him were defined by something like the immediate equivalent of this witnessing experience. That is, the apparent resemblance patients' drug-induced behaviour bore to what one assumed (without specific knowledge) had been their behaviour and competences before the onset of their illnesses, validated the technique. It is this apperception that the film seems dedicated to conveying.

The explanation Bleckwenn gave for these results was that amytal had effects on the mind that fell loosely under the category of relaxation. Amytal made people shoulder their psychical burdens more lightly: paranoid schizophrenics were less concerned with concealing the content of their fears, and more able, therefore, to relate them to the psychiatrist. Applying this principle to a variety of patients, Bleckwenn found that his catatonic patients, assumed to be submerged in the coma because they had withdrawn from the world, were able to re-engage and interact with others. The problems of individuals diagnosed with neurosis and hysteria similarly gave way under the influence of the drug, and they were able to discuss their worries to a far greater extent under its influence than previously.

Given that the main use to which these psychiatrists were putting amytal and similar barbiturates was that of gaining access to beliefs that, however confidently held by the individuals concerned, were patently false, one might have assumed that this in itself would have disqualified the technique from a forensic application. That is, one might have assumed that such techniques could not be relied upon to recover a record of events in the manner in which House expected them to be. But in fact several psychiatrists, including Bleckwenn's close colleague W. F. Lorenz (also of the Wisconsin Psychiatric Institute), saw forensic potential in them.

Lorenz referred to a range of states of consciousness correlating with dosages and types of drugs, and on the basis of this distanced himself from the notion of a truth serum:

> [D]epending on the drug and the dose used, all degrees of disturbance of consciousness can be produced ranging from a vague sense of slight bewilderment to complete disorientation or unconsciousness. Some drugs produce a delirium with vivid hallucinations, while others induce a state of narcosis without hallucinations. In short, there is no 'truth serum' but a state of mind that can be induced by a variety of drugs.[25]

Having said this, however, he did not reflect on the truth status of what was said in different states of mind, and did conclude that a skilled psychiatrist could use drugs to secure truthful statements:

> Observing certain mental states which seem to conform to what was required for the test of 'telling the truth', we undertook some further investigations and sought especially to compare results obtained from various drugs. From this investigation we conclude that if a delirium can be avoided as a general proposition, the data and information obtained from a subject during a state of narcosis is much more reliable . . .
>
> We aim to produce a state of mind in which the individual continues to be fairly clear, that is, he is oriented and seemingly aware of his environment, yet his higher faculties are so impaired temporarily that he cannot inhibit automatic reactions to various stimuli. In other words, his responses to questions are reactions well formulated but not subject to any critical censorship. This is the state of mind that we believe might best fit the 'truth telling' level.

[25] W. F. Lorenz, 'Criminal Confessions Under Narcosis', *Wisconsin Medical Journal* (1932), 245–50 (245).

We found, as might be expected, that in such a condition not only were impressions from without, that is, from the environment, reacted to without censorship, but also, thoughts arising from within, that is, associations formed on the basis of memories, were likewise uttered without apparent voluntary control. That is, such subjects would talk on, more or less continuously expressing, without any particular emphasis, a whole panorama of views, ideas, experiences, and impressions.

We found it difficult to produce this state of mind as a constant reaction with either hyoscine or scopolamine with or without morphine. Upon the other hand, as we became more experienced with the use of sodium amytal, we found it possible to produce a state of mind that fulfilled the purpose sought and at the same time with the additional advantage that delirium never developed. (246–7)

This remark needs to be placed in the context of US psychiatrists' bid for a role in legal processes. American psychiatrists worked hard to develop increasingly prominent roles for themselves in the courts during the 1920s and 1930s, and found the legal world increasingly open to psychiatric entrepreneurs. Psychiatric supervision of the use of amytal for a forensic purpose offered a significant opportunity for professional advancement on the part of men such as Lorenz. The two fora—the hospital-psychiatric and the legal—were, of course, distinct, populated by communities operating with different assumptions about the nature of evidence and the ways in which one interpreted it. This helps explain the fact that Lorenz and other psychiatrists did not reflect self-consciously on the relation, or rather the distinction, between the kind of evidence they assumed they were getting when they gave amytal to a diagnosed schizophrenic and the kind of evidence they thought they could get by giving amytal to criminal suspects. Another piece of the puzzle is provided by the culture of pragmatism within the psychiatric communities of this period. Psychiatrists of the 1930s suited the theory to the practice rather than the other way round. They seized upon bits of theory and explanation that suited the particular purposes they had in mind. In this case, then, a broadly 'narrative' framework would legitimize statements made by psychiatric patients under the influence of amytal.

By the 1940s this apprehension of the power of barbiturates to recover one's 'true self', the content of one's delusional thoughts, or lost memory had been given a more organized form. It was now to be known as 'narcoanalysis', the use of drugs to access repressed thoughts. The term was coined by the British psychiatrist Stephen Horsley, for whom it was a means of carrying out 'exploratory surgery' on the mind. Narcoanalysis, when he introduced it in 1936, was supposed to be the poor man's psychoanalysis, a means of tapping into the unconscious for the price of an IV. Through the use of post-hypnotic suggestion, he claimed, previously repressed mental content could be integrated back into an individual's life in a constructive manner.[26]

[26] His term, 'narcoanalysis', was amended to 'narcosynthesis' by the American physicians Roy Grinker and John Spiegel in 1942. See J. S. Horsley, *Narco-analysis* (London: Oxford University Press, 1943); see also R. Grinker and J. Spiegel, 'Brief Psychotherapy in War Neuroses', *Journal of Psychosomatic Medicine* 6 (1944), 123–31. R. Grinker and J. P. Spiegel, *Men Under Stress* (Philadelphia: Blakiston, 1977); R. Grinker and J. P. Spiegel, *War Neuroses* (Philadelphia: Blakiston, 1945).

The ultimate testing ground for the new barbiturates, both as therapy and as adjudicator, was provided by the Second World War. This war was for chemical therapies and psychiatric diagnoses what the First World War had been for psychoanalysis: a gigantic arena that established it as a favourite tool of a great many doctors, who continued to rely on it after the war. At first amytal was used, as it had been by Bleckwenn, Lindeman, and other psychiatrists, as a way of getting access to painful, repressed, or delusional thoughts. Later, though, it came to take on a role somewhere between this increasingly familiar psychiatric use and the forensic one associated with the term 'truth serum'. Specifically, it came to be used during and after the war as a way of differentiating between malingerers and those who were 'truly' suffering from psychiatric disorders such as battle trauma. A real sufferer would experience a cathartic release as he told his story (and could later be restored to the ranks as cured); a faker would endeavour to keep silent. In this context the salient truth was manifested when a patient did *not* blurt out personal narratives.

TRUTH TECHNIQUES AND THE LITERATURE OF DYSTOPIA

From the 1920s onward, and most intensely in the 1940s, truth serum and lie detection sparked the imagination of popular journalists and fiction writers not just in America, but much more lastingly in Britain and Europe, where the threat of an intrusive and abusive state power loomed, of course, much larger than it did in the United States. As an indication of the ephemeral nature of much of that literature, it is worth citing a minor play published in Holland in 1932 (and probably never performed), *Mind Products Unlimited*, whose plot laboured the notion that chemical expertise could produce personality-transforming and truth-inducing potions.

The most sustained and politically hostile literary engagement with the use of truth drugs came with the novel *Kallocain*, written by the Swedish poet and novelist Karin Boye (1900–41).[27] First published in Swedish in 1940, with the subtitle *Roman Från 2000-talet* (*Novel of the 21st Century*), *Kallocain* was translated into German and French in 1947, but did not appear in English until 1966.[28] Before *Kallocain*, Boye was known primarily as a poet and editor of poetry. In 1931 she helped to found the poetry magazine *Spektrum*, which brought to Swedish readers the work of T. S. Eliot and the Surrealists; she was also a co-translator of Eliot's *The Waste Land*. Before she committed suicide in 1940, she wrote six volumes of

[27] On Boye see R. B. Vowles, Introduction to Boye, *Kallocain*, pp. vii–xxi.; P. K. Garde, 'How to Remember Her?', *Connexions* 10 (1983), 22–3; H. Forsas Scott, 'Reading and Writing our Own Tongue: The Examples of Elin Wagner and Karin Boye', *Women's Studies International Forum* 9/4 (1986), 355–61.

[28] *Kallocain: Roman Från 2000-talet* (Stockholm: Albert Bonniers, 1940); for details of the first English translation see n. 2 above. German and French translation dates from Löwendhal Rare Books, London, Catalogue 9.

poetry and three of short stories, as well as five novels. *Kallocain* was her last, and it is thought to have voiced, and perhaps even intensified, the fears that drove her to suicide.

Boye's work expressed a craving for intellectual and emotional freedom, and an intense valuation of intimate self-revelation. It was heavily influenced by her knowledge of psychoanalysis: she moved to Berlin in the early 1930s to be analysed, and the experience made a strong impression on her written work thereafter. Her interest in contemporary theories of mind also presumably brought her into contact with the theory, if not the practice, of therapeutic and other uses of truth drugs. Her American editor, Richard Vowles, sees a sharp disparity between Boye's poetry and her novels, the former reflecting a 'lyrical inwardness' and the other 'an oracular sense of public responsibility', 'fields' that 'lay hopelessly apart'.[29] Vowles takes this discontinuity as one reason why she is not now better known among literary scholars outside Sweden. Yet, for all their differences, Boye's poetic and fictional writings come together in their fascination with the nature and the limits of self-disclosure: her poems are intensely confessional in manner, and the idea and practice of confession are preoccupations of her novels, noteworthy among them (besides *Kallocain*) a psychoanalytical case history entitled *KRIS* (1934).

Kallocain imagines a future totalitarian state, which has entirely subsumed the individual. In this 'Worldstate', the concept of the 'individual' is itself heretical. Privacy, intimacy, and individual relationships are almost entirely eliminated. Sons and daughters, for example, are separated from their parents forever at an early age. The novel is structured around the development and implementation in this world of a truth serum named Kallocain after its inventor, Leo Kall. Kall anticipates that his new drug will complete the desired erosion of individuality and personal privacy, producing a fully 'communal' social state and a collective sense of identity.

Boye uses the Kallocain interview in a satisfyingly double-edged way. It becomes a mechanism for characters within the plot to extract truthful confessions from subjects, but also a way for Boye to make characters tell the truth directly about the dystopic society in which they are trapped rather than revealing its features indirectly through readers' observations. Within the plot it functions both as a repressive mechanism and as a subversive one for the same reasons; for the characters, too, it is both a way of securing truths (of many kinds) and a demonstration of how the society itself institutionalizes forms of falseness and bad faith. Also, despite the overt claim that individuals are subsumed by the state, many features of the truth drug show the persistence of individualism and individual creativity: its very naming is a clue, perhaps, that it will not have the straightforward role of dissolving the individual mind into the collective as Kall initially proposes.

Boye scholars have assumed that she had no prior knowledge of any supposed

[29] Vowles, Introduction, p. vii.

'truth drug'. She committed suicide shortly after writing the novel (a writing process she described as nightmarish). She left few records that would enlighten literary historians as to the resources on which she drew to develop the notion, though the editor of one edition of *Kallocain* has noted her friendship with the chemist and poet, Ebbe Linde.[30] Yet the similarities between the hopes extended for Kallocain and for scopolamine, pentothal, and amytal are surely too great to be coincidental. Some of the hopes held out by Kall echo, to the very phrasing, those of Robert House for scopolamine. In the early stages of development of the truth drug, for instance, Kall reminds his superior, Edo Rissen, that 'we are both familiar with the shortcomings of the third degree'. He anticipates that Kallocain will provide an alternative with none of its fallibility or brutality: 'the criminal will confess, happily and without reservation'. He also anticipates, like House, that the drug will remove the need for third-party witnesses.[31]

The physiological action of the drug is also strikingly similar to that of scopolamine and its barbiturate successors. Like amytal and pentothal, Kallocain is an intravenous drug. Also like them, it takes effect in a few minutes. The altered state it produces seems identical: it, too, removes inhibitions and creates a relaxed state in which the affected individual has an urge to express herself or himself, and does so candidly.

Kallocain is not a replica of House's scopolamine, however. House thought of his drug specifically as a means of securing information, of opening what he understood to be the locked filing cabinet within a subject's brain. Kallocain is more like amytal or pentothal as reported by 1930s psychiatrists. As in the reports made by psychiatrists experimenting with the use of barbiturates on individuals suffering from trauma, the memories expressed under its influence are relived and embodied by the subject. That is, they are not narrated as abstracted truths but experienced again with something of their original force and flavour. For instance, consider the first experiment, and the one related in the greatest detail. Several minutes after the contents of the syringe are injected into his arm, the test subject ('No. 135') speaks 'with noticeable slowness' and eventually reports, 'I feel so awfully well. I've never felt so well in my whole life.' Then, however, after a further pause, 'The big, rough man started to weep in despair. He sank down in the chair until he hung like a rag over the armrest and emitted long, low, almost rhythmic moans.' When he eventually speaks, the man describes being intensely unhappy, miserable in a life that feels to him as if it is merely a long-drawn-out form of death. This individual's life literally is a lengthy process of dying because he is a member of the 'voluntary test [sacrificial] service', whose job it is to be subjected to scientific tests of all kinds—though of course the narrative is also an allegorical commentary on the state of everyone in the society. His statement is a confirmation of the absolute power of the drug, since as natural as a reader might

[30] Ibid. p. xvi. Vowles does not give any further information about Linde, nor have I been able to find any.

[31] Boye, *Kallocain*, 47.

expect his unhappiness to be, his feelings are outlawed in the new Worldstate. The state demands not only absolute compliance and self-denial in the actions of its subjects, but also enthusiasm and joy in this compliance and service. The feelings of individuals are as subject to policing by the state as are their outward actions. His description of his feelings is therefore an indication that the new drug is extraordinarily powerful.

The new drug promises to be an effective tool of repression for the state because, as Kall hoped, it can remove the 'last vestige of privacy from the individual'. The state can now seek not only to police traditional forms of crime, but to check that ordinary citizens' feelings do indeed match their loyal, apparently cheerful actions. No such thing as false consciousness in Worldstate: or rather, nothing but 'verified' false consciousness. One characteristic in which Kallocain differs from scopolamine, pentothal, and amytal intensifies its utility in enhancing the disciplinary power of the state: Kallocain never suppresses the memory of the drug interview, and subjects are left with an unrelentingly detailed memory of the confessions they have poured forth during the drug interview.

Kallocain has been criticized as unidimensional in comparison with better-known novels of dystopia, such as *Brave New World*; R. B. Vowles regards its characters as projected features of Boye's personality, for instance, rather than more complex and multifaceted individuals in themselves.[32] The role of the truth drug in the novel underscores this criticism, at least if one compares it to the portrayal of surveillance and other truth-procuring techniques in the more famous novel to which the modern reader inevitably compares *Kallocain*, George Orwell's *Nineteen Eighty-Four* (1949). Both novels depict a state in which individuality is not permitted or its existence even acknowledged (although there are, of course, proles in Oceania). Both imagine a near future around the same time as one another. But *Kallocain* is much more of a straightforward protest against totalitarianism. Scholars of *Nineteen Eighty-Four*, in contrast, have ascribed to it many different purposes and meanings, and a number have pointed out the futility of attempting to shoehorn it into a critique of a particular political system or ideology.

A central theme of *Nineteen Eighty-Four* is the question of whether there is an internal, inviolable truth at all, preservable in memory but not accessible to others. Despite the main characters' ultimate failure to preserve their confidence in such a notion, it is clear that Orwell believed in it himself—or, more important, that he believed it was politically crucial to believe in it. The possibility of a private memory (for instance, Winston Smith's diary) is a pivotal assertion of the novel early on, and towards the end the demonstration that Smith has been able neither to conceal from the state (or its representative, O'Brien) the truths of his personal memory, nor, finally, even to maintain them himself, is devastating. On the other hand, Orwell does preserve the idea, if not the ideal, of reliable collective memory. In *Kallocain* the state is in the process of successfully erasing histor-

[32] Vowles, Introduction, pp. x–xi.

ical memory in its members by policing every child's education. For Boye this does not matter as much as it does to Orwell because she is not concerned with the public memory of truth so much as with a romantic notion of truth—validated by intense, intimate personal sensibility, but resting on the ability to share intimate knowledge with another person. Orwell is more concerned with defending the notion of 'external' truth, and so the ongoing preservation of the past in the form of group memory is far more important. He limits the power of the state to erase or redefine memory by the existence of the proles, who embody the past and who live outside the thought-policing structures of the totalitarian state.

Kallocain and *Nineteen Eighty-Four* thus share a central concern with the importance of personal truth and memory that can withstand the state's efforts to know and police thought. Both imagine a state bent on erasing this kind of truth and the interpersonal relationships sustained by it. But in *Kallocain* the tool by which private truth is to be erased is a specific and extraneous one. In *Nineteen Eighty-Four*, such specificity would diminish our sense of the mysterious surveillance power of the state. We are never told exactly how O'Brien is so quick and so successful at figuring out what Smith is thinking; Orwell makes reference to truth drugs and truth techniques but never spells out how they might work and what their relationship might be to the wider apparatus of torture to which individuals are routinely subjected. Nor does he resort to the kind of physical 'confessional' acts that authors routinely use to depict characters giving themselves away. For instance, there is a point in the interrogation sequence when Winston observes, silently, how old and worn O'Brien appears to be. Orwell does not describe any outward signs of Winston's inner thoughts, as he might have done (a change of expression or posture for instance, or a slip of the tongue or an indirect hint in conversation). Yet O'Brien instantly knows what Winston has been thinking and is able to recite it to him almost verbatim. The difference between this and *Kallocain*'s specific and instrumental device is that between a relatively simple concept of testimonial truth—like that of House—and a far more complex concept, such that only the massive yet subtle psycholinguistics of doublethink could control it. Oceania admits of nothing other than the state. In *Kallocain* too science has been reduced to mere utilitarian pursuit of weapons and surveillance techniques (something that at this very time sociologist of science Robert Merton was describing as a fatal but essential characteristic of totalitarian science). But in Boye's novel things are still much more primitive than in Orwell's: this state has totalitarianism as an aspiration, not a satiric reality. The respective roles of truth techniques in the two novels thus underpin differences in the degree to which the state has assumed the power to determine what counts as truth. In short, the real truth technique in *Nineteen Eighty-Four* is the state itself.

CONCLUSION

The chronology sketched in this essay suggests several shifts in the attitudes of police, lawmakers, medical researchers, and writers during the early decades of the twentieth century as to the nature and social implications of the evidence yielded up by a truth technique. In the 1920s and early 1930s, police and medical researchers alike seized upon the notion of a technique that could supply truths defined as having a reality independent of the fantasy life of the individual concerned. Their implicit notion of memory—of a history embodied in the brain and accessible through chemical techniques—was akin to that of a filing cabinet, or, to use an anachronistic simile, a database. Truth serum, along with the lie detector and other new techniques for evaluating human testimony, thus became key features of the new field of police science, as it emerged as a concerted, self-conscious, and energetic activity in America for the first time in the 1920s and 1930s, and as it was subsequently emulated across Europe.

During this period, psychiatrists and psychoanalysts were developing a different model: truth drugs were a means of tapping into memories, fantasies, and emotions. This notion did not (necessarily) make a strong distinction between historical truths and narrative truths—truths that might be, for instance, memories of past fantasies or expressions of repressed emotions. One factor in the rise of this model was the rapid development and increasing authority of psychoanalysis after the First World War. Another was the growing, though fragmentary and anecdotal, evidence that subjects could continue to lie under the influence of these drugs, or that suggestions made to them during the interview could be reflected back unwittingly as supposed memory. This psychiatric model emphasized the use of pentothal and other chemicals (along with hypnosis) as a cathartic technique. As such, the specific statements that an individual made under the influence were less important than the way he or she said them.

This did not mean that truth sera could not be used as diagnostic aids in distinguishing between falsehood and truth, but the nature of the evidence changed in psychiatric contexts, and so did the kinds of falsehood it could ferret out. This model of truth-telling emphasized what one might call psychological plausibility in the behaviour of tested individuals. For instance, the earliest forensic models of truth-telling emphasized the value of consistency in testimony: if a subject gave the same testimony while drugged as he did sober, one could plausibly trust the testimony. Within the psychiatric and psychoanalytic models, in contrast, consistency could be a sign of falsehood, because it was seen to suggest that the subject was self-consciously controlling his behaviour instead of allowing himself to be affected by the action of the drug. So, in the case of aphasia and other forms of battle fatigue in the Second World War, subjects who seemed not to change their affect under the influence of pentothal, or amytal subjects who did not exhibit signs of improvement or release, were to be suspected as malingerers.

From the beginning the research into the extraction of memory involved two interwoven streams of research, two related evolving models—one focusing on the notion of historically valid truth, the other on narrative truth, though the latter continually chipped away at the confidence with which people asserted the possibility of accessing the former. In the Second World War they converged;

after the war they diverged sharply and came into conflict. Then, the psychiatric model won out and the forensic one subsided, at least for a time.

The recent debates about recovered or false memory syndrome suggest that there is a cyclical nature to the kinds of plausibility that these truth techniques can have. The notion of a 'truth technique' for recovering memory gained renewed popularity within the criminal and legal system in the mid- to late 1970s, after a few very striking cases in which hypnosis refreshed the memory of witnesses (whose statements were then confirmed by new physical evidence). Inspired by these extraordinary results, police departments throughout America founded training programmes in hypnosis, sometimes augmenting their techniques with amytal and pentothal.

From these beginnings came the well-known controversies over 'recovered memory syndrome' in the 1980s and early 1990s, in which judges presiding over civil lawsuits (and in some cases, criminal charges) admitted evidence gleaned from interviews of subjects under the influence of pentothal-like chemicals, on the grounds that this evidence related to historical events, specific details of which could be viably distinguished from other possible forms of psychical evidence (fantasies, remembered fantasies, or suggestions made by the interviewer that were reflected back in the speech of the subject). In these cases, individuals were well aware of the potential taints that could be present in drug-induced testimony, but in the cultural and political contexts in which this evidence was introduced these concerns had less force than they had in the past (and have had since) because they seemed to place victims of childhood sexual abuse at an evidentiary disadvantage with respect to their abusers. The course of the recovered memory debates was, however, to be speedier than that of previous developments in the career of chemical truth techniques—perhaps because the various conventions of evidence in play were so much more familiar than they had been to individuals in previous decades. It is striking that in these debates the options—historical truth, remembered fantasy/worry, and modern suggestion—were much the same as they had been in previous decades. Whether the similarities or the subtle differences are more explanatory of what occurred remains moot.

It appears that the most significant factor in the career of truth serum and associated techniques is not the cumulative effect of the many experiments, therapeutic trials, and court cases, but rather the sheer power of motivation that certain communities have felt at various times to gain access to and control over human memory. This motivation has produced recurrent propositions about the medical accessibility of memory, each similar in many respects to the last, despite a consistent pattern of complications and discreditations.

It should be clear that the history of truth serum is certainly not one of cumulative knowledge. But why has the long series of experimental failures not discredited the notion once and for all? At root, the answer is that behind the truth serum story lies a history of something a little different and, I think, more basic: a history of the need for memory. This other history is what my project will eventually seek to recover. It is a story of changing notions about what memory itself is, how to get at it, and what it would mean to get truth from remembered experience. It is a measure of the strength of this need that, in their quest for objective and exhaustive memories, researchers in successive generations have been so ready to override what might be regarded as devastating counter-evidence. In other words, the resilience of truth serum rests on a history of forgetting.

CHAPTER 14

Coming of Age

E. F. Keller

> The boy lay down, charmed by the quiet pool,
> And, while he slaked his thirst, another thirst
> Grew; as he drank he saw before his eyes
> A form, a face, and loved with leaping heart
> A hope unreal and thought the shape was real.
> ('Narcissus and Echo', Ovid, *Metamorphoses*)[1]

If Schroedinger's 'What is Life?' was widely read in the 1950s by scientists interested in biology, J. D. Watson's account of his and Francis Crick's epochal achievement, *The Double Helix*, has, it seems, been read by virtually everyone. Since its original publication in 1968, it has been reissued eleven times (in English alone)—including a 1980 appearance in a Norton Critical Edition; it has sold many thousands of copies, and promises to remain a strong seller for some time to come. In 2002, 34 years after its first appearance, eleven English language editions were still in print.

The Double Helix has also received its share of critical commentary. A flood of reviews (forty-three to be exact) appeared within the first year of publication, five in the second, and at least ten others since. It has also been made the subject of serious literary analysis (see especially, John Limon's '*The Double Helix* as Literature' (1986)[2]), in which it has been compared to Kingsley Amis's *Lucky Jim*, praised for its apparently novel mix of factual reportage and novelistic techniques (Limon calls it 'a factual novel'); and analysed for its rhetorical structures.[3] What more could possibly need to be said? Perhaps not very much, but one issue, I think, has not been adequately noted. That issue concerns the status of the body in (at least this account of) the early days of Molecular Biology. Here, in this chapter, I want to explore a rather remarkable parallelism emerging from Watson's 'personal account' among at least three different kinds of relationships: between models on the one hand and physical/biological reality on the other; between replication and

[1] Translated by A. D. Melville, with an introduction and notes by E. J. Kenney (Oxford: Oxford University Press, 1986), 64.

[2] *Raritan* 3 (1986), 26–47.

[3] Alan Gross, *The Rhetoric of Science* (Cambridge, Mass.: Harvard University Press, 1991).

reproduction; and between Cambridge 'popsies' and actual (especially mature) women. In each case, the embodied three-dimensional subject, bemired in the 'messy complexity' of the real, is displaced (and effaced) by a highly simplified two- (or one-) dimensional facsimile—a surface or model, lovely to look at, but impervious to touch. This relational parallelism, I suggest, functions crucially in Watson's journey from adolescence to manhood, and perhaps even in his casting of the central discovery as marking his arrival as a mature scientist. I begin, therefore, by tracing that journey.

SETTING THE STAGE

That *The Double Helix* is a 'coming of age' story is hardly a secret; indeed, it is made explicit by the author himself. The narrative opens with the hero a mere 23 years old, and closes with his *rite de passage* completed. The last lines read: 'But now I was alone, looking at the long-haired girls near St. Germain des Pres and knowing they were not for me. I was twenty-five and too old to be unusual.'[4] Two years earlier, however, Watson arrives in Cambridge, England, ogling (from afar) both the long- and the short-haired girls with all the brashness and ineptitude of a blustering adolescent, self-consciously groping for the 'unusual', and in desperate need of the right guide to manhood.

Fortuitously, that guide is provided at the start. Chapter 1 begins: 'I have never seen Francis Crick in a modest mood.' Crick's 'quick, penetrating mind' (7), his voice 'booming over the tea room' (9), his 'shattering bang' of a laugh (10), all clearly mark him for the role. Twelve years Watson's senior (even though as yet without his Ph.D.), Crick is at the prime of his life, and of his intellect. The 'quick manner in which he seized their facts' makes the stomachs of his older colleagues (those past their prime) 'sink with apprehension', lest he 'expose to the world the fuzziness of [their] minds' (10). In contrast to the 'popular conception supported by newspapers and mothers of scientists', Watson frankly divulges, these 'pedantic, middle-aged men' (11) are mostly 'cantankerous fools'—'not only narrow-minded and dull, but also just stupid' (14).

Finally, Francis also knows how to live. Married to the fun-loving (and tolerant) Odile, his home life stood in like contrast to that of his colleagues. No dinners for him of

> wives' drab mixtures of tasteless meat, boiled potatoes, colorless greens, and typical trifles. Instead, dinner was often gay, especially after the wine turned the conversation to the currently talked-about Cambridge popsies.
>
> There was no restraint in Francis' enthusiasms about young women—that is, as long as they showed some vitality and were distinctive in any way that permitted gossip and amusement. When young, he saw little of women and was only now discovering the sparkle they added to life. Odile did not mind this predilection . . . (65)

[4] James Dewey Watson *et al.*, *The Double Helix*, A Norton Critical Edition (New York: W. W. Norton and Co., 1980), 223.

And so the stage is set.

Together, the two buccaneers embark on their quest for the 'Rosetta Stone for unravelling the true secret of life' (14). Clearly, it is boldness, not timidity—especially not the timidity of Maurice Wilkins—that is required for 'seizing DNA', for handling 'dynamite like DNA' (16). And speed as well, for they are, we are told, not only in a contest with nature, but also against time. Across the ocean, Linus Pauling—hardly young, but still supremely potent—was dazzling the world with his powerful and 'uniquely beautiful' creations. Pauling was the reigning king, and he needed to be dethroned: 'The combination of his prodigious mind and his infectious grin were unbeatable. . . . Even if he were to say nonsense, his mesmerized students would never know because of his unquenchable self-confidence. A number of his colleagues quietly waited for the day when he would fall flat on his face by botching something important' (36). Watson and Crick's task was clear: 'Within a few days after my arrival, we knew what to do: imitate Linus Pauling and beat him at his own game' (48). Since it was only a matter of time before Pauling embarked on his own 'assault on DNA' (157), the challenge was to get there first.

The structure of this drama has a kind of classic simplicity. Although plausibly viewed as a 'factual novel', the very form of its narrative line—from youthful challenge to triumphant conquest—borrows from the oldest genre of fairy-tale. As such, it mandates at least one major obstacle, and that obstacle is introduced in chapter 2: 'The real problem, then, was Rosy' (20).

With the hindsight of maturity, victory, and Rosalind Franklin's premature death, Watson felt called upon in his Epilogue to offer something of an apology for his treatment of her in the main text, but the text stands. For it is not in fact Rosalind Franklin who is depicted (indeed, the apology may only serve to exacerbate the insult), but a mythical character called 'Rosy'. 'Rosy', as quickly becomes evident, is guardian of the magical Rosetta Stone.

But it is not only by her presence as local custodian of DNA that she impedes the hero's access to his quest; even more, it is by the qualities of her being. Angry, aggressive, hysterical, desexed, and emasculating, she is the phantasmatic virago, the screen on which the young hero's fears are projected. It is her stance (even the way she looks[5]) rather than her actions that threatens—that compromises his achievement of the requisite manliness.

In the first instance, however, Rosy is not Watson's problem, but rather that of Maurice Wilkins, her senior colleague at King's College. So much so, in fact, that Wilkins cannot take his mind off her: 'Not that he was at all in love with Rosy, as we called her from a distance. Just the opposite . . . She claimed that she had been

[5] Or at least, the way she looked to Watson. The distortions in his portrayal of Franklin become particularly evident when compared either with Ann Sayre's conspicuously more realistic portrait in *Rosalind Franklin and DNA* (New York: W. W. Norton, 1975), or with the excellent (and probably definitive) biography recently published by Brenda Maddox, *Rosalind Franklin: The Dark Lady of DNA* (New York: HarperCollins, 2002). My focus here, however, is neither on Franklin, nor on the injustice of Watson's portrait, but strictly on his creation of 'Rosy'.

given DNA for her own problem and would not think of herself as Maurice's assistant' (16). Furthermore, 'mere inspection suggested she would not bend. By choice she did not emphasize her feminine qualities. . . . it was quite easy to imagine her the product of an unsatisfied mother . . .' Perhaps even more to the point, 'given her belligerent moods, it would be very difficult for Maurice to maintain a dominant position that would allow him to think unhindered about DNA' (17). Clearly, dominance, unhindered, was a prerequisite for success.

When the book first appeared, Watson's portrait of Rosalind Franklin inevitably aroused the ire of feminists; especially, it aroused the ire of all those who knew the actual woman. But as a purely fictional character, it must be granted that Rosy comes from a long literary lineage. Indeed, it is in Watson's treatment of women, and especially of Rosy, that Limon finds the closest resemblance to *Lucky Jim*:

> the two Jims are even more brutal about women than they are about the senescent and stupid—women are 'popsies' in both books. The most distressing connection between Jim Dixon and Jim Watson is their treatment of intellectual women . . . In both cases it is considered wrong, nearly a sexual sin on the part of the aggressive, often hysterically neurotic academic woman that she does not take, as Watson says of 'Rosy,' 'even a mild interest in clothes.' . . . On the other hand, prettiness in women is very nearly all-in-all.[6]

Limon continues with a nice observation about 'prettiness':

> I use the term *pretty* advisedly—it is one of the key words in both books. . . . In *The Double Helix*, the adjective is omnipresent: all the desire in the memoir is divided between the search for the 'pretty truth' about the pretty double helix, and the search for pretty girls. The two esthetic objects compete for the attention of both Watson and Crick. The word 'beautiful,' or any comparably strong synonym, almost never appears in either book. It gets attached to only two people in *The Double Helix*, one of whom is a man and one of whom is Watson's sister. Indeed, for all the preoccupation with Cambridge popsies, there is surprisingly little sex or sexuality in either book. The only sexual activity explicitly referred to in *The Double Helix* takes place between bacteria, and it is far from obvious that any of the prettiness referred to results in erotic feeling. (32)

Limon is, of course, right—or very nearly right. Watson's sister is not actually 'beautiful', only 'very pretty'. On the other hand, the word 'beautiful' is invoked not only for Bertrand Foucarde, a young Frenchman visiting Cambridge to 'perfect his English' (indeed, Bertrand's only detectable function in this memoir, apart from an occasional game of tennis, seems to be his presence as 'the most beautiful male, if not person, in Cambridge' (174)), but also for a variety of 'man-made' structures: for Linus Pauling's alpha helix (36), for the buildings of Cambridge (43), and finally, for the model of the double helix he and Crick at last construct (213). None the less, Limon's main point stands: for all the talk about sex, both the book and the 'secret of life' Watson has so often 'daydreamt of discovering' reflect a remarkably asexual vision. The mechanism for procreation

[6] Limon, '*The Double Helix* as Literature', 31.

(or reproduction) he and Crick suggest—the 'base pairing' of nucleotides—is, finally, a mechanism for asexual replication: one DNA molecule is transformed into two identical molecules.

Sexuality (at least conventionally reproductive sexuality) is not, then, to the point. The developmental picture that unfolds is, if anything, pre-sexual. Or so, at least, is how the author fears it might appear from Rosy's perspective. Watson suspects her of regarding his favourite activity (playing tennis with Bertrand was the only thing he admits to finding 'more pleasing than model building' (177)) as mere child's play: 'only a genius of [Pauling's] stature could play like a ten-year-old boy and still get the right answer' (69).

Watson's task, then, is to demonstrate the efficacy and force of such a boyish proclivity. By his own account, his first incentive for learning crystallography was, 'I did not want Rosy to speak over my head' (56). Given her attitude towards model building, 'fear of a sharp retort from Rosy' made everyone around her reluctant 'even to mention models'. 'Certainly a bad way to go out into the foulness of a heavy, foggy November night was to be told by a woman to refrain from venturing an opinion about a subject for which you were not trained. It was a sure way of bringing back unpleasant memories of lower school' (70). Alas, Watson's initial efforts to master crystallography proved disastrously inadequate, and deeply humiliating. Ebullient over their first attempted structure of DNA, built out of carbon-atom models held together with copper wire, he and Crick had invited Rosy and Wilkins to witness their triumph. Rosy's reaction was cool, 'curt', and 'aggressive':

> Most annoyingly, her objections were not mere perversity: at this stage the embarrassing fact came out that my recollection of the water content of Rosy's DNA samples could not be right. The awkward truth became apparent that the correct DNA model must contain at least ten times more water than was found in our model. . . . there was no escaping the conclusion that our argument was soft. (94)

An argument so 'soft' could hardly provide cause for Rosy to be impressed by her 'fifty-mile excursion into adolescent blather' (95).

Embarrassed and humbled, many months would pass before they might try again. Crick is ordered to return to work on his thesis, and both men are made to promise they will leave DNA to their colleagues at King's College. To while away the time, Watson accepts an invitation to the home of Naomi Michison, 'sister of England's most clever and eccentric biologist, J. B. S. Haldane' (102), where the evenings are passed playing intellectual games. Once again, not his forte. 'Every time my limpid contribution was read, I wanted to sink behind my chair rather than face the condescending stares of the Michison women. . . . Much more agreeable were the hours playing "Murder" in the dark twisting recesses of the upstairs floors' (106).

But young Watson is not to be held down for long. Back at work, he tackles the problems of crystallography in earnest, even attempting to take his own X-ray pictures (not of course of DNA, still the preserve of King's College, but of the

RNA in TMV—tobacco mosaic virus). But the thought of DNA lying fallow at King's festers, and the spectre of Rosy's recalcitrance, and of her iron-clad grip on her male colleagues, takes on new dimensions. 'Rather than build helical models, she might twist the copper-wire models about his [Wilkins's] neck' (122).

By June, Watson had obtained the X-ray pattern needed to prove that the RNA in TMV is helical: 'It was, of course, clear what we should next conquer' (125). If only they could get rid of Rosy. 'The question of finding Rosy a job elsewhere had been brought to his [Wilkins's] boss, Randall, but the best to be hoped for would be a new position starting a year hence. Sacking her immediately on the basis of her acid smile just couldn't be done. Moreover, her X-ray pictures were getting prettier and prettier' (148). Nothing to be done but wait, or so it then seemed.

By late fall, however, a letter from Pauling to his son Peter arrived with the news that he, Pauling, had a model for DNA! But when the manuscript finally arrived, to their immense relief, it became clear that the astonishing Pauling had made a blunder. This meant that they still had time, but not a lot: 'When his mistake became known, Linus would not stop until he had captured the right structure' (162). Certainly not enough time for English niceties.

At this point, the climax of the memoir, a visit to King's to warn Wilkins seemed imperative. As it happens, Wilkins is otherwise occupied, and Watson goes instead to Rosy's lab. There, risking 'a full explosion', he attempts to explain to her the urgency, the need for someone more competent to interpret her data, and—now more than ever—for the construction of helical models: 'Suddenly Rosy came from behind the lab bench that separated us and began moving toward me. Fearing that in her hot anger she might strike me, I grabbed the Pauling manuscript and hastily retreated to the open door. My escape was blocked by Maurice, who, searching for me, had just then stuck his head through.' The young hero's bold offensive apparently ends, once again, in humiliation and retreat: 'While Maurice and Rosy looked at each other over my slouching figure . . . I was inching my body from between them, leaving Maurice face to face with Rosy' (166). But not for long. With Rosy's help, the incident serves not to underscore Maurice's paternal seniority, 'face to face with Rosy', but rather his identification with the young Watson. Removing 'Maurice from his uncertainty by turning around and firmly shutting the door', she cements their new fraternity by shutting both men out together, thus reminding Maurice not of his seniority or institutional authority, but only of the time, some months earlier, when 'she had made a similar lunge toward him' (167).

And so a new alliance is forged; Maurice Wilkins leaks the crucial X-ray print of the 'B' form that Rosy had obtained over the summer, and the rest, as we say, is history. Armed with the invaluable new data the X-ray pattern revealed, Watson rushes back to Cambridge. This indeed is just the data he and Crick need to reconstruct their model, and within a few weeks (and with a few other lucky breaks), they have a double helical model of DNA that works. The next time Rosy and Maurice come to Cambridge is to witness their triumphant victory.

IS THERE A BODY IN THIS TEXT?

As I'm sure is evident from the précis given above, my purpose is not to discuss the scientific history as here recorded by Watson. There is another set of issues to which the narrative speaks quite directly, and with surprising clarity: the fragility of at least this masculine ego; the bifurcated perception of adult women as either monstrous and dangerous or trivial and inconsequential;[7] and the use of scientific conquest—rather than, say, of heterosexual conquest—as the central *rite de passage* to manhood. For Watson, the couplings that count most are between himself and Crick on the one hand, and the base pairing of nucleotides on the other. Indeed, the popular logo, 'Watson-Crick', denotes both pairings at once.[8]

Limon is not alone in noting the conspicuous absence of overt sexuality, an absence perhaps especially striking for a coming-of-age story. But he is the only critic to date to develop this theme.

> When Watson says that 'important biological objects come in pairs', he would have to be thinking less of such pairs as Wilkins-Franklin, and more of such pairs as Watson-Crick. During Watson's digression on the subject of Bertrand Fourcade's asexual or transsexual beauty—he is, with Watson's sister, about the only thing in the book more attractive than a DNA molecule—he mentions 'Bertrand's perfectly proportioned face.' On Bertrand's asexual or transsexual face, Watson implies, are paired biological organs in perfect asexual symmetry and mutual adjustment. (39)

Limon goes on to suggest that 'from the idea that the prettiest symmetries in the world are masculine and asexual (quasi-sexual, let us say), from Bertrand Fourcade's face to Watson-Crick, Watson found the inspiration for both his DNA model and his novel' (41).

The DNA model, containing its very own secret of life, is, according to Limon (39), 'a child' that has been born of two men. His allusion to Shelley's *Frankenstein* is perhaps unavoidable. Yet, the connection between the two stories is more conspicuously one of contrast than of similarity: this 'child' is not monstrous but beautiful; it is not the malign fruit of a science gone wrong, but the ideal product of science at its best; it is also not alive, but rather, conspicuously lifeless and disembodied. Neither a living thing nor even a model of an actual living thing (not even of an asexually reproducing organism), the double helical structure is also not an actual molecule of DNA. What it is is a cardboard and metal construction—a look-alike surrogate for a biological/physical entity existing in its natural habitat.

[7] Naomi Mitchison, to whom the book is rather surprisingly dedicated, might at first glimpse appear to be an exception. She is introduced in the text as his friend Avrion Mitchison's mother, as a 'distinguished writer', and most importantly, as Haldane's sister. But she appears only briefly, and her principal contribution to the narrative seems to be limited to the hospitality of her large (and very cold) home and her fearfully 'condescending stares'.

[8] *Base Pairs* was the title (according to Bronowski) initially considered for Watson's book, and it reflects the same ambiguity of reference. Jacob Bronowski, 'Honest Jim and the Tinker Toy Model', *The Nation* 206 (18 March 1968), 381–2.

Andre Lwoff (another Nobel laureate in biology) also bemoans the absence of eros, and not only of eros but of affectivity in general. 'All things considered,' Lwoff writes in his review of Watson's book, 'it seems as though Jim's heart has not been nurtured and touched long enough by a loving and beloved person. Surely maturation is largely a matter of interaction.' Lwoff's remarks about Watson's 'lack of affectivity', about his unresponsiveness to the natural beauty of things (e.g. 'to Italy in general and Paestum in particular'), are somewhat mollified by what he calls Watson's 'sensitivity to people'. But sensitivity only to some people, and only on some occasions. As Lwoff observes, 'Narcissus takes pleasure in looking at his reflection in the shimmering water. Jim allows himself to be sensitive only insofar as the person involved reflects his own interests.'[9]

Lwoff's invocation of reflection as a figure for Watson's relation to the world is certainly compelling, and not just for its association with Narcissus. Reflection is also the signifier of mirror-image symmetry, and it is noteworthy that the biological forms that Watson finds most inspiring—from the asexual beauty of Bertrand Fourcade's face to the equally asexual marvel of Watson-Crick replication—all have in common the essential two-ness that characterizes symmetries of reflection. The identification of the genetic material with the sequence of nucleotide bases on a strand of DNA answered the crucial question of what genes are, and the answer gave us to understand these entities as one-dimensional formations. The identification of the actual physical configuration of DNA as a double helix held together by complementary base pairing solved an even more difficult problem: it suggested an exquisitely elegant mechanism for the generational immortality of the genetic material. Before 1953, no one would have dreamed that the secret of life could prove so elemental, but now, complementary base pairing provided a stunningly simple mechanism for guaranteeing the persistence of the one-dimensional genetic sequence from one generation to the next. Here in this mechanism is the mirror that reflects us back to ourselves, in each round of cell division, in each round of replication, nucleotide base by nucleotide base. Indeed, so seductive was the solution this simple structure provided to the 'problem of life' that it proved all but irresistible. But might not this 'solution' be yet another beloved reflection in an imaginary mirror of nature? Might not part of its satisfaction lie in the underlying aesthetics to which it conforms, in its vindication of mirror-symmetry as not only the most fundamental but also the sole principle of life?

Surely one needs to enquire of the whereabouts of the biological body in this endless hall of mirrors that is the double helical account of life. Where, in short, is the three-dimensional corporeal entity, developing, interacting, and reproducing in morphogenetic space and time? Watson and Crick's representation of the secret of life as the tautomeric coupling of two complementary one-dimensional molecules was a triumph scarcely precedented in the history of biology.

[9] Andre Lwoff, 'Truth, Truth, What Is Truth?', *Scientific American* 219 (July 1968), 133–8; rpt. in *The Double Helix*, 225–34 (232).

Sometimes called 'the mother-molecule of life', DNA is thought to depend on none of the complexities of the cellular body for either its existence or its ability to function, but is itself the mother—or, more accurately, the forefather—of them all. Its signal claim to biological pre-eminence lies in its purported power of 'self-replication', i.e. in the mechanism suggested by its double helical structure for reproduction without a maternal body—indeed, for reproduction without a material body of any kind.

Watson's personal triumph over Rosy rests on an analogous displacement of the body, though now not of the cellular, cytoplasmic, or organismic body but rather on the level of the molecular body. The contest between them is a contest between two visions of science, and perhaps even more, between two notions of the proper subject (or object) of science, what is conventionally called nature. Rosy insists on the primacy of the material thing (here the DNA), and of the need to study it in its actual configuration, whatever that may be. Hence her commitment to experiment. Watson, by contrast, has little interest in the actual properties of DNA, except in so far as he needs the data for his model building. Indeed, the DNA itself is construed as an enemy, a thing to be 'seized', 'assaulted', and 'conquered'. Far from being an object of deference or respect, he regards it (perhaps as he regards Rosy) as too explosive to be treated with caution. Early on, he explains that 'you did not move cautiously when you were holding dynamite like DNA' (16). The object of Watson's devotion is, in short, not the material entity, but the model that he and Crick will build—for him, the true (and the only beautiful) object (or subject) of science.

Readers may recognize my heading for this section, 'Is There a Body in This Text?' as a play on Mary Jacobus's elegant paper, 'Is There a Woman in This Text?', written two decades ago (1982) and itself a play on Stanley Fish's classic title (1980). Jacobus's concern is with the relation between women and theory, and more specifically, with the displacement and even sacrifice of women so often demanded by the adversarial relations commonplace between rival theorists. *The Double Helix*, in fact, provides her with a case in point. In the rivalry between Watson-Crick and Linus Pauling, Rosy must go. The real object of desire in this text is clearly not Rosy, but more to the point, it is not any woman. Rather, it is the double helix (not to mention the Nobel prize it is expected to bring), 'pursued less for itself than for being desired by another scientist'.[10]

But there is an additional worry, and that is that the object of desire in this text is not any body at all. My principal concern in this chapter is thus not so much the displacement of women as it is the displacement of the generic body, be it the body of a woman (or of a man), the body of the cell, or even the body of a DNA molecule. Of course, to the extent that the body, any body—male or female, cellular or

[10] Mary Jacobus, 'Is There a Woman in This Text?', *New Literary History* 14 (1982), 117–41 (130). She cites, as well, Francis Crick's reference to RNA and DNA in *Life Itself: Its Origin and Nature* (New York: Simon & Schuster, 1981) as 'the dumb blondes of the biomolecular world'. See also my discussion of this text in 'From Secrets of Life to Secrets of Death', in Evelyn Fox Keller, *Secrets of Life, Secrets of Death: Essays on Language, Gender, and Science* (New York: Routledge, 1992), 51–2.

molecular—is 'woman', these are not independent concerns.[11] Rosy may in fact be particularly subject to demonization, or so I suggest, not just because she is a woman scientist, but because, as the woman scientist she happens to be, she represents the body on all three levels.

The film version of this story[12] has Rosalind Franklin say, somewhat beatifically, 'I just want to look, I don't want to touch.' But here, in this book version, it is Watson who just wants to look and doesn't want to touch. And for the purposes of 'just looking'—at Bertrand's face, at 'popsies', or at models of DNA—cardboard versions may well be adequate; certainly, like the seductive image Narcissus finds reflected in the silvery pond, they offer the illusion of safety.

CODA

Like Watson in his 'Epilogue', I too need to add a demurral. It would surely be irresponsible to suggest that the author of *The Double Helix*, however influential he and his work have been, writes for scientists in general, just as it would be irresponsible to suggest that Watson's image of science speaks to more than one particular vision of science. I want to make clear, therefore, that I do not mean to imply that either 'Jim' or his vision can be taken to be representative—not of scientists as a group, and not of science as a whole. Yet the fact remains that, for all its idiosyncrasies, the influence this account of scientific discovery has had since its publication has been immense. Generations of students have read it as a heroic model (scientific as well as personal). That fact alone, it seems to me, warrants the effort of identifying and exhibiting the very particular uni-dimensionality of its underlying intellectual and emotional aesthetic.

[11] The association between 'woman' and body has been a recurrent theme in feminist criticism over the last twenty years. For a recent account, see Judith Butler's *Bodies That Matter* (New York: Routledge, 1993). Alternatively (or equivalently), one might say 'the body is mother', a formulation that would bring into view the elision not only of the maternal body as biological subject, but also of mothers as novelistic subject. With the exception of two references—one to Rosy's hypothesized 'unsatisfied mother', and the other, to the blissfully deluded mothers of 'narrow-minded and stupid' scientists—the stern and apparently unmotherly Naomi Michison is the only mother present in the story of *The Double Helix*.

[12] *The Race for the Double Helix*, dir. Mick Jackson (BBC Television, 1987; reissued by Films for the Humanities and Social Sciences, 2000), starring Jeffrey Goldblum as Watson, Tim Pigott-Smith as Crick, and Juliet Stevenson as Rosalind Franklin.

Notes on Contributors

Rachel Bowlby is Professor of English at the University of York. Her books include *Still Crazy After All These Years: Women, Writing and Psychoanalysis* (1992), *Shopping with Freud* (1993), and *Feminist Destinations and Further Essays on Virginia Woolf* (1997).

Kate Flint is Professor of English at Rutgers University, New Brunswick. Author of *The Woman Reader 1837–1914* (1993) and *The Victorians and the Visual Imagination* (2000), she is currently completing *The Transatlantic Indian, 1776–1930*, and continuing to work on Victorian and modernist perception and the senses.

Mary Jacobus is Grace II Professor of English at the University of Cambridge. Her books include *Reading Woman: Essays in Feminist Criticism* (1986), *First Things: The Maternal Imaginary in Literature, Art, and Psychoanalysis* (1996), and *Psychoanalysis and the Scene of Reading* (1999). She is also co-editor (with Evelyn Fox Keller and Sally Shuttleworth) of *Body/Politics: Women and the Discourses of Science* (1989).

Maroula Joannou is Senior Lecturer in English and Women's Studies at Anglia Polytechnic University in Cambridge. She is the author of *Ladies, Please Don't Smash These Windows: Women's Writing, Feminism and Social Change 1918–1938* (1995) and *Contemporary Women's Writing: From the Golden Notebook to the Color Purple* (2000), and editor of *Women Writers of the 1930s: Gender, History and Politics* (1998). Her doctoral thesis at Cambridge University was supervised by Gillian Beer.

Evelyn Fox Keller, Professor of History and Philosophy of Science in the Program in Science, Technology and Society at MIT, began her professional life as a theoretical physicist. Her most recent books are *The Century of the Gene* (2000) and *Making Sense of Life: Explaining Biological Development with Models, Metaphors, and Machines* (2002).

Nigel Leask is a lecturer in the English Faculty and a Fellow of Queens' College, Cambridge. He is the author of *The Politics of Imagination in Coleridge's Critical Thought* (1988), *British Romantic Writers and the East: Anxieties of Empire* (1992), and *Curiosity and the Aesthetics of Travel Writing 1770–1840: From an Antique Land* (2002).

George Levine is Kenneth Burke Professor of English at Rutgers University and Director of the Center for the Critical Analysis of Contemporary Culture. His books include *Darwin and the Novelists* (1988) and *Dying to Know* (2002).

Suzanne Raitt is Margaret L. Hamilton Professor of English at the College of William & Mary. Her publications include *Vita and Virginia: The Work and Friendship of V. Sackville-West and Virginia Woolf* (1993), and *May Sinclair: A Modern Victorian* (2000). She also edited *Volcanoes and Pearl Divers: Essays in Lesbian Feminist Studies* (1995) and co-edited *Women's Fiction and the Great War* (1997) with Trudi Tate. She is currently at work on a book on the concept of waste in British modernism.

Harriet Ritvo is the Arthur J. Conner Professor of History at MIT. She is the author of *The Animal Estate: The English and Other Creatures in the Victorian Age* (1987) and *The Platypus and the Mermaid, and Other Figments of the Classifying Imagination* (1997).

Jacqueline Rose teaches literature at Queen Mary College, University of London. Her publications include *The Haunting of Sylvia Plath* (1991), *States of Fantasy* (1996, the 1994 Clarendon Lectures), and *Albertine*, a novel (2002).

Sally Shuttleworth is Professor of English at the University of Sheffield. Her publications include *George Eliot and Nineteenth-Century Science: The Make-Believe of a Beginning* (1984) and Charlotte *Brontë and Victorian Psychology* (1996). She is the co-editor (with Jenny Taylor) of *Embodied Selves: An Anthology of Psychological Texts, 1830–1890* (1998).

Helen Small is Fellow in English at Pembroke College, Oxford, and Visiting Scholar at New York University (2001–3). She is the author of *Love's Madness: Medicine, the Novel, and Female Insanity, 1800–1865* (1996), co-editor of *The Practice and Representation of Reading* (1996), and editor of *The Public Intellectual* (2002).

Trudi Tate is a Fellow and Tutor of Clare Hall, University of Cambridge. Her books include *Modernism, History and the First World War* (1998), *Women, Men and the Great War: An Anthology of Stories* (1998), and *Women's Fiction and the Great War* (1997), co-edited with Suzanne Raitt. She has recently completed a book about Australian veterans' memories of the Vietnam War.

Alison Winter is Associate Professor of History at the University of Chicago. Her first book was *Mesmerized: Powers of Mind in Victorian Britain* (1998), and she has published essays on various topics in the history of the human sciences, and on Victorian science and medicine. She is currently working on a history of sciences of memory and truth technologies in the twentieth century.

Select Bibliography of Works by Gillian Beer

BOOKS

Meredith: A Change of Masks: A Study of the Novels (London: Athlone Press, 1970).
The Romance, Critical Idiom series (London: Methuen, 1970, 1977, 1986).
Darwin's Plots: Evolutionary Narrative in Darwin, George Eliot, and Nineteenth-Century Fiction (Rose Mary Crawshay Prize, 1984) (London: Routledge & Kegan Paul, 1983, 1985; 2nd edn. Cambridge: Cambridge University Press, 2000).
George Eliot (Brighton: Harvester, 1986).
Arguing with the Past: Essays in Narrative from Woolf to Sidney (London: Routledge, 1989).
Forging the Missing Link: Interdisciplinary Stories (Cambridge: Cambridge University Press, 1992; French edn. 1994).
Open Fields: Science in Cultural Encounter (Oxford: Clarendon Press, 1996).
Virginia Woolf: The Common Ground: Essays by Gillian Beer (Edinburgh: Edinburgh University Press, 1996).

EDITIONS

The Notebooks of George Meredith, with Margaret Harris (Salzburg: Institut für Anglistik und Amerikanistik, Universität Salzburg, 1983).
Virginia Woolf, *The Waves*, World's Classics (Oxford: Oxford University Press, 1992).
Virginia Woolf, *Between the Acts*, Penguin Twentieth-Century Classics (London: Penguin, 1992).
George Meredith, *Modern Love* (London: Syrens, 1995).
Charles Darwin, *The Origin of Species*, World's Classics (Oxford: Oxford University Press, 1996).
Jane Austen, *Persuasion*, Penguin Classics (London: Penguin, 1998).
Introduction to Sigmund Freud, *The 'Wolfman' and other Cases*, trans. Louise Adey Huish, New Penguin Freud Series (London: Penguin, 2002).

SELECTED ESSAYS

'Meredith's Revisions of *The Tragic Comedians*', *Review of English Studies* 14 (1963).
'George Meredith and *The Satirist*', *Review of English Studies* 15 (1965).
'Charles Kingsley and the Literary Image of the Countryside', *Victorian Studies* 8 (1965).
'Richardson, Milton, and the Status of Evil', *Review of English Studies* 19 (1968).
'Aesthetic Debate in Keats's Odes', *Modern Language Review* 64 (1969).
'*One of Our Conquerors*: Language and Music', in I. Fletcher (ed.), *Meredith Now: Some Critical Essays* (London: Routledge & Kegan Paul, 1971).
'Myth and the Single Consciousness: *Middlemarch* and "The Lifted Veil" ', in I. Adam

(ed.), *This Particular Web: Essays on Middlemarch* (Toronto: University of Toronto Press, 1975).

'"Coming Wonders": Uses of Theatre in the Victorian Novel', in M. Axton and R. Williams (eds.), *English Drama: Forms and Development: Essays in Honour of Muriel Clara Bradbrook* (Cambridge: Cambridge University Press, 1977).

'Plot and the Analogy with Science in Later Nineteenth-Century Novelists', in *Comparative Criticism: A Yearbook* 2 (1990).

'Carlyle and Mary Barton: Problems of Utterance', in F. Barker, J. Coombes, J. Hulme, C. Mercer, and D. Musselwhite (eds.), *1848: The Sociology of Literature: Proceedings of the Essex Conference on the Sociology of Literature, July 1977* (Colchester: University of Essex, 1978).

'Beyond Determinism: George Eliot and Virginia Woolf', in M. Jacobus (ed.), *Women Writing and Writing about Women* (London: Croom Helm, 1979).

'Negation in *A Passage to India*', *Essays in Criticism* 30 (1980).

'"Our Unnatural No-Voice": The Heroic Epistle, Pope, and Women's Gothic', *The Yearbook of English Studies* 12 (1982).

'Hume, Stephen and Elegy in *To the Lighthouse*', *Essays in Criticism* 34 (1984).

'Virginia Woolf and Pre-history', in E. Warner (ed.), *Virginia Woolf: A Centenary Perspective* (London: Macmillan, 1984).

'Darwin's Reading and the Fictions of Development', in D. Kohn (ed.), *The Darwinian Heritage* (Princeton: Princeton University Press, 1985).

'Origins and Oblivion in Victorian Narrative', in R. B. Yeazell (ed.), *Sex, Politics, and Science in the Nineteenth-Century Novel* (Baltimore: Johns Hopkins University Press, 1986).

'Language Theory and Darwinian Theory: Problems of Interchange and Autonomy', in *Essays on Creativity and Science: Papers Delivered at a Conference Held in Honolulu, Hawaii, March 23–24, 1985* (Honolulu: Hawaii Council of Teachers of English, 1986).

'The Body of the People in Virginia Woolf', in S. Roe (ed.), *Women Reading Women Writing* (Brighton: Harvester, 1987).

'Designing and Describing: A Problem in the Language of Discovery', in G. Levine (ed.), *One Culture: Essays in Science and Literature* (Madison: University of Wisconsin Press, 1987).

'Can the Native Return?', The Hilda Hulme Memorial Lecture, 8 December 1988 (London: University of London, 1989).

'The Death of the Sun: Solar Physics and Solar Myth', in J. B. Bullen (ed.), *The Sun is God: Painting, Literature and Mythology in the Nineteenth Century* (Oxford: Clarendon Press, 1989).

'Representing Women: Re-presenting the Past', in C. Belsey and Jane Moore (eds.), *The Feminist Reader: Essays in Gender and the Politics of Literary Criticism* (London: Macmillan, 1989; 2nd edn. 1997).

'Darwin and the Growth of Language Theory', in S. Shuttleworth and J. Christie (eds.), *Nature Transfigured: Science and Literature, 1700–1900* (Manchester: Manchester University Press, 1989).

'Discourses of the Island', in F. Amrine (ed.), *Literature and Science as Modes of Expression* (Dordrecht: Kluwer Academic Publishers, 1989).

Introductory essay to special number on 'Rhetoric and Science'(with Herminio Martins), *History of the Human Sciences* 3/2 (1990).

'Translation or Transformation?: The Relations of Literature and Science', lecture delivered at the Royal Society, 1989, *Notes and Records of the Royal Society of London* 44 (1990).

'The Reader's Wager: Lots, Sorts, and Futures', Bateson Lecture at the University of Oxford, *Essays in Criticism* 40 (1990).

'Parable, Professionalisation, and Literary Allusion in Victorian Scientific Writing', special number on narrative, *AUMLA: Journal of the Australasian Universities Language and Literature Association* 74 (1990).

'The Island and the Aeroplane: The Case of Virginia Woolf', in H. Bhabha (ed.), *Nation and Narration* (London: Routledge, 1991). Also in R. Bowlby (ed.), *Virginia Woolf* (London: Longman, 1992).

'Speaking for the Others: Relativism and Authority in Victorian Anthropological Literature', in R. Fraser (ed.), *Sir James Frazer and the Literary Imagination: Essays in Affinity and Influence* (Basingstoke: Macmillan, 1990).

'Helmholtz, Tyndall, Gerard Manley Hopkins: Leaps of the Prepared Imagination', *Comparative Criticism: A Yearbook* 13 (1991).

'Four Bodies on the *Beagle*: Touch, Sight, and Writing in a Darwin Letter', in J. Still and M. Worton (eds.), *Textuality and Sexuality: Reading Theories and Practices* (Manchester: Manchester University Press, 1993).

'Wave Theory and the Rise of Literary Modernism', in G. Levine (ed.), *Realism and Representation: Essays on the Problem of Realism in Relation to Science, Literature, and Culture* (Madison: University of Wisconsin Press, 1993).

'Rhyming as Comedy: Body, Ghost, and Banquet', in M. Cordner, P. Holland, and J. Kerrigan (eds.), *English Comedy* (Cambridge: Cambridge University Press, 1994).

'*Square Rounds* and Other Awkward Fits: Chemistry as Theatre', *Ambix* 41 (1994).

'George Eliot and the Novel of Ideas', in J. Richetti, J. Bender, D. David, *et al.* (eds.), *The Columbia History of the British Novel* (New York: Columbia University Press, 1994).

'Passion, Politics, Philosophy: The Work of Edith Simcox', *Women: A Cultural Review* 6/2 (1995).

'Eddington and the Idiom of Modernism', Rothschild Distinguished Lecture in History of Science at Harvard, in H. Krips, J. E. McGuire, and T. Melia (eds.), *Science, Reason and Rhetoric* (Pittsburgh: University of Pittsburgh Press; Konstanz: Universitätsverlag Konstanz, 1995).

' "Wireless": Radio, Physics, and Modernism', in F. Spufford and J. Uglow (eds.), *Cultural Babbage: Technology, Time and Invention* (London: Faber & Faber, 1996).

'Authentic Tidings of Invisible Things: Secularising the Invisible in Late Nineteenth Century Britain', in T. Brennan and M. Jay (eds.), *Vision in Context: Historical and Contemporary Perspectives on Sight* (London: Routledge, 1996).

'Travelling the Other Way', in N. Jardine, J. A. Secord, and E. C. Spary (eds.), *Cultures of Natural History* (Cambridge: Cambridge University Press, 1996; rpt. Patagonia: British Museum Press, 1998).

'Hardy and Decadence', in C. P. C. Pettit (ed.), *Celebrating Thomas Hardy: Insights and Appreciations* (Basingstoke: Macmillan, 1996).

'The Evolution of the Novel', in A. C. Fabian (ed.), *Evolution* (The 1995 Darwin College Lecture series) (Cambridge: Cambridge University Press, 1998).

'The Dissidence of Vernon Lee: *Satan the Waster* and the Will to Believe', in S. Raitt and T. Tate (eds.), *Women's Fiction and the Great War* (Oxford: Clarendon Press, 1997).

'The Making of a Cliché: "No Man is an Island" ', *European Journal of English Studies* 1/1 (1997). Also in *GRAAT: Publication des Groupes de Recherches Anglo-Américaines de l'Université François Rabelais de Tours* 16 (1997).

'Writing Darwin's Islands: England and the Insular Condition', in T. Lenoir (ed.),

Inscribing Science: Scientific Texts and the Materiality of Communication (Stanford: Stanford University Press, 1998).

'Has Nature a Future?', in E. S. Shaffer (ed.), *The Third Culture: Literature and Science* (Berlin: W. de Gruyter, 1998).

'Sylvia Townsend Warner: The Centrifugal Kick', in M. Joannou (ed.), *Women Writers of the Nineteen Thirties: Gender, Politics and History* (Edinburgh: Edinburgh University Press, 1999).

'Rhyming as Resurrection', in M. Campbell, J. M. Labbe, and S. Shuttleworth (eds.), *Memory and Memorials, 1789–1914: Literary and Cultural Perspectives* (London: Routledge, 2000).

'Wave, Atom, Dinosaur: Woolf's Science', a lecture delivered at the seventy-first general meeting of the English Literary Society of Japan, 30 May 1999 (Tokyo: English Literary Society of Japan, 2000; London: Virginia Woolf Society of Great Britain, 2000).

'Narrative Swerves: Grand Narratives and the Disciplines', *Women: A Cultural Review* 11/1–2 (2000).

'Knowing a Life: Edith Simcox—Sat est vixisse?', in S. Anger (ed.), *Knowing the Past: Victorian Literature and Culture* (Ithaca: Cornell University Press, 2001).

'Storytime and its Futures', in K. Ridderbos (ed.), *Time* (The 2000 Darwin College Lecture series) (Cambridge: Cambridge University Press, 2002).

'Credit Limit: Fiction and the Surplus of Belief', *Comparative Criticism* 24 (2002).

'The Afterlife and the Life Before', in P. Mallett (ed.), *Thomas Hardy: Texts and Contexts* (London: Palgrave Macmillan, 2003).

'Island Bounds', in Rod Edmond and Vanessa Smith (eds.), *Islands in History and Representation* (London: Routledge, 2003).

'The Academy: Europe in England', in G. Cantor and S. Shuttleworth (eds.), *Science Serialized: Representations of the Sciences in Nineteenth-Century Periodicals* (Boston: MIT Press, 2003).

Index